AF441844

DISTRIBUTIVE MODULES
AND RELATED TOPICS

ALGEBRA, LOGIC AND APPLICATIONS

A series edited by
R. Göbel
Universität Gesamthochschule, Essen, Germany
A. Macintyre
The Mathematical Institute, University of Oxford, UK

**Please see the back of this book for other titles in the Algebra, Logic and
Applications series.**

DISTRIBUTIVE MODULES AND RELATED TOPICS

Askar Tuganbaev

Moscow Power Engineering Institute, Russia

GORDON AND BREACH SCIENCE PUBLISHERS
Australia • Canada • China • France • Germany • India • Japan • Luxembourg
Malaysia • The Netherlands • Russia • Singapore • Switzerland

Amsteldijk 166
1st Floor
1079 LH Amsterdam
The Netherlands

British Library Cataloguing in Publication Data

Tuganbaev, Askar
   Distributive modules and related topics. – (Algebra, logic
   and applications ; v. 12)
   1. Rings (Algebra)   2. Modules (Algebra)
   I. Title
   512.4

ISBN 90-5699-192-2

Dedicated to the memory
of my parents, Akan and Alma Tuganbaev

Contents

Preface

All rings are assumed to be associative and (except for nil-rings and for some stipulated cases) to have a nonzero identity element. We denote by Lat(M) the lattice of all submodules of a module M.

A module is called *distributive* if the following two equivalent conditions hold.

(1) $F \bigcap (G + H) = F \bigcap G + F \bigcap H$ for all submodules F, G, and H of the module M.

(2) $F + G \bigcap H = (F + G) \bigcap (F + H)$ for all submodules F, G, and H of the module M.

The distributivity of M means that the lattice Lat(M) is distributive. A module M is *uniserial* if any two submodules of M are comparable with respect to inclusion (i.e., the lattice Lat(M) is a chain). All uniserial modules are distributive. Any quasi-cyclic Abelian p-group is a uniserial module over the ring of integers $\mathbf{Z}$. The additive group of rational numbers is a distributive $\mathbf{Z}$-module which is not uniserial. All valuation rings in division rings are uniserial.

The class of distributive rings includes all commutative Dedekind rings (for example, rings of integral algebraic numbers or commutative principal ideal rings). In particular, the ring of integers and the polynomial ring $F[x]$ over a field F are distributive rings. In addition, all strongly regular rings (for example, all factor-rings of direct products of skew fields and all commutative regular rings) are distributive.

All commutative Dedekind rings and all serial right Noetherian rings possess the following property: each finitely generated A-module is decomposed into a direct sum of distributive modules. Let A be a commutative regular ring (for example, a field), and let G be a locally cyclic group. Then the group ring $A[G]$, the polynomial ring $A[x]$, and the Laurent polynomial ring $A[x, x^{-1}]$ are distributive Bezout rings

A *semidistributive (resp. serial)* module is any direct sum of distributive (resp. uniserial) modules. All serial modules (in particular, all semisimple modules) are semidistributive. A module M is a *Be-*

zout module if every finitely generated submodule of M is cyclic. If all maximal right ideals of a ring A are ideals (for example, if A is commutative), then any Bezout A-modules is distributive.

Let F be a field, and let A be the 5-dimensional algebra over the field F generated by all (3×3)-matrices of the form

$$\begin{pmatrix} f_{11} & f_{12} & f_{13} \\ 0 & f_{22} & 0 \\ 0 & 0 & f_{33} \end{pmatrix}, \quad \text{where } f_{ij} \in F.$$

Then $M \equiv e_{11}A = e_{11}F + e_{12}F + e_{13}F$ is a local projective distributive Noetherian Artinian Bezout A-module which contains a direct sum of two simple modules. In particular, M is not uniserial. It can be verified that the algebra A is a right semidistribive left serial hereditary Artinian ring, all A-modules are semidistributive, and all injective A-modules are serial.

We highlight [24], [14], [122], [123], and [15] among the first works concerning distributive modules and rings in the noncommutative case. The systematic study of distributive modules over noncommutative rings was initiated in papers [1], [166], [34], and [26]. Distributive modules were considered in [17, Ch. 9],[41, §4.1], and [146, §2.2]. In [250], distributive modules are applied to complex analysis. In [228], distributive rings are used for a study of rings of continuous functions in topological spaces. In [247], distributive modules are applied to study rings with a duality.

In [90], [176], [183], and [182], distributive rings are used to investigate rings of weak global dimension one and hereditary rings. In [180], [182], and [189], some applications of distributive rings and modules to formal power series rings were obtained. Distributive group and semigroup rings were studied in [102], [82], [40], [179], [181], [182], [184], [186], and [195]. Distributive quaternion algebras were studied in [196], [197], and [198]. In [177], [184], [149], and [201], the modules which are distributive over their endomorphism rings were studied. Topological aspects of properties of distributive modules and rings were considered in [229], [227], and [184]. Distributive modules over incidence algebras were studied in [61]. Modules decomposed into a direct sum of distributive modules, and rings which are direct sums of distributive right ideals were studied in [65], [67], [167], [248], [99], [100], [101], [33], [20], [247], [243], [212], and [213]. Distributive graded modules were considered in [37]. Rings possessing faithful distributive modules were studied in [14] and [15]. Conditions sufficient for the distributivity of some lattices of linear subspaces were considered in [84] and [94]. Noncommutative rings whose lattices of two-sided ideals are distributive were studied in [24], [124], [18], [110], and [95]. Distributive modules and rings were addressed in surveys [19], [115], and [116].

Distributive modules are closely related to multiplication modules (a module is called *multiplication* if $M(M : N) = N$ for all $N \in \mathrm{Lat}(M)$). For example, a module M over a commutative ring A is distributive $\Longleftrightarrow$ all finitely generated submodules of M are multiplication [5]. Multiplication modules over noncommutative rings were considered in [221], [169], [222], [223], [160], and [211]. Multiplication modules over commutative rings were studied in [125], [32], [69], [126], [5], [80], [6], [156], [7], [11], [143], [144], [145], [159], [140], [50], [51], [54], [113], and [38].

In this book, as a rule, we consider distributive modules over noncommutative rings. Since the distributivity of a commutative ring A is equivalent to the fact that all localizations of the ring A w.r.t. its maximal ideals are uniserial rings [90] (in particular, all Prüfer domains are distributive), the commutative case deserves a special consideration. Here, we just highlight papers [90], [91], [155], [5], [229], and [194].

Moreover, it is worth noting that uniserial and serial modules and rings (e.g., valuation rings), which are addressed in considerable number of papers, are little touched on here. We only mention papers [43], [136], [137], [157], and [232].

Also, see [21], [22], [23], [36], [39], [44], [45], [46], [47], [48], [49], [56], [57], [58], [63], [64], [68], [79], [85], [87], [88], [96], [97], [98], [103], [104], [105], [106], [107], [114], [132], [133], [134], [135], [127], [128], [129], [130], [131], [147], [148], [150], [151], [152], [155], [158], [218], [233], [220], [224], [226], [230], [234], [235], [237], [238], [239], [240], [241], [242], [244], [245], [246].

The background required for this book (e.g., definitions of a ring, module, homomorphism etc) is standard, and may be found in most graduate level texts on algebra.

Before going to the main presentation we cite some remarks, notations and definitions. Expressions such as a "Noetherian ring" mean that the corresponding right and left conditions hold.

We denote by $\mathrm{End}(M)$, M^I, and $M^{(I)}$ the endomorphism ring, the direct product, and the direct sum of I copies of a module M, respectively.

For a ring A, we denote by $C(A)$ its centre. For a subset B of a ring A, we denote by $r_A(B)$ and $\ell_A(B)$ the right and left annihilators of B, respectively; $\mathrm{Id}(B)$ denotes the ideal of A generated by B. We can omit the subscripts if the situation is obvious. If F is a subset of a module M_A over a ring A and $G \in \mathrm{Lat}(M)$, then $(F : G)$ denotes the right ideal $\{a \in A \mid Fa \subseteq G\}$ of A. A *faithful* module is any module M_A such that $r_A(M) = 0$.

An element a of A is *right regular* (resp. *left regular, right invertible, left invertible*) in A if $r(a) = 0$ (resp. $\ell(a) = 0$, $aA = A$, $Aa = A$). A right and left regular (resp. invertible) element is called a *regular (invertible)* element. Right (resp. left) invertible elements are also called right (resp. left) *units* of A. The *group of units* of a ring A is denoted by $U(A)$.

A module M is a *simple* module if M has no nonzero proper submodules.

M_A is a simple module $\Longleftrightarrow$ each nonzero homomorphism $N_A \to M$ is an epimorphism $\Longleftrightarrow$ each nonzero homomorphism $M \to N_A$ is a monomorphism $\Longleftrightarrow$ $M = mA$ for each nonzero $m \in M \Longleftrightarrow$ given any nonzero $m, n \in M$, there exists $a \in A$ such that $ma = n$.

The sum of all simple submodules of a module M is denoted by $\mathrm{Soc}(M)$. (If M has no simple submodules, then $\mathrm{Soc}(M) = 0$ by definition.) $\mathrm{Soc}(M)$ is called *the socle* of M. A *semisimple* module is any direct sum of simple modules. $\mathrm{Soc}(M)$ is the largest semisimple submodule of M and coincides with the intersection of all essential submodules of M.

A submodule N of a module M is *maximal* (in M) if N is kernel of a nonzero homomorphism of M into some simple module (i.e. M/N is a simple module). The set of all maximal submodules of M is denoted by $\max(M)$.

A submodule of the factor module of a module M is called a *subfactor* of M. Any nonzero module has a simple subfactor.

The module M is said to be *finite-dimensional* (in the sense of Goldie) if M contains no infinite direct sums of nonzero submodules. A module M is *completely finite-dimensional* if all subfactors of M are finite-dimensional.

A module M is *semi-Artinian* if each nonzero subfactor of M is an essential extension of a semisimple module. A module M is *semi-Noetherian* if each nonzero subfactor of M has a maximal submodule.

A submodule H of M is a *complement* of $F \in \mathrm{Lat}(M)$ in M if and only if $F \bigcap H = 0$, $F \oplus H$ is essential in M, and $F \bigcap E \neq 0$ for any $E \in \mathrm{Lat}(M)$ properly containing H. We say that $F \in \mathrm{Lat}(M)$ is *complemented* provided F is a complement of some $H \in \mathrm{Lat}(M)$ in M.

A submodule N of a module M is *essential* (in M) if N has a nonzero intersection with any nonzero submodule of M. In this case, we say that M is an *essential extension* of N. If a submodule N of a module M contains an essential submodule of M, then M is an essential extension of N. All finite intersections of essential submodules of M are essential submodules of M.

A submodule $H \in \mathrm{Lat}(M)$ is *superfluous* (in M) if for any $N \in$

Lat(M), the equality $H + N = M$ implies the equality $N = M$.

A module M is *projective with respect to a module N* (or *N-projective*) if for each epimorphism $h : N \to \overline{N}$ and for any homomorphism $\overline{f} : M \to \overline{N}$ there exists a homomorphism $f : M \to N$ such that $\overline{f} = hf$. A module that is projective with respect to itself is called a *quasi-projective* (or *self-projective*) module.

A module M_A is a *projective* module if the following equivalent conditions hold.

(i) M is a direct summand of a free A-module.

(ii)   For each A-module epimorhism $f : N \to M$, $\mathrm{Ker}(f)$ is a direct summand of M.

(iii) For each A-module epimorhism $h : N \to \overline{N}$ and any $\overline{f} \in \mathrm{Hom}(M, \overline{N})$, there exists $f \in \mathrm{Hom}(M, N)$ such that $\overline{f} = hf$.

(iv) There exist a set $\{m_j\}_{j \in J}$ of elements of M and a set $\{f_j\}_{j \in J}$ of homomorphisms $f_j : M \to A_A$ such that for any $m \in M$, $m = \sum_{j \in J} m_j f_j(m)$ $(f_j(m) = 0$ for almost all subscripts $j)$.

A module M_A over a ring A is *hereditary* (*resp. semihereditary, Rickartian*) if all submodules (resp. all finitely generated submodules, all cyclic submodules) of M are projective.

A *domain* is any ring A such that each nonzero element of A is regular. A ring without nonzero nilpotent ideals (resp. elements) is called a *semiprime* (resp. *reduced*) ring. A ring A is *prime* if $IJ \neq 0$ of any nonzero ideals I, J of A. A *right primitive* ring A is any ring such that A has a faithful simple right module (i.e., A has a maximal right ideal which contains no nonzero ideals of A). A ring is *simple* if it has no nonzero proper ideals. An ideal B of a ring A is $\bigcap$-*irreducible* if any two nonzero ideals of A/B have a nonzero intersection. A ring A is $\bigcap$-*irreducible* if any two nonzero ideals of A have a nonzero intersection (i.e. 0 is $\bigcap$-irreducible ideal).

An ideal B of A is *semiprime* (resp. *prime, completely semiprime, right primitive*) if A/B is a semiprime (resp. *prime, reduced, right primitive*) ring. A proper ideal B of A is *completely prime* if A/B is a domain. A module M_A is *completely prime* if $r(m) = r(M)$ for any nonzero $m \in M$. If B is an ideal of a ring A, then the module $(A/B)_A$ is completely prime $\iff$ B is a completely prime ideal of A. A *minimal prime* ideal (resp. *minimal completely prime* ideal) in a ring A is any prime (resp. completely prime) ideal P such that P contains no properly any other prime ideal (resp. completely prime ideal) of A. A right (left) ideal I is called *right (left) nil-ideal* if all elements of I are nilpotent. A subset T of a ring A is a *multiplicative* set in A if $1 \in T$, $0 \notin T$, and T is closed under multiplication.

An *orthogonally finite* ring is any ring which contains no infinite sets of nonzero orthogonal idempotents. A ring is *normal* if all its idempotents are central. A *principal right ideal ring* is any ring A such that all right ideals of A are principal.

A module M is a *regular* module if every finitely generated submodule of M is a direct summand of M.

A ring A is a *regular* ring if for any $a \in A$, there exists $b \in A$ such that $a = aba$.

A is a regular ring $\iff A_A$ is a regular module $\iff {}_A A$ is a regular module $\iff$ the intersection of any two finitely generated right ideals of A and the right annihilator of any element of A are cyclic direct summands of $A_A \iff$ the intersection of any two finitely generated left ideals of A and the left annihilator of any element of A are cyclic direct summands of ${}_A A$.

A *semiregular* ring is any ring A such that $A/J(A)$ is a regular ring and idempotents of $A/J(A)$ can be lifted modulo $J(A)$.

A ring A is a *strongly regular* ring if for any $a \in A$, there exists $b \in A$ such that $a = a^2 b$.

A is a strongly regular ring $\iff$ for any $a \in A$, there exists $b \in A$ such that $a = ba^2 \iff$ every element of A is a product of a central idempotent and a unit $\iff A$ is a regular reduced ring $\iff A$ is a regular normal ring.

All factor rings and direct products of regular (strongly regular) rings are regular.

Symbols

Chapter 1

Basic notions

1.1   Radicals, idempotents, and nilpotents

1.1 The intersection $J(M)$ of kernels of all homomorphisms from a module M into simple modules is called the *Jacobson radical* of M. A module M is *semiprimitive* if $J(M) = 0$. We note that either $J(M) = M$ (if $\max(M) = \emptyset$) or $J(M)$ coincides with the intersection of all maximal submodules of M (if $\max(M) \neq \emptyset$).

A ring A is *semilocal* if $A/J(A)$ is an Artinian ring.

A is a semilocal ring $\iff$ the factor ring $A/J(A)$ is right Artinian $\iff A/J(A)$ is left Artinian $\iff A/J(A)$ is a semisimple Artinian ring $\iff A/J(A)$ is isomorphic to a finite direct product of matrix rings over division rings.

A *semiprimary* ring A is any ring A such that A is semilocal, and the radical $J(A)$ of A is nilpotent.

If M is a right A-module, then $\mathrm{Sing}(M)$ denotes the set of all $m \in M$ such that $r(m)$ is an essential right ideal of A. A module M is a *singular* module if $\mathrm{Sing}(M) = M$. A *nonsingular* module is any module M such that $\mathrm{Sing}(M) = 0$.

1.2 An ideal B of a ring A is *left vanishing* if the following equivalent conditions hold.

(1) For each sequence $b_1, b_2 \ldots$ of elements of B, there exists k such that $b_k b_{k-1} \cdots b_1 = 0$.

(2) $B \subseteq J(A)$, and for any $b_1, b_2, \ldots \in B$, the descending chain of principal left ideals $Ab_1 \supseteq Ab_2 b_1 \supseteq \ldots \supseteq Ab_k b_{k-1} \cdots b_1 \supseteq$ eventually terminates.

(3) For each right module N_A, the module NB is superfluous in N.

(4) $MB \neq M$ for each nonzero right module M_A.

Every nilpotent ideal is right and left vanishing, and every left or right vanishing ideal B of A is a nil-ideal.

In particular, $B \subseteq J(A)$.

1.3 An element a of a ring A is *strongly nilpotent* if all terms of any sequence $\{a_i\}_{i=0}^{\infty}$ such that $a_0 = a$, $a_{n+1} \in a_n A a_n$ are equal to zero beginning with some index.

The set $N(A)$ of all strongly nilpotent elements of a ring A is called the *prime radical* of A.

(1) The prime radical $N(A)$ containes the sum of all nilpotent ideals of A and is the least semiprime ideal of A.

(2) $N(A)$ coincides with the intersection N_1 of all prime ideals of A. Also, $N(A)$ coincides with the intersection N_2 of all semiprime ideals of the ring A.

(3) If A is a ring with the maximum condition on nilpotent ideals, then $N(A)$ is the largest nilpotent ideal of A.

1.4 Let A be a ring which can have no the identity element, B be an ideal of A, C be an ideal of the ring B, and let D be the ideal of A generated by C. Then $D^3 \subseteq C \subseteq D$.

Consequently, if C is a nilpotent ideal of B, then D is a nilpotent ideal of A.

◁ Since $D = C + AC + CA + ACA \subseteq B$, we obtain $D^3 \subseteq BDB = BCB + BACB + BCAB + BACAB = BCB \subseteq C \subseteq D$. ▷

1.5 Let e be a nonzero idempotent of a ring A, and let M_A be a right A-module. Then Me is a right eAe-module, and the following assertions hold.

(1) If Pe is a proper submodule of Me, then $Pe = (PeA)e$, and PeA is a proper submodule of M.

Therefore, if P is a maximal submodule of M, then either $Pe = Me$ or Pe is a maximal submodule of Me.

(2) If B is an ideal of A, then $B \bigcap eAe = eBe$ is an ideal of eAe, and the factor ring $(eAe)/(eBe)$ is isomorphic to $(e + B)(A/B)(e + B)$.

(3) If M_A is finitely generated and Pe is a maximal submodule of Me_{eAe}, then PeA is contained in some maximal submodule of M.

(4) $J(Me) \subseteq J(M)e$.

In addition, if M is finitely generated, then $J(Me) = J(M)e$.

(5) $eJ(A) = J(eA_A)$, $J(eAe) = eJ(A)e$, and $(eAe)/J(eAe) \cong (e + J(A))(A/J(A))(e + J(A))$.

(6) If B is a right ideal of A, then $B \bigcap Ae = Be$, and the submodule eB of eA_A is a homomorphic image of B_A.

(7) Let eXe, eYe, and eZe be three ideals of the ring eAe. Then $AeXeA$, $AeYeA$, and $AeZeA$ are ideals of A.

Also, $(eXe)(eYe) \subseteq eZe \iff (AeXeA)(AeYeA) \subseteq AeZeA$.

(8) Let B be a right ideal of A. Then $eJ(B_A) \subseteq J(eB_A)$.

If the module B_A is semisimple (resp. Noetherian, Artinian), then eB_A is semisimple (resp. Noetherian, Artinian).

(9) If $\{P_i\}_{i \in I}$ is a set of submodules of M_A, then $(\bigcap_{i \in I} P_i)e = \bigcap_{i \in I}(P_i e)$ and $(\sum_{i \in I} P_i)e = \sum_{i \in I}(P_i e)$.

(10) The rule $\psi(Ne) = NeA$ defines an injective lattice homomorphism $\psi : \mathrm{Lat}(Me_{eAe}) \to \mathrm{Lat}(M)$.

The rule $\varphi(N) = Ne$ defines a surjective lattice homomorphism $\varphi : \mathrm{Lat}(M) \to \mathrm{Lat}((Me)_{eAe})$.

If $M_A = \oplus_{i \in I} M_i$, then $Me_{eAe} = \oplus_{i \in I} M_i e$.

(11) If the module M_A is semisimple (resp. Noetherian, Artinian, distributive, semidistributive), then the module Me_{eAe} is semisimple (resp. Noetherian, Artinian, distributive, semidistributive).

(12) If the ring A is semisimple (resp. right Noetherian, right Artinian, right distributive, right semidistributive), then the ring eAe is semisimple (resp. right Noetherian, right Artinian, right distributive, right semidistributive).

(13) If A is a semilocal ring, then the module $eA/J(eA) = eA/eJ(A) = eA/(eA \cap J(A))$ is semisimple, and the ring eAe is semilocal.

(14) eXe is a nilpotent ideal of $eAe \iff AeXeA$ is a nilpotent ideal of A.

(15) If A is semiprime (resp. prime, simple), then eAe is semiprime (resp. prime, simple).

(16) If P is a prime ideal of A which contains no e, then ePe is a proper prime ideal of eAe.

(17) If P and Q are two incomparable prime ideals of A and e is not contained in $P + Q$, then the prime ideals ePe and eQe of eAe are not comparable, and $ePe + eQe \neq eAe$.

(18) If N is the prime radical of A, then eNe is the prime radical of eAe.

◁ (1) – (12) are directly verified. (13) follows from (6). (14) and (15) follows from (7).

(16) Let P be a prime ideal. Since $e \in A \setminus P \subseteq A \setminus ePe$, we obtain $ePe \neq eAe$. Assume that ePe is not a prime ideal of eAe. By (7), there exist two ideals X and Y of A such that $eXeYe \subseteq ePe \subseteq P$. Hence the ideal $(AeA)X(AeA)Y(AeA)$ is contained in the prime ideal P. Therefore, either $X \subseteq P$ or $Y \subseteq P$. Hence either $eXe \subseteq ePe$ or $eYe \subseteq ePe$.

(17) Since $e \in A \setminus (P + Q)$, we obtain $ePe + eQe \neq eAe$. By (14), ePe and eQe are prime ideals of eAe. Assume that $ePe \subseteq eQe$. Hence the ideal $(AeA)P(AeA)$ is contained in the prime ideal Q; this is a contradiction. Analogously, the inclusion $ePe \supseteq eQe$ is impossible.

(18) The prime radical N of A coincides with the set of all strongly nilpotent elements of A. Since $eNe \subseteq N$, all elements of eNe are strongly nilpotent in A. It is sufficient now to prove that if $a \in A$ and eae is strongly nilpotent in eAe, then eae is strongly nilpotent in A. This follows from the fact that if $a_1 = eae$, $a_{n+1} \in a_n A a_n$, then all the elements a_n belong to eAe. $\triangleright$

1.6 (1) If $\{e_i\}_{i \in I}$ is any set of orthogonal idempotents of a ring A, then $\sum_{i \in I} e_i A$, and $\sum_{i \in I} Ae_i$ are direct sums.

(2) A is orthogonally finite $\iff A_A$ satisfies the minimum condition on descending chains of cyclic direct summands $\iff {}_A A$ satisfies the minimum condition on descending chains of cyclic direct summands.

1.7 Let B be an ideal of a ring A, $\overline{A} \equiv A/B$, and let $\overline{a}$ denote the natural homomorphic image of $a \in A$ in the ring $\overline{A}$. We say that *idempotents of $\overline{A}$ can be lifted to idempotents of A* if given any idempotent $f \in \overline{A}$, there exists an idempotent $e \in A$ such that $f = \overline{e}$.

Let $\{f_i\}_{i \in I}$ be some set of orthogonal idempotents of $\overline{A}$. If $f_i = \overline{e}_i$ for all i, then we say that $\{f_i\}_{i \in I}$ *can be lifted to the set* $\{e_i\}_{i \in I}$ *of orthogonal idempotents of A.*

1.2 Local and semiperfect rings

1.8 A module M is a *local* module if $M/J(M)$ is a simple module, and $J(M)$ is superfluous in M.

A module M_A is local $\iff M$ is finitely generated and has precisely one maximal submodule $\iff M \neq J(M)$, and $M = mA$ for any $m \in M \setminus J(M)$.

1.9 A ring A is *local* if $A/J(A)$ is a division ring.

An idempotent $e \in A$ is called *local* if the ring eAe is local.

A is a local ring $\iff A_A$ is a local module $\iff {}_A A$ is a local module $\iff A = aA$ for all $a \in A \setminus J(A) \iff A = Aa$ for all $a \in A \setminus J(A) \iff a \in U(A)$ for all $a \in A \setminus J(A) \iff J(A)$ coincides with the set of all noninvertible elements of $A \iff$ for any $a, b \in A$ such that $a + b = 1$, at least one of elements a, b is a unit of A.

If A is a local ring, then every cyclic A-module is local, all factor rings of A are local, and all simple right A-modules are isomorphic to the module $(A/J(A))_A$.

A ring A is *matrix-local* if the factor ring $A/J(A)$ is isomorphic to a matrix ring over a division ring. Clearly, each local ring is matrix-local and every matrix-local ring is semilocal.

If A is a matrix-local ring, then all simple right (left) A-modules are isomorphic.

1.10 A ring A is a *semiperfect* ring if A is a semilocal ring, and idempotents of $A/J(A)$ can be lifted modulo $J(A)$.

A is a semiperfect ring $\Longleftrightarrow$ A_A is a direct sum of local modules $\Longleftrightarrow$ $_AA$ is a direct sum of local modules $\Longleftrightarrow$ A is a semiregular orthogonally finite ring $\Longleftrightarrow$ there exists a decomposition $1 = \sum_{i=1}^{n} e_i$ of the identity element of A into a sum of local orthogonal idempotents $\Longleftrightarrow$ there exists a decomposition of the identity element of A into a sum of primitive orthogonal idempotents $1 = \sum_{i=1}^{n} e_i$; all e_i in any such decomposition are local, and n is determined uniquely for all such decompositions.

All right or left serial rings and all local rings are semiperfect.

1.11 A ring A is a *right perfect* ring if A is a semilocal ring, and $J(A)$ is a left vanishing ideal.

A is a right perfect ring $\Longleftrightarrow$ A is ring with the minimum condition on principal left ideals $\Longleftrightarrow$ A is an orthogonally finite left semi-Artinian ring $\Longleftrightarrow$ A is a semiperfect left semi-Artinian ring $\Longleftrightarrow$ A is a semilocal ring, and every right A-module is semi-Noetherian.

1.3 Injective and finite-dimensional modules

1.12 Let N be a module. A module M is *injective with respect to N* (or *N-injective*) provided that for any submodule $\overline{N} \in \mathrm{Lat}(N)$, each homomorphism $\overline{N} \to M$ can be extended to a homomorphism $N \to M$.

A module M over a ring A is an *injective* module if M is injective with respect to everyy A-module.

A module M is *quasi-injective* (or *self-injective*) if M is an M-injective module.

M_A is an injective module $\Longleftrightarrow$ M is injective with respect to A_A.

If N is a module, then all direct summands and direct products of N-injective modules are N-injective. In particular, all direct summands and direct products of injective modules are injective.

1.13 An *injective hull* of a module M is any injective module which is an essential extension of M.

(1) Each module has at least one injective hull.

(2) Each injective module containing a module M contains at least one injective hull of M.

(3) If E_1 and E_2 are two injective hulls of a module M, then there exists an isomorphism $f : E_1 \to E_2$ which acts identically on M.

1.14 A submodule H of a module M is *closed* in M if H has no proper essential extensions within M.

Assume that N and H are two submodules of a module M, H is closed in M, and N is essential in H. Then we say that H is a *closure* of N in M.

For a module M, the following assertions hold.

(1) Every direct summand of M is closed in M.

(2) Every $N \in \mathrm{Lat}(M)$ has a closure in M.

(3) Let N_1 and N_2 be two submodules of M such that $N_1 \bigcap N_2 = 0$.

Then there exist two closed submodules M_1 and M_2 of M such that $M_1 \supseteq N_1$, $M_2 \supseteq N_2$, $M_1 \bigcap M_2 = 0$, M is an essential extension of $M_1 \oplus M_2$, and $M_1 \bigcap K \neq 0$ for any submodule K of M properly containing M_2.

1.15 A nonzero module M is *uniform* if any two nonzero submodules of M have nonzero intersection.

Any direct sum of uniform modules is called a *semiuniform* module.

M is a uniform module $\Longleftrightarrow$ any two nonzero cyclic submodules of M have nonzero intersection $\Longleftrightarrow$ every nonzero submodule of M is essential in M $\Longleftrightarrow$ every nonzero cyclic submodule of M is essential in M $\Longleftrightarrow$ M is a closure of any of its nonzero cyclic submodule $\Longleftrightarrow$ M has no nonzero proper closed submodules.

1.16 Let n be a positive integer. We say that M has *Goldie dimension* if M is an essential extension of the direct sum of n nonzero uniform submodules, and any direct sum N of nonzero uniform submodules such that N is essential in M has exactly n nonzero summands.

1.17 Let M be a nonzero module with the injective hull E. Then

M is finite-dimensional $\Longleftrightarrow$ M has finite Goldie dimension n $\Longleftrightarrow$ M is an essential extension of a finite-dimensional module $\Longleftrightarrow$ M is an essential extension of a finite direct sum of n nonzero uniform modules $\Longleftrightarrow$ M is an essential extension of a finitely generated module N which is a finite direct sum of n nonzero cyclic uniform modules $\Longleftrightarrow$ M contains no properly ascending infinite chains of closed submodules $\Longleftrightarrow$ M contains no properly descending infinite chains of closed submodules $\Longleftrightarrow$ M contains no infinite direct sum of closed submodules $\Longleftrightarrow$ E is a direct sum of n nonzero uniform modules $\Longleftrightarrow$ E is a finite direct sum of indecomposable modules.

1.18 A module M is *weakly finite-dimensional* if M contains no infinite direct sums of nonzero fully invariant closed submodules.

A module M is *weakly invariant* if each of its closed submodule is fully invariant in M.

(1) Let M be a weakly invariant module. Then

M is finite-dimensional $\Longleftrightarrow$ M is weakly finite-dimensional.

(2) A ring A is right weakly invariant $\Longleftrightarrow$ all closed right ideals of A are ideals of A.

(3) If a ring A is right weakly invariant and contains no infinite direct sums of nonzero ideals, then A is right finite-dimensional.

1.19 Let B and C be two ideals of a ring A, and let $B \subseteq C$. The ideal C is called a *bi-essential extension* of B if B has a nonzero intersection with any nonzero ideal of A which is contained in C.

An ideal B of a ring A is *bi-essential* if B has a nonzero intersection with any nonzero ideal of A (i.e. A is a bi-essential extension of B).

An ideal B of a ring A is called *bi-uniform* if any two nonzero ideals of A which are contained in B have a nonzero intersection.

It is verified directly that a ring A is prime $\Longleftrightarrow$ A is a semiprime bi-uniform ring.

1.20 Let B be a nonzero ideal of a semiprime ring A, B^* be the sum of all ideals of A which have the zero intersection with B, and let $h : A \to A/B^*$ be the natural ring epimorphism.

(1) $B \cap B^* = 0$, and $B^* = r(B) = \ell(B)$.

(2) The ideal B^* of A coincides with the sum B_1 of all right ideals of A which have the zero intersection with B.

Also, B^* coincides with the sum B_2 of all left ideals of A which have the zero intersection with B.

(3) $B \oplus B^*$ is an essential right ideal and an essential left ideal of A, B^*_A is a closed complement to B_A in A_A, and $_A B^*$ is a closed complement to $_A B$ in $_A A$.

(4) $h(B)$ is an essential right ideal and essential left ideal of $h(A)$.

(5) Let Z be an ideal of A such that $Z \supseteq B^*$ and $h(Z) = \mathrm{Sing}(h(A)_{h(A)})$.

Then $B \cap Z \subseteq \mathrm{Sing}(A_A)$.

(6) If A is right nonsingular, then A/B^* is right nonsingular.

(7) If B^* contains a prime ideal P of A, then $B^* = P$, B^* is a minimal prime ideal of A, and B is a bi-uniform ideal of A.

(8) If P is a prime ideal of A such that $P \cap B = 0$, then $P = B^*$, P is a minimal prime ideal of A, and B is a bi-uniform ideal of A.

(9) Let A be right nonsingular, and let P be a prime ideal of A such that $P \bigcap B = 0$.

Then the ideal P is nonsingularly prime, and P is a minimal prime ideal of A.

◁ (1) The right ideal $r(B)$ is an ideal, $(r(B)B)^2 = 0$, and A is semiprime. Therefore $r(B) \subseteq \ell(B)$. Analogously, $\ell(B) \subseteq r(B)$. Hence $r(B) = \ell(B)$. Since $(B \bigcap \ell(B))^2 = 0$, we have $B \bigcap \ell(B) = 0$ and $\ell(B) \subseteq B^* \subseteq B_1$. Let B_1 be a sum of right ideals C_i, where $B \bigcap C_i = 0$ for all i. Hence $C_i B \subseteq B \bigcap C_i = 0$ and $C_i \subseteq \ell(B)$ for all i. Therefore $B_1 \subseteq \ell(B)$. Hence $\ell(B) = B^* = B_1$. Analogously, $B^* = B_2$.

(2) and (3) follow from (1).

(4) Assume that a right ideal C of A properly contains B^*. Then $h(C) \neq 0$. Assume that $h(B) \bigcap h(C) = 0$. Then $h(CB) \subseteq h(B) \bigcap h(C) = 0$ and $CB \subseteq B^* = r(B)$. Therefore $BCB = 0$, whence $(BC)^2 = 0$. Hence $BC = 0$, $C \subseteq B^*$, and $h(C) = 0$; this is a contradiction. Therefore $h(B)$ is an essential right ideal of $h(A)$. Analogously, $h(B)$ is an essential left ideal.

(5) Let $b \in B \bigcap Z$, D be a right ideal of A such that $D \supseteq B^*$, and let $h(D) = h(B) \bigcap r_{h(A)}(h(b))$. Since $r_{h(A)}(h(b))$ and $h(B)$ are essential right ideals of $h(A)$, the right ideal $h(D)$ of $h(A)$ is essential. Hence D is an essential right ideal of A. Since $h(bD) = 0$, we have $bD \subseteq B \bigcap B^* = 0$. Therefore $r(b)$ is an essential right ideal of A, and $B \bigcap Z \subseteq \mathrm{Sing}(A_A)$.

(6) follows from (5).

(7) Assume that B^* properly contains the prime ideal P. Let $t : A \to A/P$ be the natural epimorphism. Then $t(B)$ and $t(B^*)$ are nonzero ideals of the prime ring $t(A)$. Also, $t(B)t(B^*) \subseteq t(B) \bigcap t(B^*) = 0$; this is a contradiction. Hence $B^* = P$. Therefore, the ideal B^* is prime and coincides with any prime ideal contained in B^*. Hence B^* is a minimal prime ideal. Since B^* is a prime ideal, B is a bi-uniform ideal.

(8) Since $P \bigcap B = 0$, we have $P \subseteq B^*$. Now our assertion follows from (7).

(9) follows from (6) and (8). ▷

1.21 Let A be a semiprime ring. By 1.20(1), the right and left annihilators of any ideal of A coincide. Therefore, any ideal of A which is a right (left) annihilator of some ideal B of A is called an annihilator ideal and is denoted by B^*. An annihilator ideal which is maximal among proper annihilator ideals is called a *maximal annihilator ideal*.

For a nonzero ideal B of a semiprime ring A, the following assertions hild.

(1) $(B^{**})_A$ is a closed essential extension of B_A, $_A(B^{**})$ is a closed essential extension of $_AB$, $B^{**} \cap B^* = 0$, and $B^{***} = B^*$.

(2) Every maximal annihilator ideal M of A is a minimal prime ideal of A.

(3) If B is a nonzero bi-uniform ideal of A, then B^* is a maximal annihilator ideal and a minimal prime ideal.

(4) If B is a bi-essential ideal of A, then B is an essential right ideal and an essential left ideal of A.

(5) Assume that B contains an ideal C of A. Then
C^* properly contains $B^* \iff B$ is not a bi-essential extension of C.

◁ (1) follows from 1.20(1),(2),(3).

(2) If A is prime, then 0 is the unique proper annihilator ideal and the unique minimal prime ideal. Assume that A is not prime, M is a maximal annihilator ideal, and $M = B^*$, where B is a nonzero ideal of A. By (1), we may assume that $B = B^{**}$. Assume that the ideal M is not prime. There exist two ideals L and N of A properly containing M such that $LN \subseteq M$. Hence $(BL)N = 0$, and N properly contains the maximal annihilator ideal M. Therefore $BL = 0$. Hence $L \subseteq M$; this is a contradiction. Therefore M is prime. By 1.20(7), M is a minimal prime ideal.

(3) By (2), it is sufficient to prove that B^* is a maximal annihilator ideal. Assume the contrary. Hence there exist two nonzero ideals C and N of A such that N properly contains B^* and $C \subseteq N^* \subseteq B^{**}$. Set $B_1 \equiv N \cap B$. By 1.20(3), $B_1 \neq 0$. Set $C_1 \equiv C \cap B$. By (1), $C_1 \neq 0$. Since B is a bi-uniform ideal, we have $B_1 \cap C_1 \neq 0$. In addition, $(B_1 \cap C_1)^2 \subseteq CN = 0$. Since A is semiprime, $B_1 \cap C_1 = 0$; this is a contradicton.

(4) follows from 1.20(3) and from the fact that $B^* = 0$.

(5) Assume that C^* properly contains B^*. By 1.20(3), $C^* \cap B \neq 0$. Assume that B is a bi-essential extension of C. Then $0 \neq C \cap C^* \cap B$ and $(C \cap C^* \cap B)^2 = 0$; this is a contradiction. Assume that B is not a bi-essential extension of C. There exists a nonzero ideal B_1 of A such that $B_1 \subseteq B$ and $C \cap B_1 = 0$. Hence $BB_1 \neq 0$ and $CB_1 = 0$. Therefore C^* properly contains B^*. ▷

1.22 For a semiprime ring A, the following conditions are equivalent.

(1) A contains no infinite direct sums of nonzero ideals.

(2) A is a ring with the maximum condition on annihilator ideals.

(3) A is a ring with the minimum condition on annihilator ideals.

(4) A is a bi-essential extension of a finite direct sum B of nonzero bi-uniform ideals $B_1, \ldots, B_n$.

(5) A is a finite subdirect product of prime rings.

(6) There exist minimal prime ideals $P_1, \ldots, P_n$ of A such that $P_1 \cap \ldots \cap P_n = 0$, and any intersection of less than n ideals P_i is not equal to zero.

$\triangleleft$ $(1) \Longleftrightarrow (2) \Longleftrightarrow (3)$ follow from 1.21(5). $(1) \Longleftrightarrow (4)$ and $(5) \Longleftrightarrow (6)$ are directly verified.

$(4) \Longrightarrow (5)$ By 1.19, we may assume that $n > 1$. Set $P_i \equiv B_i^*$. By 1.21(3), all P_i are minimal prime ideals. Set $P \equiv P_1 \cap \ldots \cap P_n \subseteq B^*$. If $P = 0$, then our assertion is proved. Assume that $P \neq 0$. Then $0 \neq B \cap P \subseteq B \cap B^*$; this is a contradiction.

$(6) \Longrightarrow (4)$ By 1.19, we may assume that $n > 1$. Let $B_i \equiv \bigcap_{j \neq i} P_j$, $C_i \equiv \sum_{j \neq i} B_j$, and let $B \equiv \sum_{i=1}^{n} B_i$. By assumption, all B_i are nonzero ideals and $B_i \cap P_i = 0$ for all i. Therefore $B_i \cap B_j = 0$ for $i \neq j$. Hence $B_i B_j = 0$ for $i \neq j$. Therefore $B_i C_i = 0$ for all i. Hence $(B_i \cap C_i)^2 = 0$ for all i. Therefore $B_i \cap C_i = 0$ for any i. Hence $B = \oplus_{i=1}^{n} B_i$. All P_i are prime ideals, all B_i are nonzero ideals, and $B_i \cap P_i = 0$ for all i. Therefore, the ideal B_i is bi-uniform for all i, and the ideal $P_i \oplus B_i$ is bi-essential. Hence $S \equiv \bigcap_{i=1}^{n} (P_i \oplus B_i)$ is a bi-essential ideal.

Assume that the ideal B is not bi-essential. There exists a nonzero ideal D of A such that $B \cap D = 0$. Then $D B_i = 0$ for all i. Hence $D(P_i \oplus B_i) = D P_i$ for all i. Since S is a bi-essential ideal of the semiprime ring A, we have $DS \neq 0$. Therefore $0 \neq (DS)^n \subseteq \prod_{i=1}^{n} D(P_i \oplus B_i) = \prod_{i=1}^{n} D P_i \subseteq P_1 \cap \ldots \cap P_n = 0$; this is a contradiction. $\triangleright$

1.23 (1) If $X, Y_1, \ldots, Y_t$ are ideals of a ring A with $A = X + Y_i$ for all i, then $A = X + Y_1 \cdot \ldots \cdot Y_t = X + \bigcap_{i=1}^{t} Y_i$.

(2) If $P_1, \ldots, P_n$ are ideals of a ring A such that $P_1 \cap \ldots \cap P_n = 0$ and $P_i + P_j = A$ for all $i \neq j$, then $A \cong \prod_{i=1}^{n} A/P_i$.

1.24 Let A be a semiprime ring which contains no infinite direct sums of nonzero ideals, and let the sum of any two distinct minimal prime ideals of A be equal to A.

Then A is a finite direct product of prime rings.

$\triangleleft$ This follows from 1.22 and 1.23(2). $\triangleright$

1.4   Nonsingular rings and modules

1.25 Let B and D be two rings. A (B, D)-*bimodule* is an Abelian group H equipped with a left B-module structure and a right D-module structure such that $(bh)d = b(hd)$ for any $b \in B$, $h \in H$, and $d \in D$. The symbol $_BH_D$ is used to denote this situation.

Let M_A be a right module over a ring A. Denote by Z the ring of integers.

(1) M is an $(\mathrm{End}(M), A)$-bimodule and a (Z, A)-bimodule.

(2) Every left A-module N is an $(A, \mathrm{End}(N))$-bimodule and an (A, Z)-bimodule.

(3) For each module L_A, the group $\mathrm{Hom}(L_A, M_A)$ can be naturally considered as an $(\mathrm{End}(M), \mathrm{End}(L))$-bimodule.

(4) For each $m \in M_A$, $\overline{m}$ denotes the homomorphism $A_A \to M$ such that $\overline{m}(a) = ma$ for all $a \in A$.

The map $m \to \overline{m}$ is a bimodule isomorphism $_{\mathrm{End}(M)}M_A \to {}_{\mathrm{End}(M)}\mathrm{Hom}(A_A, M)_A$.

(5) For each left module $_AN$, there exists a bimodule isomorphism $_AN_{\mathrm{End}(N)} \to {}_A\mathrm{Hom}(_AA, N)_{\mathrm{End}(M)}$ which similar to the isomorphism in (4).

1.26 Let M be an essential extension of a module N.

(1) If E is an essential extension of M, then E is an essential extension of N.

(2) Any submodule T of M is an essential extension of $T \bigcap N$.

(3) N contains every simple submodule T of M.

(4) If $f : Q \to M$ is a module homomorphism, then Q is an essential extension of $f^{-1}(N)$.

◁ (1), (2), and (3) are directly verified.

(4) Let $R \equiv f^{-1}(N)$, and let G be a submodule of Q such that $R \bigcap G = 0$. Then $f(G) \bigcap N = 0$. Therefore $f(G) = 0$ and $G \subseteq \mathrm{Ker}(f) \subseteq R$. Hence $G = G \bigcap R = 0$. ▷

1.27 Let L_A and M_A be two modules over a ring A, and let T be the subset of the group $\mathrm{Hom}(L_A, M_A)$ consisting of all homomorphisms t such that $\mathrm{Ker}(t)$ is essential in L.

(1) T is a sub-bimodule of the $(\mathrm{End}(M), \mathrm{End}(L))$-bimodule $\mathrm{Hom}(L_A, M_A)$.

(2) The set $\sum_{t \in T} t(L)$ is a fully invariant submodule in M_A.

◁ (1) Assume that $t, u \in T$, $f \in \mathrm{End}(L)$, and $g \in \mathrm{End}(M)$. It is sufficient to prove that $t + u \in T$, $gt \in T$, and $tf \in T$. The inclusion $t + u \in T$ follows from 1.26(2), 1.26(1), and from the fact that $\mathrm{Ker}(t + u) \supseteq \mathrm{Ker}(t) \bigcap \mathrm{Ker}(u)$. Since $\mathrm{Ker}(gt) \supseteq \mathrm{Ker}(t)$, the inclusion $gt \in T$ follows from 1.26(1). Since L is an essential extension of $\mathrm{Ker}(t)$, 1.26(4) shows us that L is an essential extension of $f^{-1}(\mathrm{Ker}(t))$. Since $\mathrm{Ker}(tf) \supseteq f^{-1}(\mathrm{Ker}(t))$, we obtain $tf \in T$, and the proof is completed.

(2) By (1), $gT \subseteq T$ for any $g \in \text{End}(M)$. $\triangleright$

1.28 Let M be a right module over a ring A. The set of all endomorphisms φ of M such that $\text{Ker}(\varphi)$ is an essential submodule of M is denoted by $\text{sg}(M)$.

(1) The set $\text{sg}(M)$ is an ideal of $\text{End}(M)$.

(2) The set $\text{Sing}(M)$ is a fully invariant submodule of M which is called the *singular submodule* of M.

(3) $\sum_{t \in \text{sg}(M)} t(M)$ is a fully invariant submodule of M.

(4) The natural isomorphism $\text{End}(A_A) \to A$ maps from the set $\text{sg}(A_A)$ onto $\text{Sing}(A_A)$.

(5) $\text{Sing}(A_A)$ and $\text{Sing}(_A A)$ are ideals of A.

These ideals are called the *right singular ideal* and the *left singular ideal* of A, respectively.

$\triangleleft$ (1) follows from 1.27(1) ($L = M$). (2) follows from 1.27(1) ($L = A_A$) and 1.25(4). (3) follows from (1). (4) is directly verified. (5) follows from (3) and (4). $\triangleright$

1.29 Let M_A be an essential extension of a module N_A. Then the module M/N is singular.

In addition, if $g : M \to L$ is a module homomorphism and $g(N) = 0$, then $g(M) \subseteq \text{Sing}(L)$.

1.30 For a submodule N of M, the following assertions hold.

(1) If M/N is nonsingular, then N is closed in M.

(2) Let M be nonsingular. Then

M/N is nonsingular $\Longleftrightarrow$ N is closed in M.

(3) If N is the kernel of a homomorphism from M into a nonsingular module, then N is closed in M.

1.31 For a closed submodule L of M, the following assertions hold.

(1) Every closed submodule K of L is closed in M.

(2) Let N be a closed submodule of M such that N is a proper submodule of L.

Then for any submodule T of L such that $N \bigcap T = 0$ (in particular, for $T = 0$), there exists a nonzero closed submodule K of M such that $T \subseteq K \subseteq L$, $K \bigcap N = 0$, and L is an essential extension of $N \oplus K$.

$\triangleleft$ (1) There exist a closed submodule K_1 of L and a closed submodule L_1 of M such that $K \bigcap K_1 = L \bigcap L_1 = 0$, and for any modules K_2 and L_2 such that $K \subseteq K_2 \subseteq L, L \subseteq L_2 \subseteq M, K_1 \bigcap K_2 = L_1 \bigcap L_2 = 0$, we have $K = K_2$ and $L = L_2$. Obviously, $K \bigcap (K_1 \oplus L_1) = 0$. There exists a closed submodule C of M such that $K \subseteq C$ and $C \bigcap (K_1 \oplus L_1) = 0$.

In addition, if $C \subseteq C_1 \subseteq M$ and $C_1 \bigcap (K_1 \oplus L_1) = 0$, then $C = C_1$. Set $D \equiv L \bigcap (C \oplus L_1)$. Then $K \subseteq D \subseteq L$ and $D \bigcap K_1 = 0$. Therefore $K = D$, whence $(C + L) \bigcap L_1 = 0$. Hence $C + L = L$, $C \subseteq L$, and $C = K$ is closed in M.

(2) Since $N \neq L$ and N is closed in M, L is not an essential extension of N. Let K be a complement to N in L. Then K is closed in L. By (1), K is closed in M. ▷

1.32 Assume that a module M_A has a submodule $N = \oplus_{i \in I} N_i$, and each module N_i is an essential submodule of a submodule H_i of M.

(1) $\sum_{i \in L} H_i$ is an essential extension of $\oplus_{i \in L} N_i$ for each subset L of the set I.

(2) $\sum_{i \in I} H_i = \oplus_{i \in I} H_i$.

◁ (1) Let $0 \neq h \in \sum_{i \in L} H_i$. There exists a finite set $T \subseteq L$ such that $h = \sum_{t \in T} h_t$, where $0 \neq h_t \in H_t$. Let $T = \{1, \ldots, m\}$. It is sufficient to prove that there exists $a \in A$ such that $0 \neq ha \in \sum_{i=1}^{m} N_i$. We use the induction on m. For $m = 1$, our assertion follows from the fact that H_1 is an essential extension of N_1. Assume that the assertion holds for $m - 1$. Let $G \equiv \sum_{i=1}^{m-1} N_i$, and let $f \equiv \sum_{i=1}^{m-1} h_i$. Then $h = f + h_m$. If $f = 0$, then $h = h_m$, and we can use the fact that H_m is an essential extension of N_m. Assume that $f \neq 0$. By induction, there exists $b \in A$ such that $0 \neq fb \in G$. If $h_m b = 0$, then $0 \neq fb = hb \in G \subseteq N$, and our assertion is proved. Assume that $h_m b \neq 0$. Since H_m is an essential extension of N_m, there exists $d \in A$ such that $0 \neq h_m bd \in N_m$. Therefore $hbd = fbd + h_m bd \in G + N_m$ and $h_m bd \neq 0$ (otherwise $0 \neq h_m bd = -fbd \in G \bigcap N_m$; this is a contradiction). Now we can set $a \equiv bd$.

(2) Let $0 = \sum_{t \in T} h_t$, where T is a finite subset of the set I and $h_t \in H_t$. It is sufficient to prove that $h_t = 0$ for each $t \in T$. Assume that $T = \{1, \ldots, m\}$. We use the induction on m. Our assertion holds for $m = 1$. Assume that our assertion holds for $m - 1$. Let $h_1 \neq 0$. Set $f \equiv \sum_{i=1}^{m-1} h_i$. Then $f \neq 0$, since otherwise $h_1 = 0$ by induction. By 1.32(1), $\sum_{i=1}^{m-1} H_i$ is an essential extension of $G \equiv \sum_{i=1}^{m-1} N_i$. Hence $0 \neq fa \in G$ for some $a \in A$. Since $-f = h_m$, we obtain $0 \neq -fa = h_m a \in H_m \bigcap G$. Hence $N_m \bigcap G \neq 0$, since H_m is an essential extension of N_m. Therefore $h_1 = 0$. Analogously, $0 = h_2 = \ldots h_m$. ▷

1.33 A ring A is a *commutative at zero* ring if $aAb = bAa = 0$ for any $a, b \in A$ such that $ab = 0$.

Every commutative ring is commutative at zero.

(1) Every reduced ring A is commutative at zero.

(2) Assume that $a_1, \ldots, a_n$ are elements of a commutative at zero ring A with $a_1 \cdot \ldots \cdot a_n = 0$.

Then $a_{s(1)} \cdot \ldots \cdot a_{s(n)} = 0$ for all permutations s of $1, \ldots, n$.

◁ (1) Let $ab = 0$ for $a, b \in A$. Then $(bAa)^2 = 0$, whence $bAa = 0$. Hence $(aAb)^2 = 0$. Therefore $aAb = 0$.

(2) Set $b \equiv a_{s(1)} \cdot \ldots \cdot a_{s(n)}$. We can choose b from $a_1 \cdot \ldots \cdot a_n$ by a finite number of steps changing two neighbouring elements in each step. Since A is commutative at zero, $b = 0$. ▷

1.34 (1) A module M_A is Rickartian $\Longleftrightarrow$ for any $m \in M$, the right ideal $r(m)$ of A is generated by an idempotent.

(2) Every Rickartian module is nonsingular.

(3) Each domain is a Rickartian ring.

(4) If A is a subring of a right Rickartian ring Q and A contains all idempotents of Q, then A is right Rickartian.

(5) If M is a finite-dimensional module which is a direct sum of semihereditary modules $M_1, \ldots, M_n$, then M is a semihereditary module.

◁ (1) follows from the fact that $mA \cong A/r(m)$. (2) follows from (1) and from the fact that any proper right ideal generated by an idempotent cannot be essential. (3) follows from (1).

(4) If $a \in A$, then $r_Q(a) = eA$, where $e = e^2 \in Q$. By assumption, $e \in A$, whence $r_A(a) = er_A(a) = e(A \bigcap r_Q(a)) = eA$.

(5) Let $h_i : M \to M_i$ be the natural projections, and let N be a nonzero finitely generated submodule of M. By assumption, all modules $h_i(N)$ are projective, whence $N = N_i \oplus N \bigcap (\oplus_{j \neq i} M_i)$, where $N_i \cong h_i(N)$ $(i = 1, \ldots, n)$. (5) can be verified now using the induction on n. ▷

1.35 For a reduced ring A, the following assertions hold.

(1) A is normal, the right annihilator $r(B)$ of any subset $B \subseteq A$ coincides with the left annihilator of B, and $r(B)$ is an ideal of A.

(2) If $a \in A$, then

$$r(a) = \ell(a) = r(\mathrm{Id}(a)) = \ell(\mathrm{Id}(a)) = \{b \in A \mid \mathrm{Id}(a) \bigcap \mathrm{Id}(b) = 0\}$$

is an ideal of A with $I \bigcap \mathrm{Id}(a) = 0$.

(3) A is a nonsingular ring.

(4) $r(a) = r(a^n)$ for all $a \in A$ and each positive integer n.

(5) Let T be a multiplicative set in A, and let $K(T) \equiv \{a \in A \mid at = 0$ for some $t \in T\}$.

Then $K(T)$ is an ideal of A, and $K(T) \bigcap T = 0$.

(6) Let T be a maximal multiplicative set in A.

Then $A \setminus T = \{a \in A \mid at = 0 \text{ for some } t \in T\}$ is a minimal completely prime ideal of A.

(7) Every minimal prime ideal N of A is completely prime, and $N = \{a \in A \mid at = 0 \text{ for some } t \in A \setminus N\}$.

(8) Every maximal right or left ideal M of A contains a completely prime ideal P which is a minimal prime ideal.

(9) If $B \bigcap D \neq 0$ for any two nonzero ideals B and D of A, then A is a domain.

Consequently, every reduced prime ring is a domain.

(10) If A is right uniform, then A is a right Ore domain.

(11) If A is right Rickartian, then A is left Rickartian, and each of its elements is a product of a central idempotent and a regular element.

(12) Assume that A contains no infinite direct sums of nonzero ideals, and a sum of any two distinct minimal completely prime ideals of A be equal to A.

Then A is a finite direct product of domains.

(13) A is a subdirect product of domains.

$\triangleleft$ (1) and (2) follow from 1.33(1). (3) follows from (2).

(4) If $a^n b = 0$, then $a^n b^n = 0$. By 1.33, $(ab)^n = 0$. Therefore $ab = 0$, whence $r(a^n) \subseteq r(a)$.

(5) follows from (2).

(6) Let $N \equiv A \setminus T$, and let $M \equiv \{a \in A \mid at = 0 \text{ for some } t \in T\}$. By (5), M is an ideal of A, and $M \subseteq N$. We have to prove that $M = N$. Assume the contrary. Let $a \in N \bigcap M$, and let D be the multiplicative subset of A generated by a and T. Since T is a maximal multiplicative set in A, we obtain $0 \in D$. There exist $d_1, \ldots, d_n \in D$ with $d_i = a$ for some i such that $d_1 \cdot \ldots \cdot d_n = 0$. Assume that there are precisely m occurences of a among d_i $(i = 1, \ldots n)$. Define an element t as follows. If $d_i = a$ for each i, then we set $t = 1$. Otherwise, let t be the product of all elements d_i such that $d_i \neq a$. Obviously, $t \in T$. By 1.33(1), A is commutative at zero. By 1.33, $a^m t = 0$, and therefore $t \in r(a^m)$. By (4), $r(a) = r(a^m)$, and we obtain $at = 0$. But then $a \in M$; this is a contradiction. Thus, $M = N$. Obviously, M is completely prime. Since T is a maximal multiplicative set, we conclude that M is a minimal completely prime ideal.

(7) Assume that S is a subset of A such that S is closed under multiplication, and S is generated by $A \setminus N$. If $0 \in S$, then it follows from (6) that there exist $a_1, \ldots, a_n \in A \setminus N$ such that $\mathrm{Id}(a_1) \cdot \ldots \cdot \mathrm{Id}(a_n) = 0$; a contradiction. Hence S is a multiplicative set. Then S is contained in a maximal multiplicative set T. The set $A \setminus T$ is contained in the minimal prime ideal N. By (6), $A \setminus T$ is a completely

prime ideal, and therefore $N = A \setminus T$ is completely prime. By (6), $N = A \setminus T = \{a \in A \mid at = 0 \text{ for some } t \in T\}$.

(8) Every right (left) primitive ideal is a prime ideal. Since M contains a right (left) primitive ideal, M contains a minimal prime ideal P. By (7), P is completely prime.

(9) follows from (2). (10) follows from (9).

(11) Let $a \in A$, and let e be an idempotent such that $r(a) = eA$. By 1.35(2), e is central, $\ell(a) = eA = Ae$, and therefore A is left Rickartian. To prove the second assertion, set $b \equiv (1 - e)a + e$. Since $(1 - e)a$ is a regular element in the ring $(1 - e)A$, the element b is regular in A. Therefore $a = (1 - e)b$.

(12) follows from (7), (9), and 1.24.

(13) Let $\{N_i\}_{i \in I}$ be the set of all minimal prime ideals of A. Since A is semiprime, $\bigcap_{i \in I} N_i = 0$. By (6), every N_i is completely prime, whence A is a subdirect product of domains A/N_i. $\triangleright$

1.5   Rings with maximum conditions

1.36 A ring A is a ring with the *maximum (minimum) condition on right annihilators* if A contains no a properly ascending (resp. descending) chain of right (left) ideals each of which equals to the right (left) annihilator of some subset of A.

(1) If B is a subset of a ring A, then $r_A(B) = r_A(\ell_A(r_A(B)))$ and $\ell_A(B) = \ell_A(r_A(\ell_A(B)))$.

If A is a subring of a ring Q, then $r_A(B) = A \bigcap r_Q(B)$.

(2) A is a ring with the maximum condition on right annihilators $\Longleftrightarrow$ A is a ring with the minimum condition on left annihilators $\Longleftrightarrow$ A is a subring of a ring with the maximum condition on right annihilators $\Longleftrightarrow$ every subset B of A contains a finite subset $\{b_1, \ldots, b_n\}$ such that $\ell(B) = \ell(\{b_1, \ldots, b_n\}) = \bigcap_{i=1}^{n} \ell(b_i)$.

(3) A is a ring with the maximum condition on left annihilators $\Longleftrightarrow$ every subset B of A contains a finite subset $\{b_1, \ldots, b_n\}$ such that $r(B) = r(\{b_1, \ldots, b_n\}) = \bigcap_{i=1}^{n} r(b_i)$.

1.37 For a ring A with the maximum condition on right annihilators, the following assertions hold.

(1) The ideal $\mathrm{Sing}(A_A)$ is nilpotent.

(2) $\ell(\mathrm{Sing}(_A A))$ is an essential left ideal of A, and $(\mathrm{Sing}(_A A) \bigcap \ell(\mathrm{Sing}(_A A)))^2 = 0$.

(3) If A is semiprime, then A is right and left nonsingular and has no nonzero right or left nil-ideals.

(4) If e is an idempotent of A, then eAe is a ring with the maximum condition on right annihilators.

(5) If B is an ideal of A such that $r(B)$ is an essential right ideal, then B is nilpotent.

$\lhd$ Let $M \equiv \mathrm{Sing}(A_A)$, and let $H \equiv \mathrm{Sing}(_A A)$.

(1) Since $r(M^i) \subseteq r(M^{i+1})$ for each i and A is a ring with the maximum condition on right annihilators, $r(M^n) = r(M^{n+1})$ for some n. Let us prove that $M^{n+1} = 0$. Assume the contrary. We can choose $a \in M$ such that $M^n a \neq 0$, and a has a maximal right annihilator in the set of all elements $m \in M$ such that $M^n m \neq 0$. Let $b \in M$. Since $r(b)$ is an essential right ideal, $r(b) \bigcap aA \neq 0$. Therefore, there exists $t \in A$ such that $at \neq 0$ and $bat = 0$. Then $r(ba) \not\subseteq r(a) \subseteq r(ba)$. By the choice of a, we obtain $M^n ba = 0$. Then $M^{n+1}a = 0$, and therefore $a \in r(M^{n+1}) = r(M^n)$; this is a contradiction. Hence M is nilpotent.

(2) First, we note that $(H \bigcap \ell(H))^2 = 0$. By 1.36(3), there exist $h_1, \ldots, h_m \in H$ such that $\ell(H) = \bigcap_{i=1}^n \ell(h_i)$. Then $\ell(H)$ is an essential left ideal (as the intersection of a finite number of essential left ideals).

(3) By (1), $M = 0$. By (2), $(H \bigcap \ell(H))^2 = 0$, whence $H \bigcap \ell(H) = 0$. Also, $\ell(H)$ is an essential left ideal. Hence $H = 0$. Assume that A contains a nonzero left nil-ideal B. Let $a \in B$ ($a \neq 0$). Choose $b \in Aa$ such that $r(b)$ is maximal among right annihilators of elements of Aa. Since A is semiprime, there is $d \in A$ such that $bdb \neq 0$. Then $db \neq 0$, and there exists a positive integer n such that $(db)^n \neq 0$ and $(db)^{n+1} = 0$. Since $r(b) \subseteq r((db)^n)$, it follows from the maximality of $r(b)$ that $r(b) = r((db)^n)$. Hence $db \in r((db)^n) = r(b)$, whence $bdb = 0$; this is a contradiction. Assume that there exists a nonzero right nil-ideal T of A. Consider nonzero $t \in T$ and $c \in A$. Then there is n such that $(tc)^n = 0$. Therefore $(ct)^{n+1} = c(tc)^n t = 0$, whence At is a left nil-ideal. According to what we have said above, $At = 0$, whence $t = 0$; this is a contradiction.

(4) follows from the fact that $r_{eAe}(B) = eAe \bigcap r_A(B)$ for any subset B of eAe.

(5) Since $r(B)$ is an essential right ideal, $\ell(r(B)) \subseteq M$. By (1), the ideal M is nilpotent. Since $B \subseteq \ell(r(B)) \subseteq M$, B is nilpotent. $\rhd$

1.38 Let M_A be a quasi-injective module.

(1) $\mathrm{End}(M)/J(\mathrm{End}(M))$ is a regular ring, and $J(\mathrm{End}(M)) = \mathrm{sg}(M)$.

(2) All idempotents of the ring $\mathrm{End}(M)/\mathrm{sg}(M)$ can be lifted to idempotents of $\mathrm{End}(M)$.

(3) If $\mathrm{End}(M)$ is orthogonally finite, then $\mathrm{End}(M)$ is semiperfect.

◁ (1) and (2) are proved in [236, 22.1]. (3) follows from (1) and (2). ▷

1.39 Let E be the injective hull of a module M. For every $f \in \mathrm{End}(M)$, let $\overline{f}$ be an endomorphism of E such that f is the restriction of $\overline{f}$ to M ($\overline{f}$ is not necessarily unique).

(1) The ring $\mathrm{End}(E)/\mathrm{sg}(E)$ is regular, and by the rule $g(f + \mathrm{sg}(M)) = \overline{f} + \mathrm{sg}(E)$, the ring monomorphism $\mathrm{End}(M)/\mathrm{sg}(M) \to \mathrm{End}(E)/\mathrm{sg}(E)$ is well defined.

(2) If M is finite-dimensional, then $\mathrm{End}(M)/\mathrm{sg}(M)$ is a ring with the maximum and minimum conditions on right annihilators and on left annihilators, $\mathrm{End}(E)/\mathrm{sg}(E)$ is a semisimple ring, and $J(\mathrm{End}(E)) = \mathrm{sg}(E)$.

◁ (1) Let $R \equiv \mathrm{End}(M)/\mathrm{sg}(M)$, and let $S \equiv \mathrm{End}(E)/\mathrm{sg}(E)$. By 1.38(1), $J(\mathrm{End}(E)) = \mathrm{sg}(E)$, and S is regular. Let us verify that the homomorphism g is well defined. Let $f_1, f_2 \in \mathrm{End}(M)$ such that $f_1 - f_2 \in \mathrm{sg}(M)$. Then M is an essential extension of $\mathrm{Ker}(f_1 - f_2)$. Since M is essential in $\overline{M}$, the module $\overline{M}$ is an essential extension of $\mathrm{Ker}(f_1 - f_2)$. Therefore $\overline{M}$ is an essential extension of $\mathrm{Ker}(\overline{f}_1 - \overline{f}_2)$, whence $\overline{f}_1 - \overline{f}_2 \in \mathrm{sg}(\overline{M})$. It can directly be verified that g is a ring monomorphism.

(2) By 1.38(3), S is semisimple. In particular, S is a ring with the maximum and minimum conditions on right annihilators and on left annihilators. By (1), R is isomorphic to a subring of S. (2) follows now from 1.36(2). ▷

1.40 Let A be a right finite-dimensional ring.

(1) $A/\mathrm{Sing}(A_A)$ is a ring with the maximum and minimum conditions on right annihilators and on left annihilators.

(2) If a is a right regular element of A, then aA is an essential right ideal of A.

(3) If a is a right regular element of A and A is right nonsingular, then a is a regular element in A.

(4) If A is a domain, then A is right uniform.

◁ (1) follows from 1.39(2) (for $M = A$).

(2) Let B be a right ideal of A such that $B \bigcap aA = 0$. Let us prove that for every n, the sum of right ideals $B, aB, \dots, a^n B$ is direct. Assume that $\sum_{i=0}^{n} a^i b_i = 0$, where $b_0, \dots, b_n \in B$. Then $b_0 = -a \sum_{i=1}^{n} a^{i-1} b_i \in B \bigcap aA = 0$, since $\sum_{i=1}^{n} a^{i-1} b_i \in r(a) = 0$. Analogously, $b_1 = \dots = b_n = 0$. Therefore, the sum of $B, \dots, a^n B$ is direct. By 1.17, this contradicts to the assumption. Thus, $B = 0$, whence aA is an essential right ideal.

(3) By (2), aA is an essential right ideal, and therefore $\ell(a) \subseteq \mathrm{Sing}(A_A) = 0$.

(4) Note that in this case, (2) shows us that every nonzero right ideal of A is essential. $\triangleright$

1.41 If B is an ideal of a right Noetherian ring A such that $\ell(B)$ is an essential left ideal of A, then the ideal B is nilpotent.

$\triangleleft$ Since A is a ring with the maximum condition on ideals, there exists a positive integer n such that $\ell(B^n) = \ell(B^{n+i})$ for any positive integer i. Since $\ell(B) \subseteq \ell(B^n)$, $\ell(B^n)$ is an essential left ideal. Assume that the ideal B is not nilpotent. Then $\ell(B^n) \neq A$. Set $R \equiv A/\ell(B^n)$. Since R is a right Noetherian nonzero ring, there exists $x \in A$ such that $r_R(x + \ell(B^n))$ is maximal in the set of right ideals of R which are right annihilators of nonzero elements of R. If $xB^n \subseteq \ell(B^n)$, then $x \in \ell(B^{2n}) = \ell(B^n)$ and $x + \ell(B^n) = 0$; this is a contradiction. Therefore, there exists $b \in B^n$ such that $0 \neq xb \notin \ell(B^n)$. Since $\ell(B^n)$ is an essential left ideal, there exists $y \in A$ such that $0 \neq yxb \in \ell(B^n)$. Since $yxb \in \ell(B^n)$ and $xb \notin \ell(B^n)$, then $r(x + \ell(B^n))$ is properly contained in $r(yx + \ell(B^n))$. In addition, $yx + \ell(B^n) \neq 0$, since otherwise $yxb = 0$; this is a contradiction to the choice of x. $\triangleright$

1.42 If A is a right Noetherian ring, then ideals $\mathrm{Sing}(A_A)$ and $\mathrm{Sing}(_AA)$ are nilpotent.

$\triangleleft$ By 1.37(1), the ideal $\mathrm{Sing}(A_A)$ is nilpotent. Set $B \equiv \mathrm{Sing}(_AA)$. By 1.37(2), $\ell(B)$ is an essential left ideal of A. By 1.41, the ideal B is nilpotent. $\triangleright$

1.6 Classical localizations

1.43 Let T be a set of elements in a ring A. The set T is *right permutable* if for any $a \in A$ and $t \in T$, there exist $b \in A$ and $u \in T$ such that $au = tb$. The set T is *right reversible* (resp. *weakly right reversible*) if given any $a \in A$ and $t \in T$ with $ta = 0$ (resp. with $a^2 = 0$ and $ta = 0$), there exists $u \in T$ such that $au = 0$. If I is any proper ideal in A, then the set of all elements $a \in A$ such that $a + I$ is a regular element of A/I is denoted by $c(I)$. In particular, $c(0)$ is the set of all regular elements of A.

For a subset T of a ring A, the following assertions hold.

(1) If tA is an ideal of A for any $t \in T$, then T is right permutable.

(2) If A is right invariant, then every subset of A is right permutable.

(3) Every central subset of A is right permutable.

(4) Every central subset of A is reversible.

In particular, all subsets of commutative rings are reversible.

(5) Every subset of a reduced ring is reversible.

(6) If each square-zero element $a \in A$ commutes with every element of T, then the set T is weakly reversible.

(7) For each proper ideal P of A, the set $c(P)$ is multiplicative.

(8) The intersection of any chain of prime (completely prime) ideals is a prime (completely prime) ideal.

(9) Every prime (completely prime) ideal contains a minimal prime (minimal completely prime) ideal.

(10) Every right or left primitive ideal is prime.

(11) If a maximal right ideal M of A is an ideal, then A/M is a division ring, and M is a completely prime ideal.

1.44 A multiplicative subset T in a ring A is a *right Ore set* if the following two equivalent conditions hold.

(i)   There exists a ring A_T containing A as a unitary subring such that $T \subseteq U(A_T)$ and $A_T = \{at^{-1} \mid a \in A, t \in T\}$.

(ii) T is a right permutable set, and all elements of T are regular in A.

In this case, A_T is called the *right ring of quotients of A with respect to T*. A *right Ore* ring is any ring A in which $c(0)$ is a right Ore set. In this case, the right ring of quotients of A with respect to the set $c(0)$ of all regular elements is called the *classical right ring of quotients for A* and is denoted by $Q_{\mathrm{cl}}(A)$. The ring A is also called a *right order in $Q_{\mathrm{cl}}(A)$*. The *left ring of quotients $_T A$ for A with respect to a left Ore set T* and the *classical left ring of quotients $_{\mathrm{cl}}Q(A)$* are defined symmetrically.

A multiplicative subset T of a ring A is a *two-sided Ore set* in A if the following equivalent conditions hold.

(1) There exists a ring $_T A_T$ containing A as a unitary subring such that $T \subseteq U(_T A_T)$ and $_T A_T = \{at^{-1} \mid a \in A, t \in T\} = \{t^{-1}a \mid a \in A, t \in T\}$.

(2) T is a right and left permutable set such that every element of T is regular in A.

If these conditions hold, then $_T A_T$ is the *two-sided ring of quotients of A with respect to T*.

If $c(0)$ is a two-sided Ore set in A, then A is called a *two-sided Ore ring*. In this case, $_{c(0)}A_{c(0)}$ is called the *classical (two-sided) ring of quotients of A* and is denoted by $_{\mathrm{cl}}Q_{\mathrm{cl}}(A)$. Equivalently, A is an *order in $_{\mathrm{cl}}Q_{\mathrm{cl}}(A)$*.

1.45 A is a right Ore domain $\Longleftrightarrow$ A is a right order in a division ring $\Longleftrightarrow$ A is a right uniform domain.

1.46 Let T be a right Ore set in a ring A, and let Q be the right ring of quotients of A with respect to T. For each right ideal B of A, let us denote by BQ the right ideal of Q generated by B.

(1) Let $0 \neq q = at^{-1} \in Q$, where $0 \neq a \in A$ and $t \in T$.

Then $0 \neq a = qt \in A \cap qA$, and Q_A is an essential extension of A_A.

(2) Given any $q_1, \ldots, q_n \in Q$, there exist $t \in T$ and $a_1, \ldots, a_n \in A$ with $q_i = a_i t^{-1}$ for all $i = 1, \ldots, n$.

(3) If N is a finitely generated submodule of the left module $_AQ$, then there exists $t \in T$ such that $Nt \subseteq A$.

Consequently, the left module $_AN$ is isomorphic to the finitely generated left ideal Nt of A.

(4) If B be a right ideal of A, then $BQ = \{bt^{-1} | b \in B, t \in T\}$.

Consequently, if $q \in BQ$, then $qt \in B$ for some $t \in T$.

(5) $(B \cap D)Q = BQ \cap DQ$ and $(B + D)Q = BQ + DQ$ for any two right ideals B and D of A.

(6) If B and D are two right ideals of A such that $B \cap D = 0$, then $BQ \cap DQ = 0$.

(7) If a right ideal B of A is a direct sum of nonzero right ideals $B_1, \ldots, B_n$, then the right ideal BQ of Q is a direct sum of nonzero right ideals B_iQ.

(8) If X is a nonzero right ideal of Q, then $X \cap A$ is a nonzero right ideal of A and $X = (X \cap A)Q$.

In addition, if a right ideal Y of Q properly contains X, then the right ideal $Y \cap A$ of A properly contains $X \cap A$.

(9) If a nonzero right ideal N of Q is a direct sum of nonzero right ideals $N_1, \ldots, N_n$, then $N \cap A$ is a nonzero right ideal of A which is a direct sum of nonzero right ideals $N_i \cap A$.

(10) Let M, N, and P be three right ideals of Q, $MN \subseteq P$, and let M, N properly contain P.

Then $M \cap A$, $N \cap A$, and $P \cap A$ are right ideals of A, $(M \cap A)(N \cap A) \subseteq P \subseteq A$, and $M \cap A$, $N \cap A$ properly contain the right ideal $P \cap A$.

(11) A is a reduced ring $\Longleftrightarrow$ Q is a reduced ring.

(12) If $\mathcal{E}$ is the set of all right ideals E of A such that $E = A \cap EQ$, then $\mathcal{E}$ is a sublattice of the lattice of all right ideals of A, and $\mathcal{E}$ contains any right ideal E of A such that $E = \oplus_{i=1}^n E_i$, where the E_i are contained in $\mathcal{E}$.

(13) The lattice of all right ideals of Q is isomorphic to the lattice $\mathcal{E}$ (see (12)).

(14) If A is right distributive (resp. right Noetherian, right Artinian, right finite-dimensional, right uniform), then Q is right distributive (resp. right Noetherian, right Artinian, right finite-dimensional, right uniform).

(15) If P is an ideal of Q such that $P \bigcap A$ is a prime (semiprime) ideal of A, then P is a prime (semiprime) ideal of Q.

(16) If A is a ring with the minimum condition on principal left ideals, then the left module $_A Q$ satisfies the minimum condition on cyclic submodules.

(17) If A is strongly regular, then Q is strongly regular.

(18) B is an essential right ideal of $A \Longleftrightarrow BQ$ is an essential right ideal of Q.

(19) N is an essential right ideal of $Q \Longleftrightarrow N \bigcap A$ is an essential right ideal of A.

(20) $r_Q(G) = r_A(G)Q$ for any subset G of A.

(21) A is right finite-dimensional $\Longleftrightarrow Q$ is right finite-dimensional.

(22) A is right uniform $\Longleftrightarrow Q$ is right uniform.

(23) A is right nonsingular $\Longleftrightarrow Q$ is right nonsingular.

(24) A is a ring with the maximum (minimum) condition on right annihilators $\Longleftrightarrow Q$ is a ring with the maximum (minimum) condition on right annihilators.

(25) A is a ring with the maximum (minimum) condition on left annihilators $\Longleftrightarrow Q$ is a ring with the maximum (minimum) condition on left annihilators.

(26) If $T = A \setminus M$, where M is a right ideal of A, then Q is a local ring, and $J(Q) = MQ$.

$\triangleleft$ (1) If $0 \neq a \in A$ and $t \in T$, then $0 \neq a = (at^{-1})t \in A \bigcap (at^{-1})A$.

(2) By the induction on n, we need only to consider the case $n = 2$. Assume that $q_i = d_i t_i^{-1}$, where $d_i \in A$ and $t_i \in T$ $(i = 1, 2)$. Since T is right permutable, there exist $u_1 \in T$ and $u_2 \in A$ such that $t_1 u_1 = t_2 u_2 \equiv t$. Since $t_1, u_1 \in T$, we obtain $t \in T$. Hence $t_i^{-1} = u_i t^{-1}$ and $q_i = a_i t_i^{-1}$, where $a_i \equiv d_i u_i$.

(3) follows from (2).

(4) It is sufficient to prove that given $q \in B_T$, we have $q = f(b)f(t)^{-1}$ with $b \in B$ and $t \in T$. Since $q \in B_T$, we have $q = \sum_{i=1}^n f(b_i)q_i$, $b_i \in B$, and $q_i \in Q$. By (1), $q_i = f(a_i)f(t)^{-1}$ for some $a_i \in A, t \in T$. Denoting $b \equiv \sum_{i=1}^n b_i a_i \in B$, we obtain $q = f(b)f(t)^{-1}$.

(5) Assume that $q \in BQ \bigcap DQ$. By (2), there exist $b \in B, d \in D, t_1 \in T, t_2 \in T$ such that $q = bt_1^{-1} = dt_2^{-1}$. By (1), there exist $a_1, a_2 \in A$, and $t \in T$ such that $t_1^{-1} = a_1 t^{-1}$, $t_2^{-1} = a_2 t^{-1}$. Hence $q = ba_1 t^{-1} = da_2 t^{-1}$, $ba_1 = da_2 \in B \bigcap D$, and $q \in (B \bigcap D)Q$. Hence

$(B \cap D)Q = BQ \cap DQ$. The equality $(B + D)Q = BQ + DQ$ can directly be verified.

(6) follows from (5). (7) follows from (5) and (6). (8) follows from (1). (9) follows from (8) and (7).

(10) By (8), $M \cap A$ and $N \cap A$ properly contain the right ideal $P \cap A$. In addition, $(M \cap A)(N \cap A) \subseteq (MN) \cap A \subseteq P \cap A$.

(11) $\Longleftarrow$ follows from the fact that subrings of reduced rings are reduced.

$\Longrightarrow$ Assume that $a \in A$, $t \in T$, $q = at^{-1} \in Q$, and $q^2 = 0$. Since T is right permutable, there exist $u \in T$ and $b \in A$ such that $au = tb$. Hence $0 = q^2tu = at^{-1}au = ab$. By 1.35(2), $r(a)$ is an ideal of A, and $b \in r(a)$. Hence $0 = atb = a^2u$. By 1.35(4), $r(a) = r(a^2)$. Hence $au = 0$, $a = 0$, and $q = 0$.

(12) follows from 1.46(7),(8),(9). (13) follows from (12) and 1.46(7),(8),(9). (14) follows from (13). (15) follows from 1.46(10). (16) follows from 1.46(3).

(17) Assume that $q = at^{-1} \in Q$, where $a \in A$ and $t \in T$. There exists $b \in A$ such that $a = a^2b$. Hence $1 - ab \in r_Q(a)$. By 1.46(11), Q is reduced, whence $r_Q(a)$ is an ideal of Q by 1.35(2). Therefore $t^{-1}(1 - ab) \in r_Q(a)$. Hence $q = at^{-1}ab = (at^{-1})^2tb = q^2tb$, whence Q is strongly regular.

(18) and (19) follow from (6), (8), and (9).

(20) By (5), $(A \cap r_Q(G))Q = Q \cap r_Q(G) = r_Q(G)$. In addition, $r_A(G) = A \cap r_Q(G)$.

(21) and (22) follow from (7) and (9).

(23)

$\Longleftarrow$ follows from (19) and (20).

$\Longrightarrow$ Let $q = at^{-1} \in \mathrm{Sing}(Q_Q)$ (where $a \in A$, $t \in T$), and let $N \equiv r_Q(q)$ be an essential right ideal of Q. Since $Q = t^{-1}Q$, $t^{-1}N$ is an essential right ideal of Q. By (19), $A \cap t^{-1}N$ is an essential right ideal of A. Since $aA \cap t^{-1}N = 0$, we obtain $a = 0$. Therefore $q = 0$.

(24) $\Longrightarrow$ follows from (20).

$\Longleftarrow$ By 1.36(2), any subring of a ring with the maximum (minimum) condition on right annihilators is a ring with the maximum (minimum) condition on right annihilators.

(25) By 1.36(2), the maximum (minimum) condition on left annihilators is equivalent to the minimum (maximum) condition on right annihilators. Hence (25) follows from (24).

(26) It is sufficient to prove that any element of MQ is not a unit, and any element of $Q \setminus MQ$ is a unit. Let $q \in Q$. Assume that q is a unit, and $q \in MQ$. Hence $1 \in MQ$. By (4), there exists $t \in T$ such that $1 \cdot t = t \in M \cap T$; this is a contradiction. Assume that

$q = at^{-1} \in Q \setminus MQ$, where $a \in A$ and $t \in T$. Hence $a \in A \setminus M = T$, whence a is a unit. Therefore $q = at^{-1}$ is a unit. $\triangleright$

1.47 Let Q be a right ring of quotients of a ring A with respect to a right Ore set T. Assume that either Q is right Noetherian or T is contained in the centre of A.

(1) $(BD)Q = (BQ)(DQ)$ for any two right ideals B and D of A.

(2) Let P be an ideal of Q such that $P \cap A \neq A$. Then

$P \cap A$ is a prime (semiprime) ideal of $A \iff P$ is a prime (semiprime) ideal of Q.

(3) Let P be an ideal of Q such that $P \cap A \neq A$. Then

$P \cap A$ is a product of prime (semiprime) ideals of $A \iff P$ is a product of prime (semiprime) ideals of Q.

(4) A is semiprime (prime) $\iff Q$ is semiprime (prime).

$\triangleleft$ (1) If T is a central set, then our assertion can directly be verified. Assume that Q is right Noetherian. Let $b \in B$, $d \in D$, and let $t \in T$. It is sufficient to prove the inclusion $bt^{-1}d \in (BD)_T$. Since Q is right Noetherian, its right ideal $\sum_{i=0}^{\infty} t^{-i}dQ$ is finitely generated. Therefore, there exist a positive integer n and $q_0, \dots, q_n \in Q$ such that $t^{-n-1}d = \sum_{i=0}^{n} t^{-i}dq_i$. Hence $t^{-1}d = \sum_{i=0}^{n} t^{n-i}dq_i$ and $bt^{-1}d = \sum_{i=0}^{n} bt^{n-i}dq_i \in (BD)_T$.

(2) .

$\implies$ follows from 1.46(15).

$\impliedby$ Let B and D be two ideals of A such that $BD \subseteq P \cap A$. Then $(BD)Q \subseteq P$. By (1), $(BQ)(DQ) \subseteq P$. Since P is a prime ideal of Q, either $BQ \subseteq P$ or $DQ \subseteq P$. Let $BQ \subseteq P$. Then $B \subseteq P \cap A$. Therefore $P \cap A$ is a prime ideal of A. The case of semiprime ideals can be considered similarly.

(3) and (4) follow from (1) and (2). $\triangleright$

1.48 Let T be a multiplicative subset in a ring A, and let $K(T) \equiv \{a \in A \,|\, at = 0 \text{ for some } t \in T\}$. Then the following conditions are equivalent.

(1) There exist a ring A_T and a ring homomorphism $f_T \equiv f : A \to A_T$ such that $f(T) \subseteq U(A_T)$, $A_T = \{f(a)f(t)^{-1} \,|\, a \in A, \, t \in T\}$, and $\mathrm{Ker}(f) = K(T)$.

(2) $K(T)$ is an ideal of A, and for the natural epimorphism $h : A \to A/K(T)$, the image $\overline{T}$ of T is a right Ore subset of the factor ring $A/K(T)$.

(3) T is a right permutable right reversible set.

(4) T is a right permutable weakly right reversible set.

◁ (1)⟺(2), (3)⟹(2), and (3)⟹(4) are directly verified.

(4)⟹(3) Consider elements $a \in A$ and $t \in T$ such that $ta = 0$. Set $b \equiv at$. Then $b^2 = tb = 0$, whence $bx = 0$ for some $x \in T$ (by assumption). Set $u \equiv tx \in T$. Then $au = 0$, and T is right reversible.

(2)⟹(3) Since $K(T)$ is an ideal of A, T is right reversible. Let $a \in A$, $t \in T$. Since $h(T)$ is a right Ore set in $h(A)$, there exist $b_1 \in A$ and $u_1 \in T$ such that $h(au_1) = h(tb_1)$. Then $au_1 - tb_1 \in K(T)$. Therefore, there exists $u_2 \in T$ such that $(au_1 - tb_1)u_2 = 0$. Set $u \equiv u_1 u_2$, $b \equiv b_1 u_2$. Then $u \in T$ and $au = tb$, whence T is right permutable. ▷

1.49 A subset T of a ring A is a *right denominator set* in A if T satisfies the equivalent conditions of 1.48. In this case, A_T is called the *right ring of quotients* for A with respect to T, and f_T is the *natural homomorphism*. We will write a_T instead of $f_T(a)$. (If T is a right Ore set, then this definition of A_T is compatible with the definition in 1.44.)

Let B be a right ideal of a ring A. Then B_T denotes the right ideal of A_T generated by $f_T(B)$. Consider a right ideal M of A. If $T = A \setminus M$, then we use f_M, A_M, a_M, and B_M instead of f_T, A_T, a_T, and B_T.

A *right localizable* ring is any ring A such that the right ring of quotients A_M exists for each maximal right ideal M of A. Every commutative ring A is localizable and has a classical ring of quotients and the ring of quotients A_P with respect to any completely prime ideal P.

The left ring of quotients $_TA$ for A with respect to a left denominator set T and the natural homomorphism $_Tf : A \to {_TA}$ are defined symmetrically. If M is a left ideal of A and $T = A \setminus M$, then we use M instead of T in the subscript.

1.50 Let T be a right denominator set in a ring A.

(1) If B is a ring and $g : A \to B$ is a ring homomorphism such that $g(T) \subseteq U(B)$, then there exists the unique ring homomorphism $h : A_T \to B$ such that g is the restriction of h to A.

(2) The right ring of quotients A_T is unique up to isomorphism.

(3) If T is also a left denominator set, then $A_T \cong {_TA}$.

1.51 Let T be a right permutable multiplicative set in a ring A.

(1) If each square-zero $a \in A$ commutes with any $t \in T$, then the right ring of quotients A_T exists.

(2) If given any $a \in A$, there exists a positive integer $n = n(a)$ such that $r(a^n) = r(a^{n+1})$, then the right ring of quotients A_T exists.

(3) If A is a ring with the maximum condition on right annihilators, then the right ring of quotients A_T exists.

◁ (1) follows from the fact that T is weakly reversible in this case.

(2) Consider elements $a \in A$ and $t \in T$ such that $ta = 0$. By assumption, there exists a positive integer n such that $r(t^n) = r(t^{n+1})$. Since T is right permutable, there exist $b \in A$ and $u \in T$ such that $t^n b = au$. Since $t^{n+1} b = tau = 0$, we obtain $b \in r(t^{n+1}) = r(t^n)$, whence $au = t^n b = 0$. Therefore T is right reversible.

(3) follows from (2). ▷

1.52 If T is a right permutable multiplicative set in a reduced ring A, then the right ring of quotients A_T exists and is reduced.

◁ By 1.51(1), A_T exists. Let $f : A \to A_T$ be the canonical homomorphism. Since $A_T \cong f(A)_{f(T)}$, 1.46(11) shows us that it is sufficient to prove that $f(A)$ is reduced. Let $a \in A$, $(f(a))^2 = 0$. Then there exists $t \in T$ such that $a^2 t = 0$. Since A is reduced, 1.35(4) shows us that $at = 0$. Therefore $a \in \mathrm{Ker}(f)$ and $f(a) = 0$. ▷

1.53 Let Q be the right ring of quotients of a ring A with respect to a right denominator set T, and let $f : A \to Q$ be the canonical homomorphism. If B is any subset of A, then $\overline{B}$ denotes the set $f(B)$, and B_T denotes the right ideal $f(B)Q$ of Q.

(1) Let $0 \neq q = \overline{a}\,\overline{t}^{-1} \in Q$, where $0 \neq a \in A$, $t \in T$.

Then $0 \neq \overline{a} = q\overline{t} \in \overline{A} \cap q\overline{A}$, and $Q_{\overline{A}}$ is an essential extension of $\overline{A}_{\overline{A}}$.

(2) Given any $q_1, \ldots, q_n \in Q$, there exist $t \in T$ and $a_1, \ldots, a_n \in A$ such that $q_i = f(a_i)f(t)^{-1}$ for any $i = 1, \ldots, n$.

(3) For a finitely generated submodule N of the left module $_{\overline{A}}Q$, there exists $t \in T$ such that $N\overline{t} \subseteq \overline{A}$.

Consequently, the left module $_{\overline{A}}N$ is isomorphic to the finitely generated left ideal $N\overline{t}$ of $\overline{A}$.

(4) If B is a right ideal of A, then $\overline{B}Q = \{\overline{b}\,\overline{t}^{-1} \,|\, b \in B, t \in T\}$.

Consequently, if $q \in \overline{B}Q$, then $q\overline{t} \in \overline{B}$ for some $t \in T$.

(5) $(B \cap D)_T = B_T \cap D_T$ and $(B + D)_T = B_T + D_T$ for any two right ideals B and D of A containing $\mathrm{Ker}(f)$.

(6) If B and D are two right ideals of A such that $\overline{B} \cap \overline{D} = 0$, then the intersection of right ideals B_T and D_T of Q is equal to zero.

(7) Let B be a right ideal of A such that the right ideal $\overline{B}$ of $\overline{A}$ is a direct sum of nonzero right ideals $\overline{B}_1, \ldots, \overline{B}_n$.

Then the right ideal B_T of Q is a direct sum of nonzero right ideals $\overline{B}_i Q$.

(8) If X is a nonzero right ideal of Q, then $f^{-1}(X)$ is a nonzero right ideal of A, $X \cap \overline{A}$ is a nonzero right ideal of $\overline{A}$, and $X = (f^{-1}(X))_T = (X \cap \overline{A})Q$.

In addition, if a right ideal Y of Q properly contains X, then the right ideal $f^{-1}(Y)$ of A properly contains $f^{-1}(X)$, and the right ideal $Y \cap \overline{A}$ of $\overline{A}$ properly contains $X \cap \overline{A}$.

(9) If a nonzero right ideal N of Q is a direct sum of nonzero right ideals $N_1, \ldots, N_n$, then $N \cap \overline{A}$ is a nonzero right ideal of $\overline{A}$, and $N \cap \overline{A}$ is a direct sum of nonzero right ideals $N_i \cap \overline{A}$.

(10) Let M, N, and P be three right ideals of Q, $MN \subseteq P$, and let M and N properly contain P.

Then $M \cap \overline{A}$, $N \cap \overline{A}$, and $P \cap \overline{A}$ are right ideals of $\overline{A}$, $(M \cap \overline{A})(N \cap \overline{A}) \subseteq P \subseteq \overline{A}$, and $M \cap \overline{A}$, $N \cap \overline{A}$ properly contain $P \cap \overline{A}$.

(11) If m is a cardinal number and N is a nonzero m-generated right ideal of Q, then there exists a nonzero m-generated right ideal B of A such that $N = B_T$.

(12) Let B be a right ideal of A, $a \in A$, and let $\overline{a} \in B_T$.

Then $at \in B$ for some $t \in T$.

(13) If for any $a \in A$ and $t \in T$, either $a \in T$ or $t - a \in T$, then Q is a local ring.

(14) If $T = A \setminus M$, where M is a right ideal of A, then Q is a local ring, and $J(Q) = M_T \equiv M_M$.

(15) If A is right distributive (resp. right Noetherian, right Artinian, right finite-dimensional, right uniform), then Q is right distributive (resp. right Noetherian, right Artinian, right finite-dimensional, right uniform).

$\lhd$ (1) – (10) follow from 1.46 and 1.48.

(11) Let $N = \sum_{i \in I} \overline{a}_i \overline{t}_i^{-1} Q = \sum_{i \in I} \overline{a}_i Q$, where $a_i \in A$, and the cardinality of I is equal to m. Set $B \equiv \sum_{i \in I} a_i A$.

(12) By (4), there exist $b \in B$ and $u \in T$ such that $\overline{a} = \overline{b}\overline{u}^{-1}$. Hence there exists $x \in \text{Ker}(f)$ such that $au = b + x$. Since $x \in \text{Ker}(f)$, we obtain $xv = 0$ for some $v \in T$. Set $t \equiv uv \in T$. Then $at = (b + x)v = bv \in B$.

(13) Assume that $q = \overline{a}\overline{t}^{-1}$ is not invertible in Q, where $a \in A$ and $t \in T$. Then $\overline{a}$ is not a unit. Therefore $a \in A \setminus T$. By assumption, $t - a \in T$. Hence $\overline{t} - \overline{a} \in U(Q)$, $1 - q = (\overline{t} - \overline{a})\overline{t}^{-1} \in U(Q)$. Therefore Q is local.

(14) and (15) See 1.48 and 1.46(26),(14). $\rhd$

1.54 Let A be a right invariant ring, and let T be a multiplicative subset of A.

(1) If N is a maximal right ideal of A or N is a prime ideal of A, then N is a completely prime ideal, and $A \setminus N$ is a right permutable multiplicative subset.

(2) T is weakly right reversible $\Longleftrightarrow$ T is a right denominator set.

(3) If each square-zero element $a \in A$ commutes with every element of T, then T is a right denominator set.

(4) Assume that for any $a \in A$, there exists a positive integer $n = n(a)$ such that $r(a^n) = r(a^{n+1})$.

Then T is a right denominator set, A is right localizable, and the right ring of quotients A_N exists for any prime ideal N of A.

(5) If A is a ring with the maximum condition on right annihilators, then T is a right denominator set, A is right localizable, and for any prime ideal N of A, the right ring of quotients A_N exists.

(6) If A is a semiprime ring, then A is a right localizable reduced ring, and for any prime ideal N of A, the right ring of quotients A_N exists and is reduced.

(7) If each square-zero element $a \in A$ is contained in the centre of A, then A is right localizable, and for any prime ideal N of A, the right ring of quotients A_N exists.

$\lhd$ (1) If N is a maximal right ideal of the right invariant ring A, then N is an ideal, A/N is a division ring, and the ideal N is completely prime. If N is a prime ideal of the right invariant ring A and a, b are elements of A such that $ab \in N$, then $aAb \in N$ and $a \in N$ or $b \in N$. Therefore, we may assume that N is completely prime, and the set $A \setminus N$ is multiplicative. By 1.43(2), $A \setminus N$ is right permutable.

(2) follows from 1.48 and from the fact that T is right permutable by 1.43(2). (3) follows from (1) and from the fact that T is right permutable by 1.43(2). (4) follows from (1) and 1.51(2). (5) follows from (4).

(6) Since A is a right invariant semiprime ring, A is reduced. Let N be either a maximal right ideal or a prime ideal of A. By (1), N is a completely prime ideal, and $A \setminus N$ is a right permutable multiplicative set. By 1.52, the right ring of quotients A_N exists and is reduced.

(7) follows from (1) and 1.51(1). $\rhd$

1.55 Let Q be the right ring of quotients of a ring A with respect to a right denominator set T, and let $f : A \to Q$ be the canonical homomorphism. If B is any subset of A, then $\overline{B}$ denotes the set $f(B)$, and B_T denotes the right ideal $f(B)Q$ of Q. Denote by $\mathcal{E}$ the set of all right ideals E of A such that $E = A \bigcap f^{-1}(E_T \bigcap \overline{A})$.

(1) The set $\mathcal{E}$ is a sublattice of the lattice of all right ideals of A, and $\mathcal{E}$ contains any right ideal E of A such that $\overline{E} = \oplus_{i=1}^{n} \overline{E}_i$, where the E_i are contained in $\mathcal{E}$.

(2) The lattice of all right ideals of Q is isomorphic to the lattice $\mathcal{E}$ (see (1)).

(3) If the ring A is right Noetherian (resp. right Artinian, right finite-dimensional, right uniform, right distributive), then the ring Q is right Noetherian (resp. right Artinian, right finite-dimensional, right uniform, right distributive).

(4) If P is an ideal of Q such that $P \cap \overline{A}$ is a prime (semiprime) ideal of $\overline{A}$, then $\overline{P}$ is a prime (semiprime) ideal of Q.

(5) If $\overline{A}$ is a ring with the minimum condition on principal left ideals, then the left module $_{\overline{A}}Q$ satisfies the minimum condition on cyclic submodules.

(6) If A is strongly regular, then Q is strongly regular.

(7) Let B be a right ideal of A. Then
$\overline{B}$ is an essential right ideal of $\overline{A} \iff B_T$ is an essential right ideal of Q.

(8) N is an essential right ideal of $Q \iff N \cap \overline{A}$ is an essential right ideal of $\overline{A}$.

(9) $r_Q(\overline{G}) = r_{\overline{A}}(G)Q$ for any subset G of A.

(10) $\overline{A}$ is right finite-dimensional $\iff Q$ is right finite-dimensional.

(11) $\overline{A}$ is right uniform $\iff Q$ is right uniform.

(12) $\overline{A}$ is right nonsingular $\iff Q$ is right nonsingular.

(13) $\overline{A}$ is a ring with the maximum (minimum) condition on right annihilators $\iff Q$ is a ring with the maximum (minimum) condition on right annihilators.

(14) $\overline{A}$ is a ring with the maximum (minimum) condition on left annihilators $\iff Q$ is a ring with the maximum (minimum) condition on left annihilators.

(15) $\overline{A}$ is reduced $\iff Q$ is reduced.

◁ (1) – (14) follow from 1.46, and 1.48. (15) follows from 1.46(11), 1.48, and 1.52. ▷

1.56 Let Q be the right ring of quotients of a ring A with respect to a right denominator set T, and let $f : A \to Q$ be the canonical homomorphism. Assume that either Q is right Noetherian or T is contained in the centre of A. For any subset B in A, $\overline{B}$ denotes $f(B)$, and B_T denotes the right ideal $f(B)Q$ of Q.

(1) $(BD)_T = (B_T)(D_T)$ for any two right ideals B and D of A.

(2) Let P be an ideal of Q such that $P \cap A \neq A$. Then
$f^{-1}(P)$ is a prime (semiprime) ideal of $A \iff P$ is a prime (semiprime) ideal of Q.

(3) Let P be an ideal of Q such that $P \cap A \neq A$. Then
$f^{-1}(P)$ is a product of prime (semiprime) ideals of $A \iff P$ is a product of prime (semiprime) ideals of Q.

(4) $\overline{A}$ is semiprime (prime) $\iff Q$ is semiprime (prime).

◁ Note that $A/B \cong \overline{A}/\overline{B}$ for any ideal B of A. Therefore, our assertion follows from 1.47 and 1.48. ▷

1.57 Let T be a right permutable multiplicative subset in a ring A, and let $K(T) \equiv \{a \in A \,|\, at = 0 \text{ for some } t \in T\}$. Then $K(T)$ is an ideal of A. Denote by h the natural ring epimorphism $A \to A/K(T)$.

(1) All elements of $h(T)$ are left regular in $h(A)$.

(2) T is a right denominator set in $A \iff$ all elements of $h(T)$ are right regular in $h(A)$.

(3) If all left regular elements of $h(A)$ are right regular in $h(A)$, then T is a right denominator set in A.

(4) If all left zero-divisors of $h(A)$ are nilpotent, then T is a right denominator set in A.

(5) If $h(A)$ is a right uniform ring with the maximum conditions on right annihilators, then T is a right denominator set in A.

(6) If all square-zero elements of $h(A)$ are central in $h(A)$, then T is a right denominator set in A.

1.58 Let T be a multiplicative subset in a ring A, and let $K(T) \equiv \{a \in A \,|\, at = 0, \, ua = 0 \text{ for some } t, u \in T\}$. The set T is called a *right and left denominator set* if the following equivalent conditions hold.

(1) There exists a ring $_T A_T$ and a ring homomorphism $f : A \to {}_T A_T$ such that $f(T) \subseteq U(_T A_T)$, $_T A_T = \{f(a)f(t)^{-1} \,|\, a \in A, \, t \in T\} = \{f(t)^{-1}f(a) \,|\, a \in A, \, t \in T\}$, and $\operatorname{Ker}(f) = K(T)$.

(2) $K(T)$ is an ideal of A, and the image $\overline{T}$ of the set T for a natural epimorphism $h : A \to A/K(T)$ is a two-sided Ore set in $A/K(T)$.

(3) T is a right and left permutable right and left reversible set.

(4) T is a right and left permutable weakly right and left reversible set.

◁ The assertion is directly verified with the use of 1.48. ▷

1.59 Under the equivalent conditions of 1.58, $_T A_T$ is called a *two-sided ring of quotients* of A with respect to T, and f is called the *natural homomorphism*. (If T is a two-sided Ore set, then this definition of $_T A_T$ is compatible with the definition from 1.44.)

For each ideal B of A, the ideal of the ring $_T A_T$ generated by $f(B)$ is denoted by $_T B_T$. If $T = A \setminus M$ and M is an ideal of A, then we write $_M A_M$ and $_M B_M$ instead of $_T A_T$ and $_T B_T$.

The ring $_T A_T$ exists if and only if A_T and $_T A$ exist. In this case, we have $_T A_T \cong A_T \cong {}_T A$, A_T (or, equivalently, $_T A$) is a two-sided ring of quotients of A with respect to T, and the natural homomorphism $A \to A_T$ $(A \to {}_T A)$ can be identified with the natural homomorphism $A \to {}_T A_T$.

◁ The assertions can be verified using 1.48, 1.50, and 1.58. ▷

1.60 Let A be a right localizable ring, and let $\max(A)$ be the set of all maximal right ideals of A.

(1) Assume that B is a right ideal of A, $a \in A$, and $a_M \in B_M$ for each $M \in \max(A_A)$.

Then $a \in B$.

(2) If a is an element of A with $a_M = 0$ for all $M \in \max(A_A)$, then $a = 0$.

(3) If B and D are two right ideals of A and $B_M = D_M$ for all $M \in \max(A_A)$, then $B = D$.

(4) A is reduced $\iff$ A_M is reduced for all $M \in \max(A_A)$.

(5) If B is a right ideal of A such that B_M is an ideal of A_M for each $M \in \max(A_A)$, then B is an ideal of A.

(6) If A_M is right invariant for each $M \in \max(A_A)$, then A is right invariant.

(7) Every maximal right ideal N of A is a completely prime ideal of A, and therefore A/N is a division ring.

◁ (1) Set $H \equiv \{h \in A \mid ah \in B\}$. Let us prove that $H = A$. Assume the contrary. Then H is contained in a maximal right ideal M. Consider $T \equiv A \setminus M$. By assumption, $a_M \in B_M$, whence (by 1.53(2)) there exists $t \in T$ such that $at \in B$. Hence $t \in T \bigcap M = \emptyset$; this is a contradiction.

(2) and (3) follows from (1). (4) follows from (2) and 1.52. (5) follows from (3) and from the fact that $B_M \subseteq (AB)_M = A_M B_M = B_M$. (6) follows from (5).

(7) It is sufficient to prove that N is an ideal. By (5), we need prove only that N_M is an ideal of A_M for each $M \in \max(A_A)$. If $M = N$, we obtain (by 1.53(14)) that $M_M = J(A_M)$ is an ideal of A_M. Assume now that $M \neq N$. Then $M + N = A$ and $A_M = M_M + N_M = J(A_M) + N_M$. Hence $N_M = A_M$ is an ideal of A_M. ▷

1.61 Let T be a right denominator set of a ring A, and let $A_T \equiv Q$ be a right Noetherian ring.

(1) If B is an ideal of A, then B_T is an ideal of Q.

(2) If A is right invariant, then Q is right invariant.

(3) If P is a completely prime ideal of A, then P_T is a completely prime ideal of Q.

(4) If N is a product of some completely prime ideals of A, then N_T is a product of some completely prime ideals of A_T.

◁ (1) follows from 1.56(1) and 1.53(11). (2) follows from (1) and 1.53(11).

(3) Assume the contrary. By (1) and 1.53(11), there exist two ideals B and D of A such that B and D properly contain the ideal P, $B_T D_T \subseteq P_T$, and B_T, D_T properly contain P_T. Let $b \in B \setminus P$, $d \in D \setminus P$. By 1.53(12), $bdt \in P$ for some $t \in T$, and $bd \in A \setminus P$. Therefore $t \in P$, $P_T = Q \supseteq B$; this is a contradiction.

(4) follows from (3) and 1.56(1). $\triangleright$

1.62 Let T be a multiplicative subset in a right Noetherian right invariant ring A.

(1) The right ring of quotients A_T exists and is a right invariant right Noetherian ring.

(2) If P is a prime ideal of A, then P_T is a completely prime ideal of A_T.

(3) If N is a product of some prime ideals of A, then N_T is a product of completely prime ideals of A_T.

$\triangleleft$ (1) By 1.54(5), the ring A_T exists. By 1.55(3), A_T is right Noetherian. By 1.61(2), A_T is right invariant.

(2) and (3) are verified using (1) and 1.61. $\triangleright$

1.63 Let A be a ring without the identity element, $\mathbf{Z}$ be the ring of integers, and let A^1 be the direct product of additive groups of the rings A and $\mathbf{Z}$. Define multiplication of pairs $(a_1, z_1), (a_2, z_2) \in A^1$ by the rule $(a_1, z_1) \cdot (a_2, z_2) = (a_1 a_2 + z_2 a_1 + z_1 a_2, z_1 z_2)$. The group A^1 is a ring with the identity element $(0, 1)$.

It is directly verified that the ring A is isomorphic to the subring $(A, 0)$ of A^1, $(A, 0)$ is an ideal in A^1, and every right A-module M can be considered as a right module over the ring A^1 with the identity element if we define multiplication $m \cdot (a, z)$ by the rule $m \cdot (a, z) = ma + zm$.

(1) For each A-module M, the lattice of all A-submodules of M coincides with the lattice of all A^1-submodules of M.

Consequently, M is a distributive A-module $\Longleftrightarrow$ M is a distributive A_1-module.

(2) For any two right A-modules M and N, the set of all A-homomorphisms $M \to N$ coincides with the set of all A^1-homomorphisms $M \to N$.

Consequently, $\text{End}(M_A) = \text{End}(M_{A^1})$.

Chapter 2

Distributive modules

2.1 Uniserial modules

2.1 A module M is a *Bezout module* if each finitely generated submodule of M is cyclic.

M_A is a Bezout module $\iff$ each 2-generated submodule of M is cyclic $\iff$ fFor any $m, n \in M$, there exist $a, b, c, d \in A$ such that $m = (ma + nb)c$ and $n = (ma + nb)d$ $\iff$ for any $m, n \in M$, there exist $a, b, c, d \in A$ such that $m(1 - ac) = nbc$ and $n(1 - bd) = mad$.

2.2 A module M is *uniserial* if any two submodules of M are comparable (with respect to inclusion).

M is a uniserial module $\iff$ the lattice $\mathrm{Lat}(M)$ is a chain $\iff$ any two cyclic submodules of M are comparable $\iff$ all factor modules of M are uniform $\iff$ M has no subfactors $S \oplus T$, where S and T are nonzero modules $\iff$ M has no subfactors $S \oplus T$, where S and T are simple modules $\iff$

Any uniserial module is a Bezout module.

Any finitely generated uniserial module is a local cyclic module.

Every right (left) uniserial ring is a local right (left) uniform ring.

2.3 A module M_A is a *distributive* module if the following equivalent conditions hold.

(1) $F \bigcap (G + H) = F \bigcap G + F \bigcap H$ for any $F, G, H \in \mathrm{Lat}(M)$ (i.e., the lattice $\mathrm{Lat}(M)$ is distributive).

(2) $(F + G) \bigcap (F + H) = F + G \bigcap H$ for any $F, G, H \in \mathrm{Lat}(M)$.

(3) $(m + n)A = mA \bigcap (m + n)A + nA \bigcap (m + n)A$ for any $m, n \in M$.

The class of distributive modules properly contains all uniserial and, in particular, all simple modules.

A module M is distributive $\Longleftrightarrow$ all subfactors of M are distributive $\Longleftrightarrow$ all 2-generated submodules of M are distributive.

2.4 [122]. *For a module M_A over a ring A, the following conditions are equivalent.*

(1) *M is distributive.*

(2) *For any m and $n \in M$, there exists $a \in A$ such that $maA + n(1-a)A \subseteq mA \cap nA$.*

(3) *For any $m, n \in M$, there exist $a, b, c, d \in A$ such that $1 = a + b$, $ma = nc$, and $nb = md$.*

(4) *$A = (m : nA) + (n : mA)$ for all $m, n \in M$.*

◁ $(2)\Longleftrightarrow(3)$ and $(3)\Longleftrightarrow(4)$ are directly verified.

$(1)\Longrightarrow(2)$ Let $f = m + n$, $T = mA \cap nA$. Since $fA = fA \cap mA + fA \cap nA$, there exist b and $d \in A$ such that $fb \in mA$, $fd \in nA$, and $f = fb + fd$. Hence $nb = fb - mb \in T$ and $md = fd - nd \in T$. Set $a \equiv 1 - b$, $z \equiv a - d = 1 - b - d$. Hence $1 = a + b$ and $fz = f - fb - fd = 0$. Therefore $ma = md + mz = md + fz - nz = md - nz$ and $nz = -mz \in T$. Hence $ma \in T$. Thus a is the required element.

$(2)\Longrightarrow(1)$ Let $F, G, H \in \mathrm{Lat}(M)$, and let $f \in F \cap (G + H)$. We have to prove the inclusion $f \in F \cap G + F \cap H$. Let $f = m + n$, $m \in G$, and let $n \in H$. By assumption, there exist a and $b \in A$ such that $1 = a + b$, $ma \in nA$, and $nb \in mA$. Hence $fb = mb + nb \in fA \cap mA \subseteq F \cap G$, $fa = ma + na \in fA \cap nA \subseteq F \cap H$, and $f = fb + fa \in F \cap G + F \cap H$. ▷

2.5 (1) Let M_A be a module over a ring A, and let B be a unitary subring B of A.

If every 2-generated submodule of M_A is a distributive B-module, then M is a distributive B-module.

(2) Assume that each 2-generated unitary subring of a ring A is contained in some unitary right distributive subring of A.

Then A is right distributive.

(3) Let M_i be a distributive module over a ring A_i for every $i \in I$, and let A be the direct product of all the A_i.

Then the A-modules $\prod_{i \in I} M_i$ and $\oplus_{i \in I} M_i$ are distributive.

(4) All direct products and factor rings of right distributive rings are right distributive.

◁ 2.5 follows from 2.4. ▷

2.6 Let M_A be a distributive module M_A.

(1) If $m, n \in M$ and $mA \cap nA = 0$, then there exists $a \in A$ such that $ma = n(1-a) = 0$.

(2) $\mathrm{Hom}(G, H) = 0$ for any $G, H \in \mathrm{Lat}(M)$ such that $G \cap H = 0$.

(3) $\mathrm{Hom}(G/(G \cap H), H/(G \cap H)) = 0$ for any G and $H \in \mathrm{Lat}(M)$.

(4) $\mathrm{Hom}(M/H, M/G) = 0$ for any $G, H \in \mathrm{Lat}(M)$ such that $G + H = M$.

◁ (1) follows from 2.4.

(2) Let $f \in \mathrm{Hom}(G, H)$, $m \in G$, and let $n = f(m) \in H$. By (1), there exists $a \in A$ such that $ma = n(1 - a) = 0$. Hence $n = na = f(m)a = f(ma) = f(0) = 0$.

(3) follows from (2) and from the fact that all subfactors of a distributive module are distributive.

(4) Since there exist natural isomorphisms $G/(G \cap H) \cong M/H$ and $H/(G \cap H) \cong M/G$, our assertion follows from (2). ▷

2.7 [184]. *Let all simple subfactors of a module M_A be isomorphic (this is the case if A is a matrix-local ring). Then*

$$M \text{ is distributive} \iff M \text{ is uniserial.}$$

◁ We may assume that M is distributive. By 2.6(3), M has no subfactors of the form $S \oplus T$, where S and T are isomorphic nonzero modules. It follows by assumption that M has no subfactors $S \oplus T$, where S and T are simple modules. Hence M is uniserial (see 2.2). ▷

2.8 [5], [166]. *For a module M_A over a local ring A, the following conditions are equivalent.*

(1) *M is distributive.*

(2) *M is a Bezout module.*

(3) *M is uniserial.*

◁ (1)$\iff$(3) follows from 2.7. (3)$\implies$(2) is obvious.

(2)$\implies$(3) Let $m, n \in M$. It is sufficient to prove that either $m \in nA$ or $n \in mA$. There exist $a, b, c, d \in A$ such that $m(1 - ac) = nbc$ and $n(1 - bd) = mad$. Since A is local, either $1 - ac \in U(A)$ or $ac \in U(A)$. In the first case, $m = nbc(1 - ac)^{-1} \in nA$. In the second case, $a, c \in A \setminus J(A)$. Since A is local, $a, c \in U(A)$. If $d \in U(A)$, then $ad \in U(A)$ and $m = n(1 - bd)(ad)^{-1} \in nA$. If $d \in A \setminus U(A) = J(A)$, then $1 - bd \in U(A)$ and $n = mad(1 - bd)^{-1} \in mA$. ▷

2.9 Let F be a field, and let A be the 5-dimensional algebra A over the field F generated by all (3×3)-matrices of the form

$$\begin{pmatrix} f_{11} & f_{12} & f_{13} \\ 0 & f_{22} & 0 \\ 0 & 0 & f_{33} \end{pmatrix}, \qquad \text{where } f_{ij} \in F.$$

Let e_{ij} be the matrix whose element with the subscript ij is equal to 1 and other elements are equal to zero. The matrices $\{e_{11}, e_{12}, e_{13}, e_{22}, e_{33}\}$ form an F-basis of the F-algebra A.

(1) $1 = e_{11} + e_{22} + e_{33}$, e_{11}, e_{22}, e_{33} are local orthogonal idempotents, $e_{12}F = e_{12}A$, $e_{13}F = e_{13}A$, $J(A) = e_{12}F + e_{13}F$, $(J(A))^2 = 0$, and the factor ring $A/J(A)$ is isomorphic to the direct product of three copies of the field F.

(2) $_AA = Ae_{11} \oplus Ae_{22} \oplus Ae_{33}$, where Ae_{11} is a simple projective left A-module, Ae_{22} and Ae_{33} are uniserial left A-modules which contain the simple nonisomorphic projective modules Ae_{12} and Ae_{13} respectively.

(3) $A_A = e_{11}A \oplus e_{22}A \oplus e_{33}A$, where $e_{22}A = e_{22}F$ and $e_{33}A = e_{33}F$ are simple projective right A-modules which are isomorphic to the modules $e_{12}A$ and $e_{13}A$, respectively.

(4) $J(e_{11}A) = e_{12}F + e_{13}F$ with $e_{11}A = e_{11}F + e_{12}F + e_{13}F$ is an indecomposable projective distributive local nonsingular nonuniserial Bezout A-module, and every proper nonzero submodule of $e_{11}A$ coincides either with the projective module $e_{12}A \oplus e_{13}A$ or with one of the simple projective nonisomorphic modules $e_{12}A$ and $e_{13}A$.

(5) All right (left) ideals of A are projective right (left) A-modules.

◁ 2.9 is directly verified. ▷

2.10 [213]. *There exists a local projective distributive nonsingular Artinian and Noetherian Bezout module M which contains a direct sum of two simple modules. In particular, M is not uniserial.*

◁ 2.10 follows from 2.9(4). ▷

2.2　Countably distributive modules

2.11 Let t be a cardinal. A module M is *t-distributive* if the following three equivalent conditions hold.

(i)　M has no subfactors which are direct sums of t nonzero isomorphic modules.

(ii)　M has no subfactors which are direct sums of t isomorphic simple modules.

(iii) Each homogeneous component of any semisimple subfactor of M is a direct sum of s simple modules, where $s < t$.

An $\aleph_0$-distributive module M is called a *countably distributive* module. A module is *completely finite-dimensional* if all its subfactors are finite-dimensional.

(1) If M is any module of cardinality s, then M is t-distributive for each cardinal $t > s$.

(2) A module M is t-distributive $\iff$ M is m-distributive for any cardinal $m \geq t$.

(3) A module M is countably distributive $\iff$ each homogeneous component of any semisimple subfactor of M is finite-dimensional.

(4) A module M is completely finite-dimensional $\iff$ each semisimple subfactor of M is finite-dimensional.

(5) Any completely finite-dimensional module is countably distributive.

(6) All Artinian or Noetherian modules are countably distributive.

◁ 2.11 is directly verified. ▷

2.12 (1) A distributive module M is t-distributive for any cardinal $t \geq 2$. (In particular, M is countably distributive).

(2) If a module M is an essential extension of a submodule $\sum_{i=1}^{n} M_i$ and all M_i are completely finite-dimensional, then M is a finite-dimensional module.

◁ (1) follows from 2.6(3). (2) can be proved by the induction on N using the fact that an essential extenaion of a finite-dimensional module is finite-dimensional. ▷

2.13 Let the set of all nonisomorphic simple subfactors of a module M over a ring A be finite. (This is the case if the ring A is semilocal).

(1) M is countably distributive $\iff$ M is completely finite-dimensional.

(2) If M is an essential extension of a finite sum of countably distributive submodules, then M is finite-dimensional.

◁ (1) follows from 2.11(3),(4). (2) follows from (1) and 2.12(2). ▷

2.14 [208]. *Let M_A be a semi-Artinian countably distributive module over a ring A, and let the set of all pair-wise nonisomorphic simple subfactors of M be finite. (This is the case if the ring A is semilocal).*
Then M is Artinian.
In addition, if $J(M)$ is superfluous in M, then M is a finitely generated Artinian module.

◁ Since any completely finite-dimensional semi-Artinian module is Artinian, it follows from 2.13(1) that M is Artinian. Hence $M/J(M)$ is a semisimple finitely generated module. By assumption, $J(M)$ is superfluous in M. Therefore, M is finitely generated. ▷

2.15 Let M be a right module over a ring A.

(1) Let t be a cardinal, and let $\overline{F}$ be a t-generated subfactor of M. Then there exists some t-generated submodule F of M such that $\overline{F}$ is isomorphic to a factor module of F.

(2) Assume that there exists a positive integer n such that each semisimple factor module of any $(n + 1)$-generated submodule of M is a direct sum of no more than n simple modules.

Then each subfactor of M contains no direct sums of $n + 1$ nonzero modules.

(3) Let the ring A be semilocal, and let k be the number of minimal right ideals in a decomposition of the semisimple ring $A/J(A)$ into a direct sum of minimal right ideals.

If there exists a positive integer m such that all finitely generated submodules of M are m-generated, then each subfactor of M contains no direct sums of $km + 1$ nonzero modules.

◁ (1) Suppose $\overline{F} = G/H = \sum_{i \in I}(f_i + H)A$, where $G \in \mathrm{Lat}(M)$, $H \in \mathrm{Lat}(G)$, and $f_i \in G$. Let $\mathrm{card}(I) = t$, and let $F \equiv \sum_{i \in I} f_i A \in \mathrm{Lat}(M)$. Then F is the required module.

(2) Assume the contrary. Since each nonzero module has a simple subfactor, there exists a subfactor $\overline{F}$ of M such that $\overline{F} = \oplus_{i=1}^{n+1} T_i$, where the T_i are simple modules. By (1), there exists an $(n + 1)$-generated submodule F of M such that $F \cong \oplus_{i=1}^{n+1} T_i$; this is a contradiction.

(3) Every cyclic semisimple A-module is a direct sum of no more than k simple modules. Therefore, each semisimple factor module of any finitely generated submodule of M is a direct sum of no more than km simple modules. Now our assertion follows from (2). ▷

2.16 Any Bezout module M over a semilocal ring is completely finite-dimensional. In particular, M is countably distributive.

◁ 2.16 follows from 2.13(1) and 2.15(3). ▷

2.17 Every module M is an essential extension of a direct sum of cyclic modules.

◁ Let $\mathcal{E}$ be the set of all submodules of M which are direct sums of cyclic modules. Define a partial order on $\mathcal{E}$ as follows: $F \leq G \iff G = F \oplus H$ for some $H \in \mathcal{E}$. The poset $(\mathcal{E}, \leq)$ contains the unions of all the ascending chains. By Zorn's Lemma, $\mathcal{E}$ contains a maximal element E. Hence E has the nonzero intersection with every nonzero cyclic submodule of M. Therefore, M is an essential extension of E. ▷

2.18 If A is a semilocal right Noetherian ring and $J(A)$ is a nil-ideal, then A is right Artinian.

2.19 Let A be a ring. For each cardinal t, let us denote by $F(t, A)$ the class of all right A-modules which are direct sums of modules whose cardinalities do not exceed t.

(1) Let t be a cardinal, H be a direct sum of all nonisomorphic cyclic right A-modules, N be a direct sum of t copies of H, and let E be the injective hull of N.

Then any t-distributive module M_A is isomorphic to a submodule of E.

(2) For each cardinal t, there exists a cardinal $m = m(t, A)$ such that the cardinality of any t-distributive A-module does not exceed m.

(3) A is right Noetherian $\iff$ every injective right A-module is a direct sum of indecomposable modules $\iff$ there exists a cardinal t such that all injective right A-modules belong to $F(t, A)$ $\iff$ there exists a cardinal t such that any right A-module is isomorphic to a submodule of a module from $F(t, A)$.

(4) A is right Noetherian $\iff$ there exists a cardinal t such that each injective right A-module is a direct sum of t-distributive modules $\iff$ there exists a cardinal t such that each right A-module is isomorphic to a submodule of a direct sum of t-distributive modules.

(5) If there exists a cardinal t such that any direct product of copies of the module A_A belongs to $F(t, A)$, then A is right perfect.

(6) If there exists a cardinal t such that any direct product of copies of the module A_A is a direct sum of t-distributive modules, then A is right perfect.

(7) If there exists a cardinal t such that all injective right A-modules and all direct products of projective right A-modules belong to $F(t, A)$, then A is right Artinian.

(8) If there exists a cardinal t such that all injective right A-modules and all direct products of projective right A-modules are direct sums of t-distributive modules, then A is right Artinian.

◁ (1) By 2.17, M is an essential extension of a direct sum L of cyclic modules. Since M is t-distributive, there exists a monomorphism $g : L \to N$. Since E is injective, g is extended to a homomorphism $f : M \to E$. Since $L \cap \operatorname{Ker}(f) = 0$, f is a monomorphism.

(2) follows from (1). (3) is proved in [60], 20.6, 20.9, 20.19. (4) follows from (2), (3) and 2.11(3). (5) is proved in [60], 20.21. (6) follows from (2) and (5). (7) By (3) and (5), A is a right Noetherian right perfect ring. By 2.18, A is right Artinian. (8) follows from (2) and (7). ▷

2.20 [208]. *Let M be a finite direct sum of countably distributive right modules over a ring A.*

(1) *If A is left perfect, then M is Artinian.*

(2) *If A is semilocal and $M/MJ(A)$ is superfluous in M, then M is finitely generated.*

(3) *If A is right perfect, then M is Noetherian.*

(4) *If A is a perfect, then M has a composition series.*

◁ Since finite direct sums of Artinian (Noetherian) modules are Artinian (Noetherian) modules, we may assume that M is countably distributive.

(1) follows from 2.14.

(2) By 2.13(1), M is completely finite-dimensional. The semisimple finite-dimensional module $M/MJ(A)$ is finitely generated. Since $MJ(A)$ is superfluous in M, M is finitely generated.

(3) Since A is right perfect, $J(A)$ is left vanishing. Hence $NJ(A)$ is superfluous in N for any submodule N of M. By (3), N is finitely generated. Therefore M is Noetherian.

(4) follows from (1) and (3). ▷

2.21 Let a ring A be a direct sum of countably distributive right ideals. Then the following conditions are equivalent.

(1) A is right or left perfect.

(2) A is semiprimary.

(3) A is right Artinian.

◁ (3)$\Longrightarrow$(2) and (2)$\Longrightarrow$(1) are directly verified.

(1)$\Longrightarrow$(3). If A is left perfect, then 2.20(1) shows us that A is right Artinian. Let A be right perfect. By 2.20(3), A is right Noetherian. By 2.18, A is right Artinian. ▷

2.22 [208]. *For a ring A, the following conditions are equivalent.*

(1) *Each right A-module is a direct sum of countably distributive modules.*

(2) *Each right A-module is a direct sum of finitely generated modules.*

(3) *Each right A-module is a direct sum of completely finite-dimensional modules.*

(4) *A is right Artinian, and each right A-module is a direct sum of modules with a finite composition series.*

◁ (4)$\Longrightarrow$(3), (4)$\Longrightarrow$(2), and (3)$\Longrightarrow$(1) are obvious. (1)$\Longrightarrow$(4) follows from 2.19(8) and 2.20(4). (2)$\Longleftrightarrow$(4) follows from 2.19(7). ▷

2.3 Bezout modules

2.23 Let $D = \oplus_{i=1}^{n} D_i$ be a submodule of a distributive module M.

(1) If E is a module and there exist epimorphisms $f_i : E \to D_i$ for all $i = 1, \ldots, n$, then $f_1 + \ldots + f_n : E \to D$ is an epimorphism of E onto D.

(2) If t is a cardinal and all D_i are t-generated modules, then D is t-generated.

(3) [166]. If all D_i are cyclic, then D is cyclic.

(4) Each finitely generated semisimple subfactor of M is cyclic.

$\triangleleft$ (1) By induction, we may assume that $n = 2$ and $D = D_1 \oplus D_2$. Let $H \equiv (f_1 + f_2)(E)$. Let h_1 and h_2 be the natural projections of D onto D_1 and D_2, respectively. Since $H = H \cap D = H \cap D_1 \oplus H \cap D_2$, we have $H \supseteq h_1(H) \oplus h_2(H) = L$, and $(f_1 + f_2)$ is an epimorphism.

(2) follows from (1) (E is a free module with rank t). (3) follows from (2). (4) follows from (3). $\triangleright$

2.24 Let N be a superfluous submodule of a module M.

(1) M is cyclic $\Longleftrightarrow$ M/N is cyclic.

(2) If M is a distributive module, then
M is cyclic $\Longleftrightarrow$ M/N is a finite direct sum of cyclic modules.

$\triangleleft$ (1) is directly verified.

(2) By 2.23(3), M/N is cyclic. Hence (2) follows from (1). $\triangleright$

2.25 Let E be a right module over a ring A. Then
E_A is a Bezout module $\Longleftrightarrow$ for any 2-generated submodule $M \in \mathrm{Lat}(E)$, the 2-generated module $M/J(M)$ is cyclic $\Longleftrightarrow$ for any 2-generated submodule $M \in \mathrm{Lat}(E)$, the 2-generated $A/J(A)$-module $M/MJ(A)$ is cyclic.

$\triangleleft$ Let M be a 2-generated submodule of E. Then $J(M)$ and $MJ(A)$ are superfluous in M. The assertion follows now from 2.24(1). $\triangleright$

2.26 For a distributive module E_A, the following conditions are equivalent.

(1) E is a Bezout module.

(2) For each 2-generated submodule M of E, the 2-generated module $M/J(M)$ is a direct sum of cyclic modules.

(3) For each 2-generated submodule M of E, the 2-generated $A/J(A)$-module $M/MJ(A)$ is a direct sum of cyclic modules.

$\triangleleft$ Let M be a 2-generated submodule of E. Then $J(M)$ and $MJ(A)$ are superfluous in M. Now our assertion follows from 2.24(2). $\triangleright$

2.27 [225]. *A distributive Artinian module E is a Bezout module.*

◁ Let M be a 2-generated submodule of E. The Artinian semiprimitive module $M/J(M)$ is a finite direct sum of simple modules. Now our assertion follows from 2.26. ▷

2.28 Let M_A be a distributive module over a ring A, and let each distributive finitely generated right $A/J(A)$-module be a direct sum of cyclic modules.
Then M is a Bezout module.

◁ 2.28 follows from 2.26. ▷

2.29 [167]. *Every distributive module over a semilocal ring is a completely finite-dimensional Bezout module.*

◁ 2.29 follows from 2.28 and 2.13(1). ▷

2.30 A module M is *quasi-invariant (resp. invariant)* if all its maximal submodules (resp. all its submodules) are fully invariant in M.

A ring A is right quasi-invariant (right invariant) $\Longleftrightarrow$ all maximal right ideals (all right ideals) of A are ideals $\Longleftrightarrow$ each cyclic right A-module is quasi-invariant (invariant).

2.31 Let M be a right module over a ring A. Then
M is distributive $\Longleftrightarrow$ for any subfactor $\overline{M}$ of M and for any elements $\overline{m}, \overline{n} \in \overline{M}$ such that $\overline{m}A \bigcap \overline{n}A = 0$, there exist $a, b \in A$ such that $1 = a + b$ and $\overline{m}a = \overline{n}b = 0$.

◁ $\Longrightarrow$ follows from 2.6(1).
$\Longleftarrow$ Let $m, n \in M$, $\overline{M} \equiv M/(mA \bigcap nA)$, and let $\overline{m}$, $\overline{n}$ be the natural images of m,n in $\overline{M}$. Then $\overline{m}A \bigcap \overline{n}A = 0$, $(m : n) = r(\overline{m})$, and $(n : m) = r(\overline{n})$. By assumption, $A = r(m) + r(n) = (m : n) + (n : m)$. By 2.4, M is distributive. ▷

2.32 [166], [34]. *For a module M_A, the following conditions are equivalent.*
(1) *M is distributive.*
(2) *M has no subfactors $S \oplus T$, where T is a nonzero homomorphic image of S.*
(3) *M has no subfactors which are isomorphic to $T \oplus T$, where T is a simple module.*
(4) *M is 2-distributive.*
(5) *M has no 2-generated submodules which have factor modules which are isomorphic to $T \oplus T$, where T is a simple module.*

◁ $(1)\Longrightarrow(2)$ follows from 2.6(3). $(2)\Longrightarrow(3)$ and $(4)\Longleftrightarrow(5)$ are directly verified.

$(3)\Longrightarrow(5)$ Assume the contrary. There exist an epimorphism $h : M \to \overline{M}$ and cyclic submodules F,G of M such that $h(F),h(G)$ are isomorphic simple modules and $h(F)\bigcap h(G) = 0$. Then $F + G$ is a 2-generated submodule of M, and $(F+G)/(F+G)\bigcap \mathrm{Ker}(h)$ is a direct sum of two isomorphic simple modules; this contradicts to (4).

$(5)\Longrightarrow(1)$ Assume that M is not distributive. By 2.31, there exist a subfactor $\overline{M}$ of M and $m,n \in \overline{M}$ such that $mA\bigcap nA = 0$ and $r(m) + r(n) \neq A$. Hence $r(m) + r(n) \subseteq B$, where B is a maximal right ideal of A. Therefore mA and nA have factor modules which are isomorphic to the simple module A/B. Hence M has a subfactor isomorphic to $A/B \oplus A/B$; this is a contradiction. ▷

2.33 If B,D are two ideals of a ring A and $(A/B)_A \cong (A/D)_A$, then $B = D$.

◁ Let $f : (A/B)_A \to (A/D)_A$ be an isomorphism, and let h_B, h_D be natural epimorphisms of A_A onto $(A/B)_A$ and $(A/D)_A$, respectively. Since A_A is projective, there exists $g \in \mathrm{End}(A_A)$ such that $fh_B = h_D g$. Since f is an isomorphism, $\mathrm{Ker}(fh_B) = \mathrm{Ker}(h_B) = B$. Let $d \in D$, and let $m \equiv g(1)$. Then $md \in D$, since D is an ideal. Hence $(fh_B)(d) = (h_D g)(d) = h_D(md) \subseteq h_D(D) = 0$, $d \in \mathrm{Ker}(fh_B) = B$, and $D \subseteq B$. Analogously, $B \subseteq D$. ▷

2.34 [184]. *For a module M_A over a right quasi-invariant ring A, the following conditions are equivalent.*

(1) M is distributive.

(2) Any 2-generated semisimple factor module H of an arbitrary 2-generated submodule N of M is cyclic.

(3) Any 2-generated semisimple factor module H of an arbitrary 2-generated submodule N of M is cyclic $A/J(A)$-module.

◁ $(2)\Longleftrightarrow(3)$ is directly verified. $(1)\Longrightarrow(2)$ follows from 2.23(4).

$(2)\Longrightarrow(1)$ By 2.32, it is sufficient to prove that if N is any 2-generated submodule of M, then N cannot have a factor module H which is a direct sum of two isomorphic simple modules. Assume the contrary. By assumption, H is cyclic. Therefore A has distinct maximal right ideals B and D such that $(A/B)_A \cong (A/D)_A$. By 2.33, $B = D$; this is a contradiction. ▷

2.35 [167]. *Every right Bezout module over a right quasi-invariant ring is distributive.*

◁ 2.35 follows from 2.34. ▷

2.36 Assume that a ring A has an ideal B such that $B \subseteq J(A)$, and A/B is a right quasi-invariant ring.

Then each right A-Bezout module is distributive.

◁ Let $h : A \to A/B$ be the natural epimorphism. By 2.35, it is sufficient to prove that any maximal right ideal M of A is an ideal. Since $h(A)$ is right quasi-invariant, $h(AM) \subseteq h(M)$. Hence $AM \subseteq M + B \subseteq M \subseteq J(A) = M$. ▷

2.37 Let M_A be a module over a ring A, and let $A/J(A)$ be a finite direct product of division rings. Then

M is distributive $\Longleftrightarrow$ M is a Bezout module.

◁ $\Longrightarrow$ follows from 2.29.

$\Longleftarrow$ follows from 2.36 $(B \equiv J(A))$ and from the fact that finite direct products of division rings are right invariant. ▷

2.4   Modules over semiperfect rings

2.38 Let A be a semiperfect ring. There exists the decomposition $1 = \sum_{i=1}^{n} e_i$ of the identity element of A into a sum of local orthogonal nonzero idempotents e_i. Without loss of generality, we may assume that $\{e_i\}_{i=1}^{t}$ $(t \leq n)$ is the set of all pair-wise nonisomorphic modules in the set $\{\{e_i A\}_{i=1}^{n}\}$. The idempotent $e \equiv e_1 + \ldots + e_t$ is called a *basis idempotent* of the semiperfect ring A. The ring eAe is a *basis ring* of A. The module eA is called a *basis right module* of A.

A semiperfect ring A is said to be a *self-basis* ring if the following three equivalent conditions hold.

(i) $A/J(A)$ is a finite direct product of division rings.

(ii) e coincides with the identity element of A.

(iii) A coincides with its basis ring eAe.

For the semiperfect ring A, the following assertions hold.

(1) The basis ring eAe is always self-basis.

(2) There exists a positive integer m such that the free cyclic module A_A is isomorphic to a direct summand of the direct sum of m copies of the basis module eA.

(3) $e_i = e_i e = e e_i$ for $i = 1, \ldots, t$.

(4) $Ne \neq 0$ for any nonzero right A-module N.

(5) If M is a right A-module, then by the rule $\varphi(N) = Ne$, the lattice isomorphism $\varphi : \mathrm{Lat}(M_A) \to \mathrm{Lat}(Me_{eAe})$ is well defined.

(6) A module M_A is distributive (resp. Noetherian, Artinian, semisimple, semidistributive) $\Longleftrightarrow$ the module Me_{eAe} is distributive (resp. Noetherian, Artinian, semisimple, semidistributive).

$\lhd$ (1), (2), and (3) are directly verified.

(4) The nonzero module N_A has a simple subfactor $T \cong e_i A/e_i J$, where $i \leq t$ and $J \equiv J(A)$. Assume that $Ne = 0$. Then $Te = 0$. Hence $e_i e \in e_i J$. By (3), $e_i = e_i e \in J$; this is impossible.

(5) By 1.5(10), φ is a surjective lattice homomorphism. Let $F, G \in$ Lat(M), and let $N_A \equiv (F + G)/(F \bigcap G)$. Assume that $\varphi(F) = \varphi(G)$. Then $Fe = Ge$, whence $Ne = 0$. By (4), $N = 0$. Hence $F + G = F \bigcap G$, $F = G$, and φ is an isomorphism.

(6) follows from (5). $\rhd$

2.39 A module M is a *completely cyclic* module if that all submodules of M are cyclic.

M is a completely cyclic module $\Longleftrightarrow$ M is a Noetherian Bezout module.

2.40 [187]. *Let M be a right module over a semiperfect ring A, and let e be a basis idempotent of A.*

(1) M_A is distributive $\Longleftrightarrow$ Me_{eAe} is a Bezout module.

(2) Let A be left perfect. Then

M_A is distributive $\Longleftrightarrow$ Me_{eAe} is an Artinian Bezout module.

(3) Let A be right perfect. Then

M_A is distributive $\Longleftrightarrow$ Me_{eAe} is a completely cyclic module.

(4) Let A be perfect. Then

M_A is distributive $\Longleftrightarrow$ Me_{eAe} is a completely cyclic module with a composition series.

$\lhd$ (1) By 2.38(1), $eAe/J(eAe)$ is a finite direct product of division rings. By 2.37, Me_{eAe} is distributive $\Longleftrightarrow$ Me_{eAe} is a Bezout module. Now we apply 2.38(6).

(2) follows from (1) and 2.20(1). (3) follows from (1), 2.39, and 2.20(3). (4) follows from (2) and (3). $\rhd$

2.41 Let $1 = e_1 + \ldots + e_n$ be a decomposition of the identity element of A into a sum of nonzero orthogonal idempotents, P be the prime radical of A, N be a right ideal of A, and let M_A be a right A-module.

(1) If N is a right nil-ideal of A, then $e_i N e_i$ is a right nil-ideal of the ring $e_i A e_i$ for all i.

(2) If $e_i N e_i = 0$ for all i, then $N^n = 0$.

(3) If a right ideal $e_i N e_i$ of $e_i A e_i$ is contained in the prime radical P_i of $e_i A e_i$ for all i, then N is contained in P.

(4) If the prime radical of $e_i A e_i$ contains all right nil-ideals (resp. all left nil-ideals) of $e_i A e_i$ for all i, then P contains all right nil-ideals (resp.a ll left nil-ideals) of A.

(5) If the ring $e_i A e_i$ is semiprime for all i, then $P^n = 0$.

(6) If $i \in \{1, \ldots, n\}$ and $ae_i \in Ae_i$, then

$ae_i \in \mathrm{Sing}(A_A) \iff$ there exists an essential submodule B_i of $e_i A$ such that $ae_i B_i = 0$.

(7) If a is an element of A, then

$a \in \mathrm{Sing}(A_A) \iff$ for any e_i, there exists an essential submodule B_i of $e_i A$ such that $ae_i B_i = 0$.

(8) A is right nonsingular $\iff$ for any idempotent e_i, we have that $aB \neq 0$ for each nonzero $a \in Ae_i$ and for any essential submodule B of $e_i A$.

(9) Each right A-module X_A is a group direct sum of right $e_i A e_i$-modules Xe_i.

(10) If F and G are two submodules of M_A, then $F + G = \sum_{i=1}^{n}(Fe_i + Ge_i)$ and $F \cap G = \sum_{i=1}^{n} Fe_i \cap Ge_i$.

(11) M_A is distributive $\iff$ all $e_i A e_i$-modules Me_i are distributive.

$\lhd$ (1) is directly verified.

(2) Since $e_i N e_i = 0$ for all i and N is a right or left ideal, $N e_i N \subseteq N$ and $e_j N e_i N e_j = 0$ for all i, j. Hence $N = \sum_{i \neq j} e_i N e_j$ and $N^n = (\sum_{i \neq j} e_i N e_j)^n = 0$.

(3) Let $h : A \to A/P$ be the natural epimorphism. By 1.5(2),(18) and by assumption, $h(e_i)h(N)h(e_i) \subseteq h(P_i) = h(e_i)h(P)h(e_i) = 0$ for all i. Apply (2) to $h(A)$ and $h(N)$. Then $h(N)$ is a nilpotent right ideal of the semiprime ring $h(A)$. Therefore $h(N) \subseteq h(P)$ and $N \subseteq P$.

(4) follows from (3) and its left-side analog. (5) follows from (2) and 1.5(18).

(6) $\Longrightarrow$ Set $B_i \equiv r(ae_i) \cap e_i A$. If $ae_i \in \mathrm{Sing}(A_A)$, then B_i is essential in $e_i A$ and $ae_i B_i = 0$.

$\Longleftarrow$ By assumption, there exists an essential submodule B_i of $e_i A$ such that $B_i \subseteq r(ae_i)$. Set $D \equiv B_i \oplus \sum_{j \neq i}$. Then D is an essential right ideal of A which is contained in $r(ae_i)$. Therefore $ae_i \in S$.

(7) follows from (6) and from the fact that $a = \sum_{i=1}^{n} ae_i$. (8) follows from (7). (9) is directly verified. (10) follows from (9). (11) follows from (10). $\rhd$

2.42 Each finitely generated projective module over a semiperfect ring A, is a finite direct sum of indecomposable direct summands of A_A, and every such indecomposable summand is isomorphic to some $e_i A$, where e_i is some local idempotent of A.

2.5 Closed submodules

2.43 Let $\{G_i\}_{i\in I}$ be a set of submodules of a distributive module M, and let $G \equiv \sum_{i\in I} G_i$.

(1) $X \bigcap \sum_{i\in I} G_i = \sum_{i\in I}(X \bigcap G_i)$ for any submodule X of M.

(2) If X is a submodule of M such that $X \bigcap \sum_{i\in I} G_i \neq 0$, then $X \bigcap G_i \neq 0$ for some $i \in I$.

(3) If H is a submodule of M such that $H \bigcap G_i = 0$ for all $i \in I$, then $H \bigcap \sum_{i\in I} G_i = 0$.

(4) If $\{N_i\}_{i\in I}$ is a set of submodules of M such that G_i is an essential extension of N_i for any $i \in I$, then $\sum_{i\in I} G_i$ is an essential extension of $\sum_{i\in I} N_i$.

(5) If all modules G_i are essential extensions of a module N, then $\sum_{i\in I} G_i$ is an essential extension of N.

◁ (1) Let $L \equiv \sum_{i\in I}(X \bigcap G_i)$, and let $t \in X \bigcap \sum_{i\in I} G_i$. The assertion follows from the fact that there exists a finite subset $D \subseteq I$ such that $t \in X \bigcap \sum_{i\in D} G_i = \sum_{i\in D}(X \bigcap G_i)$.

(2) and (3) follows from (1).

(4) Let X be a nonzero submodule of $\sum_{i\in I} N_i$. By (2), $X \bigcap G_i \neq 0$ for some $i \in I$. Therefore $X \bigcap \sum_{i\in I} G_i \neq 0$.

(5) follows from (4). ▷

2.44 Every submodule N of a distributive module M has the unique closure G in M and the unique complement H in M.

Also, G coincides with the sum of all essential extensions of N in M, and H coincides with the sum of all submodules of M which have the zero intersection with N.

◁ Let G be the sum of all essential extensions of N in M, and let H be the sum of all submodules of M having zero intersection with N. By 2.43(5), G is the largest essential extension of N in M. Therefore G is the unique closure of N in M. By 2.43(3), H is the largest submodule of M which has zero intersection with N. Therefore H is the unique complement to N in M. ▷

2.45 Let N be a submodule of a distributive module M.

(1) If G is a submodule of M such that N is an essential extension of $G \bigcap N$, then $G + N$ is an essential extension of G.

(2) If G is a closed submodule of M and N is an essential extension of $G \bigcap N$, then $N \subseteq G$.

(3) If G is a closed submodule of M and N is uniform, then either $N \subseteq G$ or $N \bigcap G = 0$.

(4) The intersection of any two distinct closed uniform submodules of M is equal to zero.

◁ (1) follows from 2.43(4) and from the fact that $G = G + N \bigcap G$. (2) follows from (1). (3) follows from (2). (4) follows from (3). ▷

2.46 Let M be a distributive module M.

(1) If $\{G_i\}_{i \in I}$ is the set of all distinct closed uniform submodules of M and $G \equiv \sum_{i \in I} G_i$, then $G = \oplus_{i \in I} G_i$ and G contains all uniform submodules of M.

(2) [184] M has a semiuniform submodule containing all uniform submodules of M.

(3) If $M = \oplus_{i \in I} M_i = \oplus_{j \in J} N_j$, where the M_i and N_j are nonzero indecomposable modules, then there exists a bijection $s : I \to J$ such that M_i coincides with $N_{s(i)}$.

(4) If M is generated by its nonzero uniform submodules, then M is the unique direct sum of nonzero uniform modules.

(5) M is weakly invariant, and $N + f(N)$ is an essential extension of N for any $N \in \mathrm{Lat}(M)$ and each $f \in \mathrm{End}(M)$.

(6) M is uniform $\Longleftrightarrow$ M has no nonzero proper fully invariant closed submodules $\Longleftrightarrow$ any two nonzero fully invariant closed submodules of M have a nonzero intersection.

(7) M is finite-dimensional $\Longleftrightarrow$ each submodule of M is an essential extension of a cyclic module which is a finite direct sum of cyclic uniform modules $\Longleftrightarrow$ M contains no infinite direct sums of nonzero fully invariant closed submodules $\Longleftrightarrow$ M satisfies the maximum condition on nonzero fully invariant closed submodules $\Longleftrightarrow$ M satisfies the minimum condition on nonzero fully invariant closed submodules.

◁ (1) follows from 2.45(4) and 2.43(3). (2) follows from (1). (3) follows from 2.43(1). (4) follows from (2) and (3).

(5) Let $N \in \mathrm{Lat}(M)$, $f \in \mathrm{End}(M)$, and let H be a submodule of $N + f(N)$ such that $H \bigcap N = 0$. Then $H = H \bigcap (N + f(N)) = H \bigcap N + H \bigcap f(N) = H \bigcap f(N) \subseteq f(N)$. Set $G \equiv f^{-1}(H) \in \mathrm{Lat}(N)$. Then $H = f(G)$ and $G \bigcap f(G) = 0$. By 2.6(2), $\mathrm{Hom}(G, H) = 0$. Hence $H = f(G) = 0$, and $N + f(N)$ is an essential extension of N. If N is closed in M, then $f(N) \subseteq N$, since $N + f(N)$ is an essential extension of N.

(6) follows from (5). (7) follows from (5) and 2.23(3). ▷

2.47 Let M_A be a direct sum of distributive modules M_i, where $i \in I$. Then the following conditions are equivalent.

(1) M is distributive.

(2) $h_i(m) \in mA$ for any $m \in M$ and for every natural projection $h_i : M \to M_i$.

(3) $A = r(f) + r(g)$ for all $f \in M_i$, $g \in \sum_{j \in I \setminus i} M_j$ and for each subscript $i \in I$.

◁ $(2)\Longrightarrow(1)$ is directly verified. $(1)\Longrightarrow(3)$ follows from 2.31.

$(3)\Longrightarrow(2)$ Set $f \equiv h_i(m) \in M_i$, $g \equiv m - f \in \sum_{j\in I\setminus i} M_j$. By assumption, there exist $a, b \in A$ such that $1 = a + b$ and $fa = gb = 0$. Therefore $f = f(a + b) + gb = (f + g)b = mb \in mA$. ▷

2.48 Let M_A be a module over a right invariant ring A, and let $M_A = \oplus_{i\in I} M_i$. Then

M is distributive $\Longleftrightarrow$ all M_i are distributive, and $A = r(m_i) + r(m_j)$ for all distinct subscripts $i, j \in I$ and for all $m_i \in M_i$, $m_j \in M_j$.

◁ $\Longrightarrow$ follows from 2.31.

$\Longleftarrow$: Let $i \in I$, $f \in M_i$, and let $g \in \oplus_{j\in I\setminus i} M_j$. We may assume that $g = \sum_{i=1}^n g_i$, where $1, \ldots, n \in I$. By assumption, $A = r(f) + r(g_i)$ for $i = 1, \ldots, n$. Since all $r(g_i)$ are ideals, $A = r(f) + \bigcap_{i=1}^n r(g_i) = r(f) + r(g)$. By 2.47, M is distributive. ▷

2.49 Let M_A be a module and let M be a direct sum of cyclic modules $m_i A$, where $i \in I$.

(1) If all modules $m_i A$ are distributive and $r(m_i)$ are ideals of A for all $i \in I$, then

M is distributive $\Longleftrightarrow$ $A = r(m_i) + r(m_j)$ for all distinct subscripts $i, j \in I$.

(2) If A is a right distributive right invariant ring, then

M is distributive $\Longleftrightarrow$ $A = r(m_i) + r(m_j)$ for all distinct subscripts $i, j \in I$.

◁ (1)

$\Longrightarrow$ follows from 2.31.

$\Longleftarrow$ Let $f \in M_i$, and let $g \in \sum_{j\in R} M_j$, where $R = \{1, \ldots, n\}$ is a finite subset of the set $I \setminus i$. Set $B \equiv \bigcap_{j=1}^n r(m_j)$. Since all $r(m_j)$ are ideals of A and $A = r(m_i) + r(m_j)$ for $1 \leq j \leq n$, we have $A = r(m_i) + B$, $B \subseteq r(g)$, and $r(m_i) \subseteq r(f)$. Hence $A = r(f) + r(g)$. By 2.47, M is distributive.

(2) follows from (1) and from the fact that each cyclic right module over a right distributive ring is distributive. ▷

2.6 Endomorphisms

2.50 (1) The endomorphism ring of a module M is normal $\Longleftrightarrow$ $\mathrm{Hom}(G, H) = 0$ for any direct decomposition $M = G \oplus H$ $\Longleftrightarrow$ each direct summand of M is fully invariant in M.

(2) If a module M is distributive, then the ring $\mathrm{End}(M)$ is normal and each direct summand of M is fully invariant in M.

(3) Every right or left distributive ring is normal.

◁ (1) is directly verified. (2) follows from (1) and 2.6(2). (3) follows from (2). ▷

2.51 The endomorphism rings of all factor modules of a module M are normal $\Longleftrightarrow \mathrm{Hom}(G/(G \cap H), H/(G \cap H)) = 0$ for all submodules $G, H \in \mathrm{Lat}(M)$ such that $G + H = M \Longleftrightarrow \mathrm{Hom}(M/H, M/G) = 0$ for all submodules $G, H \in \mathrm{Lat}(M)$ such that $G + H = M$.

◁ If $G, H \in \mathrm{Lat}(M)$ and $G + H = M$, then $G/(G \cap H) \cong M/H$ and $H/(G \cap H) \cong M/G$. We can now apply 2.50(1). ▷

2.52 If M is a quasi-invariant module, then the ring $\mathrm{End}(M/J(M))$ is isomorphic to a subring of a direct product of division rings, and $\mathrm{End}(M/J(M))$ is a reduced ring.

In particular, if a ring A is right quasi-invariant, then $A/J(A)$ is reduced.

◁ Each maximal submodule N of $\overline{M} \equiv M/J(M)$ is a fully invariant submodule, and $\mathrm{End}(\overline{M}/N)$ is a division ring. Therefore, there exists a natural ring homomorphism $f_N : \mathrm{End}(\overline{M}) \to \mathrm{End}(\overline{M}/N)$. The intersection of kernels of all homomorphisms f_N is equal to zero. Therefore, the ring $\mathrm{End}(M/J(M))$ is isomorphic to a subring of a direct product of division rings. Hence $\mathrm{End}(M/J(M))$ is reduced. ▷

2.53 Let the endomorphism rings of all factor modules of a module M be normal (by 2.50(2) this is the case if M is distributive), $f \in \mathrm{End}(M)$, and let $N \in \mathrm{Lat}(M)$.

(1) If $f^{-1}(N) + N = M$, then $f(M) \subseteq N$.

(2) If $f(M) \subseteq N + f(N)$, then $f(M) \subseteq N$.

(3) If $M = N + f(N)$, then $M = N$.

(4) M is quasi-invariant, $\mathrm{End}(M/J(M))$ is a reduced ring, and $\mathrm{End}(M/J(M))$ is isomorphic to a subring of a direct product of division rings.

◁ (1) By the rule $g(m + f^{-1}(N)) = f(m) + N$, the monomorphism $g : M/f^{-1}(N) \to M/N$ is well defined. By 2.51, $g \equiv 0$. Hence $f^{-1}(N) = M$, $f(M) \subseteq N$.

(2) Let $m \in M$. By assumption, there exist $n, t \in N$ such that $f(m) = n + f(t)$. Hence $m - t \in f^{-1}(N)$, $m = (m - t) + t \in f^{-1}(N) + N$, $M = f^{-1}(N) + N$. By (1), $f(M) \subseteq N$.

(3) follows from (2) and the inclusion $f(N) \subseteq f(M)$. (4) follows from (3) and 2.52. ▷

2.54 [166] [207]. Every subfactor N of a distributive module M is a quasi-invariant distributive module, the ring $\mathrm{End}(N)$ is normal, and the ring $\mathrm{End}(N/J(N))$ is reduced.

◁ The module N is distributive. By 2.50(2), $\mathrm{End}(N)$ is normal. By 2.53(4), N is quasi-invariant, and $\mathrm{End}(N/J(N))$ is reduced. ▷

2.55 Let L_A and M_A be two modules over a ring A, and let T be a subset of the group $\mathrm{Hom}(L_A, M_A)$ such that T is generated by all homomorphisms whose images are superfluous in L. The set of all endomorphisms $f \in \mathrm{End}(M)$ such that $f(M)$ is superfluous in M is denoted by $\mathrm{gs}(M)$.

(1) T is a subbimodule of $(\mathrm{End}(M), \mathrm{End}(L))$-bimodule $\mathrm{Hom}(L_A, M_A)$.

(2) $\sum_{t \in T} t(L)$ is a fully invariant submodule in M_A.

(3) $\mathrm{gs}(M)$ is an ideal of $\mathrm{End}(M)$.

(4) The natural isomorphism $\mathrm{End}(A_A) \to A$ maps from the ideal $\mathrm{gs}(A_A)$ of $\mathrm{End}(A_A)$ onto $J(A)$.

2.56 Let the endomorphism rings of all factor modules of a module M be normal (by 2.50(2), this is the case if M is distributive).

(1) Assume that $f \in \mathrm{End}(M)$, n is a positive integer, and $f^n(M)$ is superfluous in M.

Then $f(M)$ is superfluous in M.

(2) $\mathrm{End}(M)/\mathrm{gs}(M)$ is a reduced ring.

(3) Images of all nilpotent endomorphisms of the module M are superfluous in M.

◁ (1) It is sufficient to prove that if m is a positive integer and $f^{m+1}(M)$ is superfluous in M, then $f^m(M)$ is superfluous in M. Let B be a submodule of M such that $M = B + f^m(M)$. Then $f(B) + f^{m+1}(M) = f(M) \supseteq f^m(M)$. Therefore $M = B + f(B) + f^{m+1}(M)$. Since $f^{m+1}(M)$ is superfluous in M, we obtain $M = B + f(B)$. By 2.53(3) applied to B, $M = B$. Therefore $f^m(M)$ is superfluous in M.

(2) follows from (1). (3) follows from (2). ▷

2.57 An endomorphism f of a module M is called *locally nilpotent* if for any $m \in M$, there exists a positive integer $n = n(m)$ such that $f^n(m) = 0$. A module M is *locally Noetherian* if each of its finitely generated submodule is Noetherian.

(1) A module M is locally Noetherian $\Longleftrightarrow$ each cyclic submodule of M is Noetherian.

(2) Every right module over a right Noetherian ring is locally Noetherian.

(3) If f is an endomorphism of a locally Noetherian module M and $\mathrm{Ker}(f)$ is an essential submodule of M, then f is locally nilpotent.

◁ (1) follows from the fact that an extension of a Noetherian module by a Noetherian module is Noetherian.

(2) follows from (1) and from the fact that each cyclic right module over a right Noetherian ring is Noetherian.

(3) Let D be a cyclic submodule of M, and let $N_i = D \cap \mathrm{Ker}(f^i)$. Since $N_i \subseteq N_{i+1}$ and D is Noetherian by assumption, there exists a positive integer n such that $N_{n+1} = N_n$. Hence a homomorphism f^n induces an isomorphism $h : D/N_n \to f^n(D)$. Let $H = \mathrm{Ker}(f) \cap f^n(D)$, and let $B/N_n = h^{-1}(H)$, where $N_n \subseteq B \subseteq D$. If $f^n(D) = 0$, then f is locally nilpotent. Assume that $f^n(D) \neq 0$. Since M is an essential extension of $\mathrm{Ker}(f)$, the module B properly contains N^n, and $f^{n+1}(B) = f(H) = 0$. Hence $B \subseteq N_{n+1} = N_n$; we have a contradiction. ▷

2.58 Let the endomorphism rings of all factor modules of a module M be normal (by 2.50(2), this is the case if M is distributive), $f \in \mathrm{End}(M)$, and let $N \in \mathrm{Lat}(M)$.

If $M = \sum_{i=0}^{n} f^i(N)$ for some positive integer n, then $M = N$.

◁ For $n = 1$, our assertion follows from 2.53(3). Assume that $n > 1$, the assertion holds for $n - 1$, and $D = \sum_{i=0}^{n-1} f^i(N)$. Then $M = D + f(D)$. By 2.53(3), $M = D$. Our assertion follows now from the induction hypothesis. ▷

2.59 Let an element a of a ring A be a root of a polynomial $f(x)$ with coefficients from the centre $C(A)$ of A. If the leading coefficient of $f(x)$ is a unit of A, then the element a is called *integral* over $C(A)$. If the leading coefficient of $f(x)$ is invertible in A, then a is *algebraic* over $C(A)$. A ring A is *integral (algebraic)* over its centre if all its elements are integral (algebraic) over $C(A)$.

2.60 Let M be a distributive module.

(1) [166] Let $f \in \mathrm{End}(M)$, $N \in \mathrm{Lat}(M)$, and let $f^{n+1}(N) \subseteq \sum_{i=0}^{n} f^i(N)$ for some positive integer n.

Then $f(N) \subseteq N$.

(2) [166] Let an endomorphism $t \in M$ be integral over the centre of $\mathrm{End}(M)$.

Then t maps from every endomorphic image of M into itself.

(3) If the ring $\mathrm{End}(M)$ is integral over its centre, then each endomorphic image of M is fully invariant in M.

◁ (1) Set $T \equiv \sum_{i=0}^{n} f^i(N)$. Then $f(T) \subseteq T$. Therefore f induces $g \in \mathrm{End}(T)$, $T = \sum_{i=0}^{n} g^i(N)$, and endomorphism rings of all subfactors of T are normal. By 2.58 applied to T, we obtain $f(N) = g(N) \subseteq N$.

(2) Let $h \in \mathrm{End}(M)$, and let $N \equiv h(M)$. It is sufficient to prove that $t(N) \subseteq N$. Let $t^{n+1} = \sum_{i=0}^{n} t^i g_i$, where all g_i are contained in the centre of $\mathrm{End}(M)$, and let $D = \sum_{i=0}^{n} t^i(N)$. Since $g_i h = h g_i$, we obtain $g_i(N) \subseteq N$. Therefore $t^{n+1}(N) \subseteq D$. Hence t induces an endomorphism f of D, and $D = \sum_{i=0}^{n} f^i(N)$. By 2.58 applied to D, $t(N) = f(N) \subseteq N$.

(3) follows from (2). ▷

2.61 Let A be a unitary subring of a ring B, and let the natural module B_A be distributive.

(1) If $N \in \mathrm{Lat}(B_A)$ and $B = \sum_{i=0}^{n} b^i N$ for some positive integer n, then $B = N$.

(2) Let $b \in B$, $N \in \mathrm{Lat}(B_A)$, and let $b^{n+1} N = \sum_{i=0}^{n} b_i N$ for some positive integer n.

Then $bN \subseteq N$.

(3) Assume that for $b \in B$, there exist $a_0, \ldots, a_n \in A \bigcap C(B)$ such that $b^{n+1} = \sum_{i=0}^{n} b^i a_i$ (e.g., this is the case if b is either an idempotent or a nilpotent element).

Then $b \in A$ and $bN \subseteq N$ for any $N \in \mathrm{Lat}(B_A)$.

(4) B is right distributive, and all idempotents and all nilpotent elements of B are contained in A.

(5) The prime radical E of B is contained in the prime radical D of the ring A.

If the ring B/E is reduced, then $D = E$, A/D is a reduced ring, and D coincides with the set of all nilpotent elements of B.

◁ (1) For each $b \in B$, the rule $f_b(x) = bx$ defines an endomorphism f_b of the distributive module B_A. Since A is a unitary subring in B, the inclusion $b \in A$ is equivalent to inclusion $f_b(A) \subseteq A$. Therefore, (1) follows from 2.58.

(2) follows from (1). (3) follows from (2), since A is a unitary subring of B.

(4) Since B_A is distributive, A_A is distributive. Therefore, (4) follows from (2).

(5) By (4), all nilpotent elements of B are contained in A. Therefore $E \subseteq D$. If B/E is reduced, then A/E is a reduced ring and all nilpotent elements of B are contained in E. Hence $D = E$. ▷

2.62 [176] *If a right distributive ring A is integral over its centre, then A is a right invariant ring.*

◁ 2.62 follows from 2.60(3). ▷

2.63 A module M is a *nilpotently invariant* module if each of its nilpotent endomorphism maps from every submodule of M into itself.

If M is a distributive module, then each locally nilpotent endomorphism f of M maps from every submodule of M into itself.

In particular, M is nilpotently invariant.

◁ (1) It is sufficient to prove that $f(N) \subseteq N$ for any cyclic submodule $N \in \mathrm{Lat}(M)$. Then $f^n(N) = 0$ for some n. Further, we use 2.60(1).

(2) follows from (1). ▷

2.64 [213]. Let M be a nilpotently invariant module, N be a submodule of M, and let T be the set of all nilpotent endomorphisms of M.

(1) The ideal $\mathrm{sg}(M)$ of $\mathrm{End}(M)$ contains all nilpotent endomorphisms of M.

(2) The image of each nilpotent endomorphism of M is contained in $\mathrm{Sing}(M)$.

(3) If M is nonsingular, then the ring $\mathrm{End}(M)$ is reduced.

(4) If M is nonsingular and $\mathrm{End}(M)$ is prime, then $\mathrm{End}(M)$ is a domain.

(5) If f_i and g_j are nilpotent endomorphisms of M with $1 \leq i, j \leq n$ and $(f_1 f_2 \ldots f_n)(N) = 0$, then $(f_1 g_1 f_2 g_2 \ldots f_n g_n)(N) = 0$.

(6) If f and g are two nilpotent endomorphisms of M and $f^m(N) = g^n(N) = 0$, then $(f + g)^{m+n}(N) = 0$.

(7) Let f be a nilpotent endomorphism of M, T_1 be the subring of $\mathrm{End}(M)$ which is generated by $\{1\} \bigcup T$, and let $f^n(N) = 0$.

Then $(fT_1)^n(N) = 0$.

(8) T is a nilsubring of $\mathrm{End}(M)$.

(9) The nil-ring T coincides with the sum of its nilpotent ideals.

(10) The prime radical P of $\mathrm{End}(M)$ contains all right nil-ideals and all left nil-ideals of $\mathrm{End}(M)$.

◁ (1) Let f be a nilpotent endomorphism of M. It is sufficient to prove that $\mathrm{Ker}(f^{n+1})$ is an essential extension of $\mathrm{Ker}(f^n)$ for any n. Let $U = \mathrm{Ker}(f^n)$, and let V be a submodule of $\mathrm{Ker}(f^{n+1})$ such that $U \bigcap V = 0$. Since $f^{n+1}(V) = 0$, $f(V) \subseteq U$. Since M is nilpotently invariant, $f(V) \subseteq V \bigcap U = 0$. Hence $V \subseteq \mathrm{Ker}(f) \subseteq U$. Therefore $V = V \bigcap U = 0$. Hence $\mathrm{Ker}(f^{n+1})$ is an essential extension of $\mathrm{Ker}(f^n)$.

(2) follows from (1) and 1.29. (3) follows from (2). (4) follows from (3) and from the fact that by 1.35(2), a reduced prime ring is a domain. (5) follows from the fact that nilpotent endomorphisms f_i, g_j

of the nilpotently invariant module M maps from every submodule of M into itself. (6) and (7) follows from (5). (8) follows from (5) and (6). (9) follows from (5) and (7).

(10) Let $R \equiv \mathrm{End}(M)$, and let D be an ideal of R generated by an arbitrary right nil-ideal E of R. The equality $(xy)^{n+1} = x(yx)^n y$ implies that D is a nil-ideal of R. By (6), $D = \sum_{i \in I} T_i$, where $\{T_i\}_{i \in I}$ is some set of nilpotent ideals of the nil-ring D. Let D_i be the ideal of R generated by T_i. By 1.4, D_i is a nilpotent ideal. Therefore D_i is contained in P. Since $D = \sum_{i \in I} T_i$ and D is an ideal of R, $D = \sum_{i \in I} D_i \subseteq P$. Hence $E \subseteq D \subseteq P$. Analogously, P contains all left nil-ideals of R. $\triangleright$

2.65 If M is a module, then $\mathrm{sg}(M) \bigcap \mathrm{gs}(M) \subseteq J(\mathrm{End}(M))$.

$\triangleleft$ Let $R \equiv \mathrm{End}(M)$, $f \in \mathrm{sg}(M) \bigcap \mathrm{gs}(M) \equiv E$, and let $g \equiv 1-f \in R$. Then $\mathrm{Ker}(f) \bigcap \mathrm{Ker}(g) = 0$ and $f(M) + g(M) = M$. Since $f \in E$, $\mathrm{Ker}(g) = 0$ and $g(M) = M$. Therefore g is a unit of R. In addition, E is an ideal of R. Hence $E \subseteq J(R)$. $\triangleright$

2.66 If M is a nilpotently invariant module and f is a nilpotent endomorphism of M, then $f(M)$ is a superfluous submodule of M.

$\triangleleft$ Assume the contrary. Then $M = N + f(M)$ for some proper submodule $N \in \mathrm{Lat}(M)$. Let $h : M \to M/(N \bigcap f(M))$ be the natural epimorphism. Since $M \neq N$ and $M = N + f(M)$, we obtain $h(f(M)) \neq 0$. Since f is nilpotent, $M \neq f(M)$. Therefore $h(N) \neq 0$. Since f is nilpotent and $h(f(M)) \neq 0$, there exists a positive integer n such that $h(f^n(M)) \neq 0$ and $h(f^{n+1}(M)) = 0$. By assumption, $f(N) \subseteq N \bigcap f(M)$. Therefore $h(f(N)) = 0$. Since M is nilpotently invariant, $h(N \bigcap f(M)) \subseteq N \bigcap f(M)$. Therefore, the equality $h(f(N)) = 0$ implies the equality $h(f^n(N)) = 0$. Hence $0 \neq h(f^n(M)) = h(f^n(N + f(M))) \subseteq h(f^n(N)) + h(f^{n+1}(M)) = 0$; this is a contradiction. $\triangleright$

2.67 Let M be a nilpotently invariant module (by 2.63, this is the case if M is distributive), T be the set of all nilpotent endomorphisms of M, U be the automorphism group of M, $R \equiv \mathrm{End}(M)$, and let P be the prime radical of R.

(1) T is a nilsubring of the ideal $\mathrm{sg}(M) \bigcap \mathrm{gs}(M) \bigcap J(R)$ of R.

(2) $UT = TU$ is an ideal of the ring $J(R)$ without the identity element.

(3) P is the largest right nil-ideal and the largest left nil-ideal of R.

(4) If the nil-ring T is right or left vanishing, then $T = P$, and in particular, the ring R/P is reduced.

(5) If the nil-ring T is left vanishing and P_R is a uniserial module, then $T = P$ is a nilpotent ideal of R.

◁ (1) By 2.64(1), $T \subseteq \mathrm{sg}(M)$. By 2.66, $T \subseteq \mathrm{gs}(M)$. By 2.65, $\mathrm{sg}(M) \bigcap \mathrm{gs}(M) \subseteq J(\mathrm{End}(M))$. Therefore $T \subseteq \mathrm{sg}(M) \bigcap \mathrm{gs}(M) \bigcap J(\mathrm{End}(M))$. By 2.64(8), T is a nil-ring.

(2) Let $\alpha \in U$, $t \in T$, and let D be a cyclic submodule of M. Then $t^n(\alpha^{-1}(D)) = 0$ for some positive integer n. Therefore $(\alpha t \alpha^{-1})^n(D) = (\alpha t^n \alpha^{-1})(D) = 0$. Hence $\alpha t \alpha^{-1} \in T$. Analogously, $\alpha^{-1} t \alpha \in T$. Therefore $UT = TU$. By (1), $T \subseteq J(R)$. Hence $UT \subseteq J(R)$. Let $a \in U$, $t \in T$, and let $b \in J(R)$. Then $1 + b \in U$, $bat = (1 + b)at - at \in UT$, and $tab = ta(1 + b) - ta \in TU = UT$.

(3) Since P is a nil-ideal, our assertion follows from 2.64(10).

(4) Let the nil-ring T be right vanishing, $\alpha \in U$, $f \in T$, and let $f_i \equiv \alpha^i f \alpha^{-i}$ for $i = 1, 2 \ldots$. Then $(f_1 \ldots f_n)(M) = 0$ for some n. Therefore $(\alpha f)^n(M) = (f_1 \ldots f_n \alpha^n)(M) = 0$. Hence $T = UT$. By (2), T is an ideal of $J(R)$. By 2.64(9), the nil-ring T is a sum of its nilpotent ideals. Therefore, 1.4 implies the equality $T = P$. If the nil-ring T is left vanishing, then the equality $T = P$ can be proved in similar manner.

(5) By (4), $T = P$. We may assume that $P \neq 0$. A nonzero ideal P is left vanishing. Hence $P \neq P^2$. Let $x \in P \setminus P^2$. Since P_R is uniserial, $P^2 \subseteq xR$. Since $x \in P$, xR is a nilpotent right ideal. Hence P^2 is a nilpotent ideal. Therefore, P is nilpotent. ▷

2.68 Assume that either endomorphism rings of all submodules of a module M are normal (by 2.50(2), this is the case if M is distributive), or M is weakly invariant. Let $f \in \mathrm{End}(M)$.

(1) If $N \in \mathrm{Lat}(M)$ and $N \bigcap f(N) = 0$, then $f(N) = 0$.

(2) $\mathrm{Ker}(f^{n+1})$ is an essential extension of $\mathrm{Ker}(f^n)$ for any positive integer n.

(3) $\bigcup_{i=0}^{\infty} \mathrm{Ker}(f^i)$ is an essential extension of $\mathrm{Ker}(f)$.

(4) If f is locally nilpotent, then M is an essential extension of $\mathrm{Ker}(f)$ and $f(M) \subseteq \mathrm{Sing}(M)$.

(5) $\mathrm{End}(M)/\mathrm{sg}(M)$ is a reduced ring, and the ideal $\mathrm{sg}(M)$ of $\mathrm{End}(M)$ contains all locally nilpotent endomorphisms of M.

(6) If M is nonsingular, then $\mathrm{End}(M)$ is a reduced ring.

◁ (1) If endomorphism rings of all submodules of M are normal, then (1) follows from 2.50(1). Let M be weakly invariant, and let G be a closure of N in M. Then $f(G) \subseteq G$. Therefore $f(N) \subseteq G$. Since $N \bigcap f(N) = 0$, we obtain $G \bigcap f(N) = 0$. Therefore $f(N) \subseteq G \bigcap f(N) = 0$.

(2) Let $T = \mathrm{Ker}(f^n)$, and let N be a submodule of $\mathrm{Ker}(f^{n+1})$ such that $N \bigcap T = 0$. Since $f^{n+1}(N) = 0$, $f(N) \subseteq T$. By (1), $N \subseteq \mathrm{Ker}(f) \subseteq T$. Hence $N = N \bigcap T = 0$.

(3) Since the union of any ascending chain of essential extensions of $\mathrm{Ker}(f)$ is an essential extension of $\mathrm{Ker}(f)$, our assertion follows from (2) and from inclusions $\mathrm{Ker}(f^i) \subseteq \mathrm{Ker}(f^{i+1})$.

(4) By (3), M is an essential extension of $\mathrm{Ker}(f)$. By 1.29, $f(M) \subseteq \mathrm{Sing}(M)$.

(5) By (4), it is sufficient to prove that the ring $\mathrm{End}(M)/\mathrm{sg}(M)$ is reduced. Let $f \in \mathrm{End}(M)$, n be a positive integer, and let $f^n \in \mathrm{sg}(M)$. Then M is an essential extension of $\mathrm{Ker}(f^n)$. By (3), $\mathrm{Ker}(f^n)$ is an essential extension of $\mathrm{Ker}(f)$. Therefore M is an essential extension of $\mathrm{Ker}(f)$, and $f \in \mathrm{sg}(M)$.

(6) By 1.29, $\mathrm{sg}(M) = 0$. Our assertion follows now from (5). $\triangleright$

2.69 Let M be a distributive module M.

(1) If f is a locally nilpotent endomorphism of M, then $\mathrm{Ker}(f)$ is an essential submodule of M, and f maps from every submodule of M into itself.

(2) If M is locally Noetherian, then the ideal $\mathrm{sg}(M)$ of $\mathrm{End}(M)$ coincides with the set of all locally nilpotent endomorphisms of M, and each endomorphism $f \in \mathrm{sg}(M)$ maps from every submodule of M into itself.

$\triangleleft$ (1) By 2.68(4), $f \in \mathrm{sg}(M)$. It is sufficient to prove that $f(N) \subseteq N$ for any cyclic $N \in \mathrm{Lat}(M)$. There exists n such that $f^n(N) = 0$. Our assertion follows now from 2.60(1).

(2) follows from (1), 2.57(3) and 2.68(4). $\triangleright$

2.70 [201],[207],[213]. *Assume that each locally nilpotent endomorphism of a module M maps from any submodule of M into itself (by 2.69(1), this is the case if M is distributive). Let H and T be the set of all locally nilpotent endomorphisms and the set of all nilpotent endomorphisms of M, respectively, and let $N \in \mathrm{Lat}(M)$.*

(1) *The ideal $\mathrm{sg}(M)$ of $\mathrm{End}(M)$ contains all locally nilpotent endomorphisms of M.*

(2) *The image of each locally nilpotent endomorphism of M is contained in $\mathrm{Sing}(M)$.*

(3) *If f_i and g_j are locally nilpotent endomorphisms of M ($1 \leq i, j \leq n$) and $(f_1 f_2 \ldots f_n)(N) = 0$, then $(f_1 g_1 f_2 g_2 \ldots f_n g_n)(N) = 0$.*

(4) *If f and g are two locally nilpotent endomorphisms of M and $f^m(N) = g^n(N) = 0$, then $(f + g)^{m+n}(N) = 0$.*

(5) *H is a subring (without the identity element) of $\mathrm{End}(M)$.*

(6) *Let f be a locally nilpotent endomorphism of M, H_1 be the subring of $\mathrm{End}(M)$ generated by the identity element and by the subset H, and let $f^n(N) = 0$.*

Then $(fH_1)^n(N) = 0$.

(7) *T is a nil-ideal of the ring (without the identity element) H and coincides with the sum of all nilpotent ideals of H.*

(8) *If U is the automorphism group of M, then $UH = HU$.*

(9) *If M is locally Noetherian, then the ideal $\mathrm{sg}(M)$ of $\mathrm{End}(M)$ coincides with the set H of all locally nilpotent endomorphisms of M, and each endomorphism $f \in \mathrm{sg}(M)$ maps from every submodule of M into itself.*

◁ (1) Let g be a locally nilpotent endomorphism of M, and let Q be a nonzero cyclic submodule of M. By assumption, g induces a nilpotent endomorphism f of the cyclic module Q. It is sufficient to prove that $\mathrm{Ker}(f^{n+1})$ is an essential extension of $\mathrm{Ker}(f^n)$ for any n. Let $U = \mathrm{Ker}(f^n)$, and let V be a submodule of $\mathrm{Ker}(f^{n+1})$ such that $U \bigcap V = 0$. Since $f^{n+1}(V) = 0$, we obtain $f(V) \subseteq U$. By assumption, $f(V) \subseteq V \bigcap U = 0$. Hence $V \subseteq \mathrm{Ker}(f) \subseteq U$. Therefore $V = V \bigcap U = 0$. Hence $\mathrm{Ker}(f^{n+1})$ is an essential extension of $\mathrm{Ker}(f^n)$.

(2) follows from (1) and 1.29.

(3) By assumption, locally nilpotent endomorphisms f_i, g_j of M map every submodule of M into itself. Therefore, (3) holds.

(4) follows from (3). (5) follows from (3) and (4). (6) follows from (1). (7) follows from (3) and (6).

(8) Let $\alpha \in U$, $h \in H$, and let D be a cyclic submodule of M. Then $h^n(\alpha^{-1}(D)) = 0$ for some positive integer n. Therefore $(\alpha h \alpha^{-1})^n(D) = (\alpha h^n \alpha^{-1})(D) = 0$. Hence $\alpha h \alpha^{-1} \in H$. Analogously, $\alpha^{-1} h \alpha \in H$. Therefore $UH = HU$.

(9) By 2.57(3), $\mathrm{sg}(M) \subseteq H$. Therefore, each endomorphism $f \in \mathrm{sg}(M)$ maps from every submodule of M into itself. By (1), $H \subseteq \mathrm{sg}(M)$. ▷

2.71 (1) [208] *If each subdirectly indecomposable factor module of a distributive module M is semi-Noetherian, then M is an invariant module.*

(2) [166] *If M is a distributive semi-Artinian module, then M is an invariant module.*

◁ (1) Let $M \neq N \in \mathrm{Lat}(M)$, $N \subseteq H \in \mathrm{Lat}(M)$, and let M/H be subdirectly indecomposable. Since M/N is a subdirect product of subdirectly indecomposable modules and any intersection of fully invariant submodules of M is fully invariant in M, it is sufficient to prove that H is fully invariant in M. It is sufficient to prove that H coincides with the intersection F of all fully invariant submodules of M containing H. Assume the contrary. By assumption, M/H is semi-Noetherian. Therefore, the nonzero module F/H has a maximal submodule G/H. Hence G is a maximal submodule of F containing H. By 2.54, G is fully invariant in F. Also, F is fully invariant in M. Therefore G is fully invariant in M and $H \subseteq G \subset F$; this is a contradiction to the choice of F.

(2) Let $f \in \mathrm{End}(M)$, $P \in \mathrm{Lat}(M)$, and let Q be the largest submodule of P such that $f(Q) \subseteq Q$. If $P = Q$, then $f(P) \subseteq P$. Assume that $P \neq Q$. Let $\overline{M} \equiv M/Q$. Since $f(Q) \subseteq Q$, f induces an endomorphism $\overline{f}$ of $\overline{M}$. Let $h : M \to \overline{M}$ be the natural epimorphism. The nonzero semi-Artinian module $h(P)$ has a simple submodule $h(S)$, where $S \in \mathrm{Lat}(P)$ and $Q \subset S$. By 2.32, $\overline{f}(h(S)) \subseteq h(S)$. Hence $f(S) \subseteq S$. Since Q is the largest submodule of P such that $f(Q) \subseteq Q$, we have $S \subseteq P$. Hence $h(S) = 0$; this is a contradiction. $\triangleright$

2.72 [208]. If A is a right distributive ring and all subdirectly indecomposable cyclic right A-modules are semi-Noetherian, then A is right invariant.

$\triangleleft$ 2.72 follows from 2.71(1). $\triangleright$

2.7 Distributivity and spectra

2.73 The sets $\{N \in \mathrm{Lat}(M) \mid M/N \text{ is a nonzero uniform module}\}$ and $\{N \in \mathrm{Lat}(M) \mid M/N \text{ is a subdirectly indecomposable module}\}$ are denoted by $\mathrm{Sp}^1(M)$ and $\mathrm{Sp}^2(M)$, respectively.

Let N be a submodule of a module M, $C^i(N) \equiv \{S \in \mathrm{Sp}^i(M) \mid N \subseteq S\}$, $O^i(N) \equiv \mathrm{Sp}^i(M) \setminus C^i(N) = \{S \in \mathrm{Sp}^i(M) \mid N \nsubseteq S\}$, $\mathrm{Close}^i(M) \equiv \{C^i(N) \mid N \in \mathrm{Lat}(M)\}$, $\mathrm{Open}^i(M) \equiv \{O^i(N) \mid N \in \mathrm{Lat}(M)\}$, and let $\varphi_i(N) \equiv$ be the intersection of all modules from $C^i(N)$ $(i = 1, 2)$.

Obviously, $\mathrm{Sp}^2(M) \subseteq \mathrm{Sp}^1(M)$.

(1) $X = \varphi_1(X) = \varphi_2(X)$ for every submodule X of a module M.

(2) If X and Y are submodules of a module M, then
$X = Y \iff \varphi_1(X) = \varphi_1(Y) \iff \varphi_2(X) = \varphi_2(Y)$.

(3) If X and Y are submodules of a module M, then
$X = Y \iff C^1(X) = C^1(Y) \iff C^2(X) = C^2(Y) \iff O^1(X) = O^1(Y) \iff O^2(X) = O^2(Y)$.

(4) If $\{X_j\}_{j \in J}$ is a set of submodules of a module M, then
$M = \sum_{j \in J} X_j \iff \bigcap_{j \in J} C^1(X_j) = \emptyset \iff \bigcup_{j \in J} O^1(X_j) = \mathrm{Sp}^1(M)$
$\iff \bigcap_{j \in J} C^2(X_j) = \emptyset \iff \bigcup_{j \in J} O^2(X_j) = \mathrm{Sp}^2(M)$.

(5) $C^i(X) \bigcup C^i(Y) \subseteq C^i(X \bigcap Y)$, $O^i(X) \bigcap O^i(Y) \subseteq O^i(X \bigcap Y)$ for all submodules X and Y of a module M ($i = 1, 2$).

(6) $\bigcap_{j \in J} C^i(X_j) = C^i(\sum_{j \in J} X_j)$, $\bigcup_{j \in J} O^i(X_j) = O^i(\sum_{j \in J} X_j)$ for every set $\{X_j\}_{j \in J}$ of submodules of a module M ($i = 1, 2$).

◁ (1) The equality $X = \varphi_2(X)$ follows from the fact that every nonzero module is a subdirect product of subdirectly indecomposable modules. The equality $X = \varphi_1(X)$ follows from the equality $X = \varphi_2(X)$ and from the fact that every subdirectly indecomposable module is uniform.

(2) follows from (1).

(3) Equivalences $C^1(X) = C^1(Y) \Longleftrightarrow O^1(X) = O^1(Y)$, $C^2(X) = C^2(Y) \Longleftrightarrow O^2(X) = O^2(Y)$ and implications $X = Y \Longrightarrow C^1(X) = C^1(Y)$, $X = Y \Longrightarrow C^2(X) = C^2(Y)$ are obvious.

Implications $C^1(X) = C^1(Y) \Longrightarrow X = Y$ and $C^2(X) = C^2(Y) \Longrightarrow X = Y$ follow from (2).

(4) follows from (3) and from the fact that $C^1(M) = C^2(M) = \emptyset$ and $O^i(M) = \mathrm{Sp}^i(M)$. (5) and (6) can be directly verified. ▷

2.74 Let M be a module such that $C^i(X) \bigcup C^i(Y) = C^i(X \bigcap Y)$ for all submodules X and Y of the module M.

Then $O^i(X) \bigcap O^i(Y) = O^i(X \bigcap Y)$ for all submodules X, Y of the module M, and the set $\mathrm{Sp}^i(M)$ can be made into a topological space by taking the closed (open) sets in $\mathrm{Sp}^i(M)$ to be all sets from the set $\mathrm{Close}^i(M)$ (the set $\mathrm{Open}^i(M)$).

Taking into account 2.73(6), we can directly verify the assertion. In this situation, the module M is said to be *i-spectral* and the topological space $\mathrm{Sp}^i(M)$ is said to be the *i-spectrum* of the module M ($i = 1, 2$).

2.75 A topological space is a T_0-*space* if for any two its points, at least one of these points has an open neighborhood which does not contain the second point.

Let M be a distributive module, and let i be one of the numbers $1, 2$.

(1) M is an i-spectral module, $\mathrm{Sp}^i(M)$ is a topological T_0-space, and by the rules $\alpha(X) = O^i(X)$ and $\beta(X) = C^i(X)$, the lattice isomorphism $\alpha : \mathrm{Lat}(M) \to \mathrm{Open}^i(M)$ and the lattice anti-isomorphism $\beta : \mathrm{Lat}(M) \to \mathrm{Close}^i(M)$ are defined.

(2) If t is an infinite cardinal number, then

M is a t-generated module $\Longleftrightarrow$ every open covering of the space $\mathrm{Sp}^i(M)$ contains a subcovering whose cardinality does not exceed t.

(3) M is a finitely generated module $\Longleftrightarrow$ $\mathrm{Sp}^i(M)$ is a compact T_0-space.

◁ (1) Let X and Y be two submodules of the module M, and let $Z \in C^i(X \cap Y)$. Since $X \cap Y \subseteq Z$ and M is a distributive module, $Z = Z + X \cap Y = (Z + X) \cap (Z + Y)$. Therefore, natural images modules $Z + X$ and $Z + Y$ in the uniform module M/Z have the zero intersection. Then either $Z + X = Z$ or $Z + Y = Y$. Therefore $Z \in C^i(X) \cup C^i(Y)$. Taking into account 2.73(5), we obtain that $C^i(X) \cup C^i(Y) = C^i(X \cap Y)$. Then $O^i(X) \cap O^i(Y) = O^i(X \cap Y)$. It follows from the above and from 2.73(3),(6) that α is a lattice anti-isomorphism and β is a lattice isomorphism. Let $F, G \in \mathrm{Sp}^i(M)$ with $F \neq G$. Then either F does not contain G or G does not contain F. Assume that F does not contain G. Then $F \in O^i(G)$. Therefore $O^i(G)$ is an open set in $\mathrm{Sp}^i(M)$ which contains F and does not contain G. Then $\mathrm{Sp}^i(M)$ is a T_0-space.

(2) We will assume that $i = 1$ for definiteness.

$\Longrightarrow$ Let $M = \sum_{j \in J} M_j$, where all M_j cyclic modules and $\mathrm{card}(J) = t$. In addition, let $\mathrm{Sp}^1(M) = \bigcup_{g \in G} O^1(N_g)$, where N_g be a submodule of the module M. By 2.73(4), $M = \sum_{g \in G} N_g$. For every cyclic module M_j, there exists a finite subset $F(j) \subseteq G$ such that $M_j \subseteq \sum_{g \in F(j)} N_g$. Let $F \equiv \bigcup_{j \in J} F(j) \subseteq G$. Then $\mathrm{card}(F) = t$, and $M = \sum_{g \in F} N_g$. By 2.73(4), $\mathrm{Sp}^1(M) = \bigcup_{g \in F} O^1(N_g)$ is the required subcovering of open covering $\bigcup_{g \in G} O^1(N_g)$ of the space $\mathrm{Sp}^1(M)$.

$\Longleftarrow$ Let $\{N_g\}_{g \in G}$ be the set of all cyclic submodules of the module M. Then $\mathrm{Sp}^1(M) = \bigcup_{g \in G} O^1(N_g)$. By assumption, there exists a subset J of the set G such that $\mathrm{card}(J) = t$ and $\mathrm{Sp}^1(M) = \bigcup_{j \in J} O^1(N_j)$. By 2.73(4), $M = \sum_{j \in J} N_j$, whence the module M is t-generated.

(3) The proof of (3) is analogous to the proof of (2). ▷

2.76 *M is a distributive module $\Longleftrightarrow$*
the lattice $\mathrm{Lat}(M)$ *of all submodules of the module M is isomorphic to the lattice of all open subsets of a topological T_0-space.*

In addition, the lattice of all submodules of a finitely generated distributive module is isomorphic to the lattice of all open subsets of a compact topological T_0-space.

◁ 2.76 follows from 2.75 and from the fact that the lattice of all open subsets of a topological space is always distributive. ▷

2.77 *For a module M, the following conditions are equivalent.*
(1) M is a distributive module.
(2) The module M is 1-spectral.
(3) The module M is 2-spectral.

◁ By 2.32, condition (1) is equivalent to the following condition (4): M has no subfactors which are direct sums of two isomorphic simple modules.

$(1){\Longrightarrow}(2)$ follows from 2.75. $(2){\Longrightarrow}(3)$ can be directly verified.

$(3){\Longrightarrow}(4)$ Assume the contrary. Then there exist submodules N, P, S, and T of M such that $N = S \bigcap T$, $S + T = P$, S/N and T/N are simple modules, and there exists an isomorphism $f : S/N \to T/N$. Let $h : M \to M/N$ be the natural epimorphism, and let $h(Y) \equiv \{x + f(x) \mid x \in h(S)\} \subseteq h(S) \oplus h(T)$, where $N \subset Y \in \mathrm{Lat}(M)$. Then

$$h(S) \bigcap h(Y) = 0 \text{ and } h(S) \oplus h(T) = h(S) \oplus h(Y) = h(T) \oplus h(Y).$$

Let $\mathcal{E}$ be the non-empty set of all submodules of the module $h(M)$ that contain the simple module $h(Y)$ and have the zero intersection with the simple module $h(S)$. By Zorn's Lemma, the set $\mathcal{E}$ has a maximal element $h(E)$, where $Y \subseteq E \in \mathrm{Lat}(M)$. Then M/E is an essential extension of the simple module $(S + E)/E$. Therefore M/E is a subdirectly indecomposable module. Then $E \in C^2(S \bigcap T) = C^2(S) \bigcup C^2(T)$. Then $E \in C^2(T)$, since E does not contain S. Therefore $T \subseteq E$, whence $h(T) \subseteq h(E)$. Then $h(T) \subset h(S) \oplus h(Y) = h(T) \oplus h(Y) \subseteq h(E)$; this is a contradiction to the equality $h(S) \bigcap h(E) = 0$. $\triangleright$

2.78 Let A be a ring without the identity element, and let A^1 be the ring with the identity element constructed in 1.63.

(1) If M is a right A-module, then

M is a distributive A-module $\Longleftrightarrow$ for any two elements $m_1, m_2 \in M$, there exist three integers z, y_1, y_2 and three elements $a, b_1, b_2 \in A$ such that $m_1 a + z m_1 = m_2 b_2 + y_2 m_2$ and $m_2 - m_2 a - z m_2 = m_1 b_1 + y_1 m_1$.

(2) A^1 is a right distributive ring $\Longleftrightarrow$ A^1 is a distributive right A-module $\Longleftrightarrow$ for any two elements $a_1, a_2 \in A$ and for any two integers x_1 and x_2, there exist five integers z, x_1, x_2, y_1, y_2 and three elements $a, b_1, b_2 \in A$ such that $a_1 a + x_1 a + z a_1 = a_2 b_2 + x_2 b_2 + y_2 a_2$, $a_2 - a_2 a - x_2 a - z a_2 = a_1 b_1 + x_1 b_1 + y_1 a_1$, $z x_1 = y_2 x_2$, and $x_2 - z x_2 = y_1 x_1$.

(3) If B is a distributive right ideal of A and M is a maximal submodule in B_A, then M is an ideal in the ring B without the identity element.

$\triangleleft$ (1) follows from 1.63(1) and 2.4. (2) follows from (1) and 1.63(1).

(3) For each $b \in B$, we have an endomorphism $f_b \in \mathrm{End}(B_A)$ such that $f_b(x) = bx$ for all $x \in B$. By 2.54 and 1.63, M is fully invariant in B_A. Hence $BM \subseteq M$ and M is an ideal in B. $\triangleright$

Chapter 3

Semidistributive modules and rings

3.1 Distributive Goldie rings

3.1 If M is a distributive module which contains no infinite direct sums of nonzero fully invariant submodules, then M is finite-dimensional, and $\mathrm{End}(M)/\mathrm{sg}(M)$ is a reduced ring with maximum conditions on right annihilators and on left annihilators.

◁ 3.1 follows from 2.46(7), 1.39(3), and 2.68(5). ▷

3.2 Let a ring A be either right weakly invariant or right distributive.

(1) A is right nonsingular $\Longleftrightarrow$ A is reduced.

(2) A contains no infinite direct sums of nonzero ideals $\Longleftrightarrow$ A is right finite-dimensional.

(3) If A is right nonsingular and contains no infinite direct sums of nonzero ideals, then A is a right finite-dimensional reduced ring and is a right order in a finite direct product of division rings.

◁ By 2.46(5), right distributive rings are right weakly invariant. Therefore, we may assume that A is right weakly invariant.

(1)

$\Longleftarrow$ follows from 1.35(4).

$\Longrightarrow$ Let $a \in A$, $a^2 = 0$. There exists a closed right ideal H of A such that $H \bigcap aA = 0$ and $aA \oplus H$ is an essential right ideal. Since H is an ideal by assumption, $aH \subseteq H \bigcap aA = 0$. In addition, $a^2 A = 0$.

Hence $aA \oplus H \subseteq r(a)$. Therefore $r(a)$ is an essential right ideal, whence $a = 0$.

(2) We have only to prove $\Longrightarrow$. By assumption, A contains no infinite direct sums of closed right ideals. By 1.17, A is right finite-dimensional.

(3) By (1) and (2), A is a right finite-dimensional reduced ring. Hence A is a right order in a finite direct product of division rings. $\triangleright$

3.3 If A is a right distributive ring A, then

A contains no infinite direct sums of nonzero ideals $\Longleftrightarrow$ each right ideal of A is an essential extension of a principal right ideal I such that I is a finite direct sum of principal right ideals.

$\triangleleft$ 3.3 follows from 2.46(7). $\triangleright$

3.4 If m and n are two nonzero elements of a distributive module M_A with $mA \cap nA = 0$, then $mA \oplus nA = (m + n)A$ and $r(m + n) = r(m) \cap r(n) \neq r(m)$.

$\triangleleft$ By 2.6(1), there exists $a \in A$ such that $ma = n(1 - a) = 0$. Hence $(m + n)a = n \neq 0$ and $(m + n)(1 - a) = m$. Therefore $mA \oplus nA = (m + n)A$ and $a \in r(m) \setminus r(m + n)$. If $x \in r(m + n)$, then $mx = -nx \in mA \cap nA = 0$ and $x \in r(m) \cap r(n)$. Hence $r(m + n) = r(m) \cap r(n)$, and $r(m)$ properly contains $r(m + n)$. $\triangleright$

3.5 Let M be a distributive right module over a ring A.

(1) If M is not finite-dimensional, then there exists a set $\{m_i\}_{i=1}^{\infty}$ of nonzero $m_i \in M$ such that the sum of all cyclic modules $m_i A$ is a direct sum, the cyclic module $(\sum_{i=1}^{n+1} m_i)A$ properly contains the cyclic module $(\sum_{i=1}^{n} m_i)A$ for any n, and $r(\sum_{i=1}^{n} m_i)$ properly contains $r(\sum_{i=1}^{n+1} m_i)$ for any n.

(2) If either M is a module with the maximum condition on cyclic submodules or A is a ring with the minimum condition on right annihilators of elements of M, then M is finite-dimensional.

$\triangleleft$ (1) Since M is not finite-dimensional, there exists a set $\{m_i\}_{i=1}^{\infty}$ of nonzero $m_i \in M$ such that the sum of all cyclic modules $m_i A$ is a direct sum. Hence the cyclic module $(\sum_{i=1}^{n+1} m_i)A$ properly contains the cyclic module $(\sum_{i=1}^{n} m_i)A$ for any n. By 3.4, $r(\sum_{i=1}^{n} m_i)$ properly contains $r(\sum_{i=1}^{n+1} m_i)$ for any n.

(2) follows from (1). $\triangleright$

3.6 Let A be a right distributive ring, N be a completely prime ideal of A, and let $T \equiv A \setminus N$.

(1) For any $x \in A$ and $t \in T$, there exists $u \in T$ such that $xu \in tA$.

(2) If all elements of T are regular in A, then T is a right Ore set, A_T is a right uniserial ring, and $J(A_T) = NA_T$.

(3) Assume that either T is right weakly reversible or each square-zero element $a \in A$ commutes with every element of T or for any element $a \in A$, there exists a positive integer $n = n(a)$ such that $r(a^n) = r(a^{n+1})$ or A is a ring with the maximum condition on right annihilators.

Then T is a right denominator set, the right ring of quotients A_N is right uniserial, and $J(A_T) = f(N)A_T$, where $f : A \to A_T$ is the canonical ring homomorphism.

(4) If A is either a ring with the maximum condition on principal right ideals or a ring with the maximum condition on left annihilators, then A is right finite-dimensional.

◁ (1) By 2.4, there exist $a, b, c, d \in A$ such that $1 = a + b$, $xa = tc$, and $tb = xd$. If $a \in T$, then we may assume that $u \equiv a$. Assume that $a \in A \setminus T = N$. Then $b = 1 - a \in A \setminus N = T$, whence $xd = tb \in T$. If $d \in N$, then $tb = xd \in N \bigcap (A \setminus N)$; this is a contradiction. Therefore $d \in T$, and we may assume that $u \equiv d$.

(2) By (1), A_T exists. By 1.46(14) and 1.46(26), A_T is a local right distributive ring, and $J(A_T) = NA_T$. By 2.8, A_T is right uniserial.

(3) By (1), T is a right permutable set. By 1.48 and 1.51, T is a right denominator set. (3) follows now from (2) and 1.48.

(4) Note that any ring with the maximum condition on left annihilators is a ring with the minimum condition on right annihilators. Hence our assertion follows from 3.5(2). ▷

3.7 (Goldie's Theorem) A *right Goldie ring* is any right finite-dimensional ring with the maximum condition on right annihilators. Every right Noetherian ring or every right Ore domain is right Goldie.

A ring A is a right order in a semisimple Artinian ring $Q \iff A$ is a semiprime right Goldie ring $\iff A$ is a semiprime right nonsingular right finite-dimensional ring $\iff A$ is a semiprime right finite-dimensional ring with the maximum condition on left annihilators $\iff$ for each right ideal D of A, D is essential if and only if D contains a regular element.

3.8 [185]. *For a right distributive ring A, the following conditions are equivalent.*

(1) A is a semiprime ring with the maximum condition on right annihilators.

(2) A is a semiprime ring with the maximum condition on left annihilators.

(3) *A is a right nonsingular ring which contains no infinite direct sums of nonzero ideals.*

(4) *A is a reduced right finite-dimensional ring.*

(5) *A is a right order in a semisimple ring.*

(6) *A is a right order in a finite direct product of division rings.*

(7) *A is a finite direct product of right Ore domains.*

◁ $(1)\Longrightarrow(2)$ By 1.37(3), A is right nonsingular. By 1.35(4), A is reduced. By 1.35(2), right annihilators of subsets of a reduced ring A coincide with their left annihilators. Hence the maximum condition on right annihilators implies the maximum condition on left annihilators.

$(2)\Longrightarrow(3)$ By 1.37(3), A is right nonsingular. By 3.5, A is right finite-dimensional.

$(3)\Longrightarrow(4)$ follows from 3.2(3).

$(4)\Longrightarrow(7)$ By 3.3, there exists $b \in A$ such that $bA = \oplus_{i=1}^{n} B_i$ is an essential right ideal, and all B_i are nonzero uniform right ideals. By 1.35(4), A is nonsingular. Therefore $\ell(b) = 0$. By 1.35(2), $r(b) = 0$. Hence $A_A \cong bA$. Therefore A is a direct sum of nonzero uniform right ideals $A_1, \ldots, A_n$. By 2.50(2), all right ideals A_i are ideals. Therefore $A = \prod_{i=1}^{n} A_i$ is a ring direct product of uniform right nonsingular rings A_i. It can be directly proved that each right nonsingular right uniform ring A_i is a domain. Hence A_i are right Ore domains.

$(7)\Longrightarrow(6)$ and $(6)\Longrightarrow(5)$ can be directly verified. $(5)\Longrightarrow(1)$ follows from 3.7. ▷

3.9 Let A be a right distributive ring which either contains no infinite direct sums of nonzero ideals or is a ring with the maximum condition on left annihilators.

Then A is right finite-dimensional, and the factor ring $A/\mathrm{Sing}(A_A)$ is a finite direct product of right distributive right uniform domains.

◁ By 3.6(4), A is right finite-dimensional. By 3.1, $A/\mathrm{Sing}(A_A)$ is a right distributive reduced ring with the maximum condition on right annihilators. By 3.8, $A/\mathrm{Sing}(A_A)$ is a finite direct product of right distributive right uniform domains. ▷

3.10 If N is the prime radical of a normal ring A, then
A is an indecomposable ring $\Longleftrightarrow A/N$ is an indecomposable ring.

◁ Since A is normal and N is a nil-ideal, our assertion follows from the fact that idempotents of A/N can be lifted to the idempotents of A. ▷

3.11 Let A be a right distributive indecomposable ring, and let N be the prime radical of A. Then

A/N is a ring with the maximum condition either on left annihilators or on right annihilators $\Longleftrightarrow A/N$ is a right uniform domain $\Longleftrightarrow$ N is a completely prime ideal of A.

$\triangleleft$ 3.11 follows from 3.10 and 3.8. $\triangleright$

3.12 Let M be a nonzero right ideal of a ring A such that the module M_A is distributive.

(1) If B and D are two right ideals of A such that $B, D \subseteq M$ and $B \cap D = 0$, then $BD = DB = \mathrm{Id}(B)\mathrm{Id}(D) = 0$.

(2) If A is prime, then M_A is uniform.

$\triangleleft$ (1) Let $d \in D$. Define a homomorphism $f : B_A \to D_A)$ as follows: $f(b) = db$ for any $b \in B$. By 2.6(2), $f \equiv 0$. Therefore $BD = 0$. Hence $\mathrm{Id}(B)\mathrm{Id}(D) = 0$.

(2) follows from (1). $\triangleright$

3.13 Let A be a domain which is either a left distributive ring or a left Bezout ring or a left finite-dimensional ring.

Then A is a left uniform domain.

$\triangleleft$ If A is a left distributive domain, then we can apply 3.12(2). Assume that there exist nonzero $m, n \in A$ such that $Am \cap An = 0$. Let A be a left Bezout ring. There exist $a, b, c, d \in A$ such that $(1 - ca)m = cbn$ and $(1 - db)n = dam$. Therefore $ca = 1$, $db = 1$, and elements a, b, c, d of the domain A are units. Hence $n = (cb)^{-1}(1 - ca)m$ which contradicts $Am \cap An = 0$. Let A be a left finite-dimensional domain. Then the sum $\sum_{j=0}^{\infty} Anm^j$ cannot be a direct sum. Therefore, there exist elements $a_1, \ldots, a_t$ such that $\sum_{j=0}^{t} a_j nm^j = 0$ and $a_j \neq 0$ for some j. We may assume that $a_0 \neq 0$. Then $0 \neq a_0 n \in Am \cap An$; this is a contradiction. $\triangleright$

3.14 Let N be a completely prime ideal of a right distributive ring A, and let $N \subseteq J(A)$.

(1) $N = mN \subset mA$ for all $m \in A \setminus N$, and A has no nontrivial idempotents.

(2) N is comparable with any right ideal of A.

(3) $MN = N$ for any right ideal M of A which is not contained in N.

(4) Either $N = 0$ and A is a right uniform domain, or N is a nonzero essential right ideal of A.

(5) Either the module N_A is not uniform, or A is right uniform.

(6) If N is a finitely generated left ideal, then either $N = J(A)$, or $N = 0$ and A is a right uniform domain.

◁ (1) Let $m \in A \setminus N$, $n \in N$. By 2.4, there exist $a, b, c, d \in A$ such that $1 = a + b$, $ma = nc \in N$, and $nb = md$. Since N is a completely prime ideal and $ma, md \in N$, we obtain $a, d \in N \subseteq J(A)$. Therefore $b = 1 - a$ is a unit, $n = mdb^{-1} \in mN$. Since N contains no nontrivial idempotents and A/N is a domain, A has no nontrivial idempotents.

(2) and (3) follow from (1). (4) follows from (2) and 3.11. (5) follows from (4).

(6) Assume that N is properly contained in $J(A)$. By (3), $J(A)N = N$. By Nakayama's Lemma, $N = 0$. Therefore A is a domain. By 3.12(2), the right distributive domain A is right uniform. ▷

3.15 [176], [119], [201]. *Let A be a right distributive ring, T be the set of all nilpotent elements of A, and let P be the prime radical of A.*

(1) If an element a of A is integral over $C(A)$, then $aB \subseteq B$ for any right ideal B of A.

(2) $A/\mathrm{Sing}(A_A)$, $A/J(A)$, and $A/(J(A) \cap \mathrm{Sing}(A_A))$ are reduced rings.

(3) For each element a of A and for any positive integer n, the right ideal $r(a^n)$ is an essential extension of the right ideal $r(a)$.

(4) H is a subring (without the identity element) of the ideal $\mathrm{Sing}(A_A)$ of A, and T is a nilsubring of the ideal $\mathrm{Sing}(A_A) \cap J(A)$ of A.

(5) If U is the group of units of A, then $UT = TU$ is an ideal of the ring $J(A)$ without the identity element.

(6) P is the largest right nil-ideal of A and the largest left nil-ideal of A.

(7) If the nil-ring T is right or left vanishing, then $T = P$, and in particular, A/P is reduced.

(8) If the nil-ring T is left vanishing and P_A is a uniserial module, then $T = P$ is a nilpotent ideal of A.

◁ (1) follows from 2.61(3).

(2) By 2.68(5) and 2.53(4), $A/\mathrm{Sing}(A_A)$ and $A/J(A)$ are reduced rings. Since the ring $A/(J(A) \cap \mathrm{Sing}(A_A))$ is isomorphic to a subdirect product of reduced rings $A/\mathrm{Sing}(A_A)$ and $A/J(A)$, then $A/(J(A) \cap \mathrm{Sing}(A_A))$ is a reduced ring.

(3) follows from 2.68(3). (4), (5), (6), (7), and (8) follow from 2.67. ▷

3.16 Let A be a right distributive ring with the maximum condition on right annihilators, and let N be the prime radical of A.

(1) N is a nilpotent ideal, $N = \mathrm{Sing}(A_A)$, and N coincides with the set of all nilpotent elements of A.

(2) If N is a completely prime ideal, then N coincides with the set of all left zero-divisors of A, any element $m \in A \setminus N$ is right regular, and by the rule $\widehat{m}(x) = mx$, the automorphism $\widehat{m}$ of N_A is well defined.

◁ (1) Let H be the set of all nilpotent elements of A. By 3.15(4), $H \subseteq \mathrm{Sing}(A_A)$. By 1.37(1), $\mathrm{Sing}(A_A)$ is a nilpotent ideal. Therefore $\mathrm{Sing}(A_A) \subseteq N \subseteq H$.

(2) We may assume that A is not a right uniform domain. Let $m \in A \setminus N$. By 3.14(1), $mN = N$. Therefore $\widehat{m}$ is an epimorphism. Since A is a ring with the maximum condition on right annihilators, $r(m^n) = r(m^i)$ for some positive integer n and all $i > n$. Set $g \equiv (\widehat{m})^n \in \mathrm{End}(N_A)$. Then $g(M) = M$ and $\mathrm{Ker}(g) = \mathrm{Ker}(g^2)$. For each $x \in \mathrm{Ker}(g)$, there exists $y \in N$ such that $x = g(y)$, whence $y \in \mathrm{Ker}(g^2) = \mathrm{Ker}(g)$ and $x = 0$. Therefore g is an automorphism of N_A, and $\widehat{m}$ is an automorphism of N_A. Hence $N \bigcap r(m) = 0$, and N is an essential right ideal. Therefore m is not a left zero-divisor. ▷

3.17 (1) A is a normal semiperfect ring $\iff$ A is a finite direct product of local rings.

(2) [166] A is a semiperfect right distributive ring $\iff$ A is a finite direct product of right uniserial rings.

◁ (1) is obvious. (2) follows from (1) and 2.9. ▷

3.18 [166]. *For a ring A, the following conditions are equivalent.*
(1) A is a right or left perfect right distributive ring.
(2) A is a semiprimary right distributive ring.
(3) A is a semilocal right distributive ring, and $J(A)$ is a finitely generated right nil-ideal.
(4) A is a finite direct product of right uniserial right Artinian rings.

◁ We may assume that A is semiperfect. By 3.17(2), we may assume that A is right uniserial.

(1)$\iff$(2)$\iff$(4) follows from 2.21. (4)$\Longrightarrow$(3) is directly verified.

(3)$\Longrightarrow$(2) Set $J \equiv J(A)$. Since J_A is a finitely generated uniserial module, there exists $m \in J$ such that $J = mA$. Since m is nilpotent and mA is an ideal, J is a nilpotent ideal. ▷

3.19 Let A be an indecomposable right distributive right Goldie ring which is not a right uniform domain, and let N be the prime radical of A.

(1) N is a nonzero completely prime nilpotent ideal and an essential right ideal, N contains all left zero-divisors of A, A/N is a right uniform domain, and $tN = N$ for any $t \in A \setminus N$.

(2) If N contains all right zero-divisors of A, then A is a right order in a right uniserial right Artinian ring Q, $NQ = J(Q)$, $N = QN$ is a nilpotent left ideal of Q, and $A \setminus N$ coincides with the set of all regular elements of A.

◁ (1) By 3.16(1), the ideal N is nilpotent and coincides with $\mathrm{Sing}(A_A)$. By 3.9, A/N is a finite direct product of domains. In particular, A/N is a ring with the maximum condition on right annihilators. By 3.11, A/N is a right uniform domain, and N is a completely prime ideal. Our assertion follows now from 3.14(1), 3.14(4), and 3.16(2).

(2) By (1) and by assumption, $A \setminus N \equiv T$ is the set of all regular elements of A. By 3.6(2), A is a right order in a right uniserial ring Q, and $J(Q) = NQ$. By (1), $tN = N$ for any $t \in T$. Therefore $N = QN$ is a left ideal of Q, and by (1), there exists a positive integer n such that $N^n = (QN)^n = 0$. Hence $(J(Q))^{n+1} = (NQ)^{n+1} = N(QN)^n Q = 0$, and Q is a right uniserial semiprimary ring. By 3.18, Q is right Artinian. ▷

3.20 For a ring A, the following conditions are equivalent.

(1) A is a right distributive right Goldie ring, and the prime radical N of A is a finitely generated left ideal of A.

(2) A is a right distributive ring with the maximum conditions on right annihilators and on left annihilators, and N is a finitely generated left ideal.

(3) A is a finite direct product of Artinian right uniserial rings and right distributive right uniform domains.

◁ (3)$\Longrightarrow$(2) is directly verified. (2)$\Longrightarrow$(1) follows from 3.9.

(1)$\Longrightarrow$(3) Without loss of generality, we may assume that A is an indecomposable ring which is not a right uniform domain. By 3.19(1), $N \neq 0$, $A/N \equiv R$ is a right uniform domain, N is a completely prime nilpotent ideal, $N = \mathrm{Sing}(A_A)$, N coincides with the set of all left zero-divisors of A, and $aN = N \subset aA$ for all $a \in A \setminus N$. By the rule $f(x) = ax$, the automorphism f of N_A is well defined. Let $n + 1$ be the nilpotence index of a nonzero nilpotent ideal N, and let $M = N^n \neq 0$. Since N is a finitely generated left ideal, M is a finitely generated left ideal, and $NM = 0$. Therefore M can be considered as a finitely generated left module over a right Ore domain R with the right division ring of quotients Q. Since N contains all left zero-divisors of A, $_R M$ is a torsion-free module. Since $aN = N$ for any $a \in A \setminus N$, $M = (aN)^n \subseteq$

$aM \subseteq M$. Therefore $_RM$ is a finitely generated torsion-free R-module. Hence M can be considered as a nonzero finitely generated left module over the division ring Q. Therefore $_RQ$ is a finitely generated module. Let $Q = \sum_{i=1}^{t} Rq_i$. There exist $a_1, \ldots, a_t, b \in R$, such that $b \neq 0$ and $q_i = a_i b^{-1}$ $(i = 1, \ldots, t)$. Then $Q = Qb = \sum_{i=1}^{t} Ra_i \subseteq R$, $A/N = Q$ is a division ring, and N is a nilpotent finitely generated left ideal. Hence A is a left Artinian right distributive local ring, and $N = J(A)$. By 3.18, A is a right uniserial right Artinian ring. $\triangleright$

3.21 For a ring A with the prime radical N, the following conditions are equivalent.

(1) A is a distributive ring with the maximum condition on right annihilators, and N is a finitely generated left ideal.

(2) A is a distributive ring with the maximum condition on left annihilators, and N is a finitely generated left ideal.

(3) A is a finite direct product of uniserial Artinian rings and distributive uniform domains.

$\triangleleft$ It is sufficient to prove (1)$\Longleftrightarrow$(3).

(3)$\Longrightarrow$(1) is directly verified.

(1)$\Longrightarrow$(3) By the symmetrical analog of 3.9, A is left finite-dimensional. By 2.46(7), A is right finite-dimensional. By 3.12(2), all distributive domains are uniform. Without loss of generality, we may assume that A is an indecomposable ring with zero divisors. By 3.20, A is an Artinian right uniserial left distributive ring. By the symmetrical analog of 2.9, A is uniserial. $\triangleright$

3.2 Distributive Noetherian rings

3.22 A ring A is a *right Prüfer* ring if the following equivalent conditions hold [183], [207].

(1) *For each maximal right ideal M of A, the right ring of quotients A_M exists and is right uniserial.*

(2) *A is right distributive, and for each maximal right ideal M, the set $A \setminus M$ is weakly right reversible.*

(3) *A is a right distributive right localizable ring.*

(4) *For each maximal right ideal M, the right ring of quotients A_M exists and is right distributive.*

(5) *Each maximal right ideal of A is a completely prime ideal, and for any completely prime ideal M of A, the right ring of quotients A_M exists and is right uniserial.*

◁ $(3)\Longleftrightarrow(4)$ follows from 1.53(15). $(3)\Longrightarrow(2)$ follows from 1.48. $(2)\Longrightarrow(5)$ follows 1.60(7) and 3.6(3). $(5)\Longrightarrow(1)$ is obvious.

$(1)\Longleftrightarrow(3)$ Let F, G, and H be three right ideals of A, and let M be a maximal right ideal of A. Since A_M is right uniserial, 1.53(5) implies that $(F\cap(G+H))_M = (F\cap G+F\cap H)_M$. By 1.60(3), $F\cap(G+H) = F\cap G+F\cap H$. ▷

3.23 Let A be a right distributive domain, and let $F \equiv \max(A_A)$.

(1) A is a right order in a division ring Q, and for every $M \in F$, there exists a right uniserial subring A_M of a division ring Q such that $A \subseteq A_M$, A_M is the right ring of quotients of A with respect to M, and the natural embedding $A \to A_M$ is the canonical homomorphism.

(2) $B = \bigcap_{M\in F} B_M$ for any right ideal B of A.

(3) If $0 \neq y \in A$, then $yA = \bigcap_{M\in F}(yA_M)$ and $Ay = \bigcap_{M\in F}(A_M y)$.

(4) If x and y are two nonzero elements of the domain A and $x \in A_M y$ for all $M \in F$, then $x \in Ay$.

(5) If $0 \neq y \in A$ and the left ideal $A_M y$ of A_M is an ideal of A_M for all $M \in F$, then Ay is an ideal of A.

(6) If A_M is left invariant for all $M \in F$, then A is left invariant.

◁ (1) By 3.12(2), the domain A is right uniform. Therefore A has the classical right division ring of quotients Q. For each $M \in F$, 3.22 shows that the right ring of quotients A_M exists and is a right uniserial domain. Since A is a domain, the canonical homomorphisms $A \to A_M$ are monomorphisms. Hence we may assume that $A \subseteq A_M \subseteq Q$ for all $M \in F$. In addition, we may assume that natural embeddings $A \to A_M$ are canonical homomorphisms.

(2) follows from (1) and 1.60(3). (3) follows from the equality $A = \bigcap_{M\in F} A_M$ and from the fact that A is a domain. (4) follows from (3) and from the fact that A is a domain.

(5) Let $a \in A, M \in F$. By assumption, $ay \in yA_M$. By (4), $ay \in yA$.

(6) follows from (5). ▷

3.24 Let A be a local ring, $M \equiv J(A)$, and let $M = mA$ for some $m \in M$.

(1) For each positive integer n, the equality $M^n = m^n A$ holds, and either $M^n = 0$ or $M^n \neq M^{n+1}$.

(2) If B is a right ideal of A such that $B \supseteq M^k$ for some k, then there exists a positive integer n such that $B = m^n A = M^n$, and A/B is a right invariant right uniserial right Artinian principal right ideal ring.

◁ (1) Since M is an ideal, $M = mA = AmA$. Therefore $M^n = (mA)^n = m^n A$. If $M^n \neq 0$, then $M^n \neq M^{n+1}$ by Nakayama's Lemma.

(2) Every nonzero cyclic right module M^n/M^{n+1} over the division ring A/M is simple. Therefore B coincides with the principal right ideal M^n for some n. Hence A/B is a right invariant right uniserial right Artinian principal right ideal ring. $\triangleright$

3.25 Let A be a right uniserial domain, $M \equiv J(A) \neq 0$, and let $N \equiv \bigcap_{i=1}^{\infty} M_i$, and let $M = mA$ for some $m \in M$.

(1) $M^n = m^n A \neq M^{n+1}$ for each positive integer n.

In addition, if a right ideal B of A properly contains N, then there exists a positive integer n such that $B = m^n A = M^n$, and A/B is a right invariant right uniserial right Artinian principal right ideal ring.

(2) N is a completely prime ideal of A, and $N = MN$.

(3) If $N \neq 0$, then N is not a finitely generated left ideal of A.

(4) A/N is a right invariant principal right ideal domain, each nonzero proper right ideal of A/N coincides with some power of the ideal M/N, and each proper factor ring of the domain A/N is a right uniserial right Artinian ring.

(5) If N is a principal right ideal and M is a principal left ideal, then $N = 0$.

$\triangleleft$ (1) By 3.24(1), $M^n = m^n A \neq M^{n+1}$ for any n. Assume that B properly contains an ideal N. Since A is right uniserial, B properly contains some power of M. Every nonzero cyclic right module M^n/M^{n+1} over a division ring A/M is simple. Therefore B coincides with one of the principal right ideals M^n. Hence A is right invariant, and A/B is a right invariant right uniserial right Artinian principal right ideal ring.

(2) Assume that N is not a completely prime ideal. Then $ab \in N$ for some $a, b \in A \setminus N$. By (1), there exist integers $k, n \geq 0$ such that $a \in m^k A \setminus m^{k+1}A$ and $b \in m^n A$. Therefore, there exist $u, v \in A \setminus M = U(A)$ such that $a = m^k u$ and $b = m^n v$. Since $ab \in N \subseteq m^{n+1}A$, there exists $x \in A$ such that $m^k u m^n v = ab = m^{k+n+1}x$. Therefore $m^n = u^{-1}m^{n+1}xv^{-1} \in M^{n+1}$. By (1), $m^n A = M^n$. Hence $M^n = M^{n+1}$; this is a contradiction to (1). Therefore N is a completely prime ideal and $N = (m : N)$. In addition, $N \subseteq mA$. Hence $N = m(m : N) = mN = MN$.

(3) follows from (2) and Nakayama's Lemma. (4) follows from (1) and (2).

(5) The principal left ideal M contains the ideal N. Therefore, there exists an ideal P such that $N = PM$. Since N is completely prime, $N = NM$. The right ideal N is principal. By Nakayama's Lemma, $N = 0$. $\triangleright$

3.26 If A is a right uniserial left finite-dimensional domain, then A is a uniserial domain.

◁ Let a and b be two nonzero elements of A. It is sufficient to prove that either $a \in Ab$ or $b \in Aa$. By 3.13, A is left uniform. There exist $u, v \in A$ such that $ua = vb \neq 0$. Since A is right uniserial, either $u \in vA$ or $v \in uA$. Assume that $v = ux$, where $x \neq 0$. Then $0 \neq ua = uxb$. Therefore $a = xb \in Ab$. Analogously, if $u \in vA$, then $b \in Aa$. ▷

3.27 Let A be a right uniserial right Noetherian left finite-dimensional domain, and let $M \equiv J(A)$.

Then A is an invariant uniserial Noetherian principal ideal domain, and each nonzero proper ideal of A is a power of M.

◁ The ring A is right Noetherian and right uniserial. Therefore A is a principal right ideal ring. Assume that A is not a division ring. Then $M = mA \neq 0$. Set $N \equiv \bigcap_{n=1}^{\infty} M^n$. It follows from $M \neq 0$ that $M \neq M^2$. Therefore M properly contains N. Let us prove that $M = Am$. Assume the contrary. There exists $t \in M \setminus Am$. By 3.26, A is left uniserial. Therefore At properly contains Am. Hence $m = ut$ for some $u \in M$. Since $M = mA$, $u = mv$ for some $v \in A$. Hence $m = mvt$ and $1 = vt \in M$; this is a contradiction. Therefore, $M = Am = mA$. By 3.25(1), every power of M is a principal right ideal and a principal left ideal. By 3.25(5), $N = 0$. By 3.25(4), A is an invariant domain, and each nonzero proper right or left ideal of A is a principal ideal coinciding with some power of M. ▷

3.28 For a ring A, the following conditions are equivalent.

(1) A is a semiprime right distributive right Noetherian left finite-dimensional ring.

(2) A is a semiprime left distributive left Noetherian right finite-dimensional ring.

(3) A is a distributive semiprime right Noetherian ring.

(4) A is a distributive semiprime left Noetherian ring.

(5) A is a finite direct product of invariant distributive Noetherian domains.

◁ It is sufficient to prove (1)$\Longleftrightarrow$(3)$\Longleftrightarrow$(5). We may assume that A is indecomposable. By 3.8 and 3.13, we may assume that A is a uniform domain.

(5)$\Longrightarrow$(3) and (3)$\Longrightarrow$(1) are obvious.

(1)$\Longrightarrow$(5) By 2.72, A is a right invariant domain. It is sufficient to prove that A is left invariant. Let $M \in \max A_A$. By 3.23(1), A is a right order in a division ring Q, and there exists a right uniserial subring A_M of Q such that $A \subseteq A_M$, A_M is the right ring of quotients of A with respect to M, and the natural embedding $A \to A_M$ is the canonical homomorphism. By 3.23(5), it is sufficient to prove that the right uniserial domain A_M is left invariant. The domain A is left finite-dimensional. By 3.13, A is a left Ore domain. Hence Q is the two-sided division ring of quotients of A. Therefore A_M is a right uniserial left uniform domain. By 1.53(15), A_M is a right Noetherian domain. By 3.27, A_M is an invariant domain. $\triangleright$

3.29 Let L be a partially ordered set (poset). If $L_1, L_2 \in L$ and $L_2 \leq L_1$, then L_1/L_2 denotes the poset in L such that $N \in L_1/L_2$ if and only if $L_2 \leq N \leq L_1$. The *deviation* $\mathrm{dev}(L)$ of L can be defined by the transfinite induction as follows.

$\mathrm{dev}(L) = -1$ if and only if L contains exactly one element.

Consider an ordinal $\alpha \geq 0$; also, assume that we have already defined which posets have deviation β for ordinals $\beta < \alpha$.

Then $\mathrm{dev}(L) = \alpha$ if and only if

(i) we have not already defined $\mathrm{dev}(L) = \beta$ for some $\beta < \alpha$, and

(ii) for every (countable) descending chain $L_1 > L_2 > \ldots$ of elements of L, we have $\mathrm{dev}(L_i/L_{i+1}) < \alpha$ for all but finitely many subscripts i (this means that, for all but finitely many i, the deviation of L_i/L_{i+1} has previously been defined, and therefore, is an ordinal less than α).

We accept the convention that an assertion like $\mathrm{dev}(L) = \alpha$ is an abbreviation for "the deviation of L exists and equals to α".

Note that an arbitrary poset does not necessarilly have a deviation.

If the lattice $\mathrm{Lat}(M)$ of submodules of a module M has deviation, then this deviation is called the *Krull dimension* of M and is denoted by $\mathrm{Kdim}(M)$.

(1) $\mathrm{Kdim}(M) = -1$ if and only if $M = 0$.

(2) $\mathrm{Kdim}(M) = 0$ if and only if M is a nonzero Artinian module.

(3) Every Noetherian module M has Krull dimension.

(4) If a module has Krull dimension, then each subfactor of this module has Krull dimension and is finite-dimensional.

3.30 A is an indecomposable Noetherian right distributive ring $\Longleftrightarrow$ either A is an Artinian right uniserial ring or A is an invariant domain, and each proper factor ring of A is a finite direct product of Artinian uniserial rings.

$\triangleleft$ $\Longleftarrow$ is directly verified.

$\Longrightarrow$ By 3.20, we may assume that A is a Noetherian right distributive domain which is not an Artinian right uniserial ring. By 3.28, A is a distributive Noetherian invariant domain. Let B be a nonzero ideal of A, $0 \neq b \in B$, and let $h : A \to A/b^2A$ be the natural ring epimorphism. Assume that $h(A)$ is not an Artinian ring. In addition, $h(b)$ is a nonzero nilpotent element of $h(A)$. By 3.20, $h(A) = F \times G$, where F is a nonzero Artinian ring, and G is a nonzero reduced ring. Let $0 \neq h(g) \in G$, where $g \in G \setminus b^2 A$. Since $h(b)h(g) = 0$, there exists $a \in A$ such that $bg = b^2 a$. Then $g = ba$ and $h(g) = h(b)h(a)$. Since $(h(b))^2 = 0$ and the ring $h(A)$ is invariant, $(h(g))^2 = 0$; this is a contradiction, since $h(g)$ is a nonzero element of the reduced ring G. Hence $h(A)$ is an Artinian ring. Since A/B is a homomorphic image of $h(A)$, A/B is an Artinian distributive ring. By 3.18, A/B is a finite direct product of Artinian uniserial rings. $\triangleright$

3.31 If A is a Noetherian right distributive ring, then every right ideal of A is 2-generated, $\mathrm{Kdim}(A_A) \leq 1$, and $\mathrm{Kdim}(_A A) \leq 1$.

$\triangleleft$ 3.31 follows from 3.30. $\triangleright$

3.32 For a unitary subring A of a ring B, the following assertions hold.

(1) Assume that M_A be a submodule of B_A and there exist $m_1, \ldots, m_n \in M$, $b_1, \ldots, b_n \in B$ such that $1 = \sum_{i=1}^{n} m_i b_i$ and $b_i M \subseteq A$ for all i.

Then $M = \sum_{i=1}^{n} m_i A$ is a projective n-generated module.

(2) Let M, C, and D be two-sided A-submodules of B such that $MC = MD = A$.

Then M_A and $_A M$ are finitely generated projective A-modules, and $C = D$.

$\triangleleft$ (1) Let $f_1, \ldots, f_n : M_A \to A_A$ be homomorphisms such that $f(m) = b_i m$ for $m \in M$. Then $m = \sum_{i=1}^{n} m_i f_i(m)$ for any $m \in M$. Hence M is projective (see 1.34).

(2) Since $C = DMC = D$, we have $MC = CM = A$. Hence there exist $b_1, \ldots, b_n \in C$ such that $1 = \sum_{i=1}^{n} m_i b_i$ for some $m_1, \ldots, m_n \in M$, and $b_i M \subseteq A$ for all i. By (1), M_A is a finitely generated projective module. Analogously, $_A M$ is a finitely generated projective module. $\triangleright$

3.33 Let A be a distributive ring A which is an order in a ring Q, and let M be an ideal of A such that $M = Aa_1 + Aa_2 = a_3 A + a_4 A$, where a_1, a_2, a_3, a_4 are regular elements of A.

Then M_A and $_A M$ are projective modules.

$\triangleleft$ There exist $b_1, b_2, t \in A$ such that t is a regular element, $a_1^{-1} = t^{-1} b_1$, and $a_2^{-1} = t^{-1} b_2$. By 2.4, there exist $f_1, f_2, h_1, h_2 \in A$ such that $1 = f_1 + f_2$, $b_1 f_1 = b_2 h_2$, and $b_2 f_2 = b_1 h_1$. Hence $a_2 a_1^{-1} f_1 = a_2 t^{-1} b_1 f_1 = a_2 t^{-1} b_2 h_2 = a_2 a_2^{-1} h_2 = h_2 \in A$ and

$a_1 a_2^{-1} f_2 = a_1 t^{-1} b_2 f_2 = a_1 t^{-1} b_1 h_1 = a_1 a_1^{-1} h_1 = h_1 \in A$. Set $C \equiv a_1^{-1} f_1 A + a_2^{-1} f_2 A \subseteq Q$. Then $MC = A f_1 A + A a_1 a_2^{-1} f_2 A + A f_2 A + A a_2 a_1^{-1} f_1 A \subseteq A$. Since $1 = f_1 + f_2 \in MC$, we obtain $A \subseteq MC \subseteq A, A = MC$. Analogously, there exists a submodule D of $_A Q$ such that $A = DM$. It can directly be verified that C and D both are subbimodules of the bimodule $_A Q_A$. By 3.32(2), the modules M_A and $_A M$ are projective. $\triangleright$

3.34 [208]. *For a ring A, the following conditions are equivalent.*

(1) A is a semiprime right distributive right Noetherian left finite-dimensional ring.

(2) A is a semiprime left distributive left Noetherian right finite-dimensional ring.

(3) A is a finite direct product of invariant distributive Noetherian domains.

(4) $A = A_1 \times \cdots \times A_n$, where every A_i is an invariant domain, and each proper factor ring of A_i is a finite direct product of Artinian uniserial rings.

(5) A is a finite direct product of invariant hereditary Noetherian domains.

$\triangleleft$ (1)$\Longleftrightarrow$(2)$\Longleftrightarrow$(3) follows from 3.28. (3)$\Longleftrightarrow$(4) follows from 3.30. (5)$\Longrightarrow$(4) follows from [60],25.5.1.

(4)$\Longrightarrow$(5) We may assume that A is an invariant domain. Let M be a nonzero ideal of A. By 3.31, M is 2-generated. By 3.33, M_A and $_A M$ are projective modules. $\triangleright$

3.35 (1) [173] *A is a Noetherian right distributive ring $\Longleftrightarrow$ A is a finite direct product of Artinian right uniserial rings and invariant hereditary Noetherian domains.*

(2) [185] A is a right distributive left Noetherian ring $\Longleftrightarrow$ A is a finite direct product of Artinian right uniserial rings and right distributive left Noetherian domains.

(3) [175] A is a distributive right Noetherian ring $\Longleftrightarrow$ A is a distributive left Noetherian ring $\Longleftrightarrow$ A is a finite direct product of uniserial Artinian rings and invariant hereditary Noetherian domains.

$\triangleleft$ (1) follows from 3.30 and 3.34.

(2)

$\Longleftarrow$ is directly verified.

$\Longrightarrow$ We may assume that A is an indecomposable ring which is not a domain. Let N be the prime radical of A, and let $R \equiv A/N$. By 3.11, R is a right uniform domain, and N is a nonzero completely prime ideal. Since A is left Noetherian, N is a nilpotent ideal. Since A is left Noetherian, there exists an ideal B such that $N^2 \subseteq B \subset N$, and $\overline{N} \equiv N/B$ is a minimal nonzero ideal of the right distributive left Noetherian indecomposable ring $\overline{A} \equiv A/B$ with zero divisors. The following two cases are possible:

(*) $B = 0$; (**) $B \neq 0$.

Consider case (*). Since N is a minimal nonzero ideal, the module N_A contains no proper nonzero fully invariant submodules. By 2.46(6), N_A is uniform. By 3.14(5), A is right uniform. Hence any nonzero ideal of A contains N. Therefore $\mathrm{Sing}(A_A)$ coincides with the set of all left zero-divisors of A,and $\mathrm{Sing}(A_A)$ contains N. Let $H \equiv \mathrm{Sing}(A_A)$, and let $L \equiv r(H)$. The ring A is a ring with the maximum condition on left annihilators. By 1.37(2), L is an essential right ideal. Hence the least nonzero ideal N is contained in L. Therefore $HN = 0$. Assume that H properly contains N. By 3.14(3), $N = HN = 0$; this is a contradiction. Therefore $H = N$, N contains all left zero-divisors of A, and $N^2 = 0$. Hence N is a left torsion-free module over the right uniform domain R which has a classical right division ring of quotients Q. By 3.14(1), $N = mN$ for all $m \in A \setminus N$. Hence N can be considered as a nonzero vector space over the division ring Q. Therefore $_Q Q$ is isomorphic to a direct summand of $_Q N$. Hence $_R Q$ is isomorphic to a direct summand of $_R N$. In addition, $_R N$ is a finitely generated module. Therefore $_R Q$ is a finitely generated module. Let $Q = \sum_{i=1}^{n} Rq_i$. By 1.53(2), there exist $t, m_1, \ldots, m_n \in R$ such that $q_i = m_i t^{-1}$ for $i = 1, \ldots, n$. Therefore $Q = Qt = \sum_{i=1}^{n} Rm_i \subseteq R \subseteq Q$. Hence $R \equiv A/N$ is a division ring.

Consider case (**). Since case (*) can be applied to the ring $\overline{A}$, the factor ring $\overline{A}/\overline{N}$ is a division ring and is isomorphic to A/N. In both cases, A is a right distributive left Noetherian semiprimary local ring and $N = J(A)$. By 3.18, A is a right uniserial right Artinian left Noetherian ring, whence A is Artinian.

(3) follows from 3.35(2) and 3.34. $\triangleright$

3.3   Semidistributive modules

3.36 For a ring A, the following conditions are equivalent.

(1) A is a normal right Rickartian ring.

(2) A is a normal left Rickartian ring.

(3) A is a Rickartian reduced ring.

(4) Every element of A is a product of a central idempotent and a regular element.

◁ (3)$\Longrightarrow$(4) follows from 1.35(10). (4)$\Longrightarrow$(3) is directly verified.

(3)$\Longrightarrow$(1) and (3)$\Longrightarrow$(2) By 1.35(2), each reduced ring is normal.

(1)$\Longrightarrow$(3) and (2)$\Longrightarrow$(3) It is sufficient to prove only the first implication. Let $a \in A$, and let e be a central idempotent such that $r(a) = eA$. If $a^2 = 0$, then $a \in r(a) = eA$ and $a = ea = ae = 0$. Therefore A is reduced. By 1.35(2), $\ell(a) = r(a) = eA = Ae$. Therefore A is left Rickartian. ▷

3.37 Let e be a nonzero idempotent of a ring A.

(1) If n is a positive integer and all n-generated right ideals of A are projective, then all n-generated right ideals of eAe are projective.

In particular, if A is right semihereditary (right Rickartian), then eAe is right semihereditary (right Rickartian).

(2) If A is right nonsingular and eA is a uniform right A-module, then $r(ae) = (1 - e)A$ for any nonzero element $ae \in Ae$.

In particular, aeA_A is a projective module.

(3) If the prime radical P of A contains all right nil-ideals of A, then the prime radical of the ring eAe contains all right nil-ideals of eAe.

◁ (1) Let $B = \sum_{i=1}^{n} b_i eAe$ be any n-generated right ideal of eAe, and let $\overline{B} \equiv B + BA(1 - e)$. Then the right ideal $\overline{B} = \sum_{i=1}^{n} b_i A$ of A is n-generated. Let F_A be a direct sum of n copies of eA. Consider elements of F as rows consisting of n elements. Denote by e_i the row from F such that its i-th component is equal to e, and other components are equal to zero. There exists an epimorphism $h : F \to \overline{B}$ such that $h(e_i) = b_i$ for all i. By assumption, $\overline{B}_A$ is projective. Therefore, there exists a monomorphism $g : \overline{B} \to F$ such that hg acts identically on $\overline{B}$. Hence $F = \mathrm{Im}(g) \oplus \mathrm{Ker}(h)$. Multiply this equality by e. Then $(eAe)^{(n)} = \mathrm{Im}(g)e \oplus \mathrm{Ker}(h)e$. Since $\mathrm{Im}(g)e = B$, B_{eAe} is projective.

(2) Since $(1 - e)A \subseteq r(ae)$, $r(ae) = eB \oplus (1 - e)A$, where eB is a submodule of eA. Assume that $(1 - e)A \neq r(ae)$. Then $eB \neq 0$. Since eB is a nonzero submodule of the uniform module eA, eA is an essential extension of eB. Hence $r(ae) = eB \oplus (1 - e)A$ is an essential right ideal of A. Since A is right nonsingular, $ae = 0$; this is a contradiction.

(3) By 1.5(18) ePe is the prime radical of eAe. Let $X = eXe = eXeAe$ be a right nil-ideal of eAe, a be an element of A, and let $x = exe \in X$. Since $exeae$ belongs to the right nil-ideal X of eAe, there exists positive integer n such that $(exeae)^n = 0$. Then $(xa)^{n+2} = (exea)^{n+2} = exea(exea)^n exea = exea(exeae)^n exea = 0$. Hence xA is a right nil-ideal of A. By assumption, $xA \subseteq N$. Hence $x \in ePe$ and $X \subseteq ePe$. ▷

3.38 Let M_R be a module over a ring R, $\mathrm{End}(M) \equiv A$, P be the prime radical of A, N be a right ideal of A, and let $M_R = \oplus_{i=1}^{n} M_i$.

(1) There exists a decomposition $1 = e_1 + \ldots + e_n$ of the identity element of A into a sum of orthogonal idempotents such that $M_i = e_i(M)$, and there exists a natural ring isomorphism $\operatorname{End}(M_i) \cong e_i A e_i$ for all i.

(2) Let the natural image of the right ideal $e_i N e_i$ of $e_i A e_i$ in the ring $\operatorname{End}(M_i)$ be contained for all i in the prime radical P_i of $\operatorname{End}(M_i)$.

Then N is contained in P.

(3) Assume that the prime radical of $\operatorname{End}(M_i)$ contains all right nil-ideals (resp. all left nil-ideals) of $\operatorname{End}(M_i)$ for all i.

Then P contains all right nil-ideals (resp. all left nil-ideals) of A.

(4) If $\operatorname{End}(M_i)$ is a semiprime ring for all i, then $P^n = 0$.

(5) If A is right or left Rickartian and all the rings $\operatorname{End}(M_i)$ are normal, then $P^n = 0$.

◁ (1) is directly verified. (2), (3) and (4) follow from (1) and 2.41.

(5) By 3.37(1) and by assumption, all $\operatorname{End}(M_i)$ are right or left Rickartian normal rings. By 3.36, all $\operatorname{End}(M_i)$ are reduced. Now we can apply (4). ▷

3.39 [213]. *Let M be a finite direct sum of n nilpotently invariant modules $M_1, \ldots, M_n$ and let P be the prime radical of* $\operatorname{End}(M)$.

(1) P contains all right nil-ideals and all left nil-ideals of $\operatorname{End}(M)$.

(2) If M is nonsingular, then $P^n = 0$ and all the rings $\operatorname{End}(M_i)$ *are reduced.*

(3) If M is nonsingular and the ring $\operatorname{End}(M)$ *is prime, then all the rings* $\operatorname{End}(M_i)$ *are domains.*

◁ (1) follows from 2.64(10) and 3.38(3). (2) follows from 2.64(3) and 3.38(4). (3) follows from 2.64(4), 1.5(15) and 3.38(1). ▷

3.40 [213]. *Let the identity element of a ring A be a sum of nonzero orthogonal idempotents $e_1, \ldots, e_n$ such that all $e_i A$ are nilpotently invariant right A-modules.*

(1) The prime radical P of A contains all right nil-ideals and all left nil-ideals of A.

(2) If A is right nonsingular, then $P^n = 0$, and all the rings $e_i A e_i$ are reduced.

(3) If A is a prime right nonsingular ring, then all the rings $e_i A e_i$ are domains.

◁ 3.40 follows from 3.39. ▷

3.41 [213]. *Let M be a finite direct sum of n distributive modules $M_1, \ldots, M_n$, and let P be the prime radical of* $\operatorname{End}(M)$.

(1) *P contains all right or left nil-ideals of* $\mathrm{End}(M)$.

(2) *If either M is nonsingular or $\mathrm{End}(M)$ is right or left Rickartian ring, then $P^n = 0$.*

◁ By 2.63, all M_i are nilpotently invariant modules.

(1) follows from 3.39(1).

(2) If M is nonsingular, then the assertion follows from 3.39(2). Let $\mathrm{End}(M)$ be right or left Rickartian ring. By 3.39(2), all the rings $\mathrm{End}(M_i)$ are normal. By 3.38(5), $P^n = 0$. ▷

3.42 Let E be a uniform submodule of a module $M = \oplus_{i \in I} M_i$. Then for some $i \in I$, there exists a monomorphism $E \to M_i$.

In particular, if all modules M_i are distributive (resp. uniserial, t-distributive), then the module E is distributive (resp. uniserial, t-distributive).

◁ The uniform module E is an essential extension of a cyclic module N. There exists a finite subset $W \subseteq I$ such that $N \subseteq \sum_{i \in W} M_i \equiv F$. Assume that $W = \{1, \dots, n\}$. Let $G \equiv \sum_{i \in I \setminus W} M_i$, $h : M \to F$ be the projection with kernel G, and let $Q \equiv h(E)$. Since $N \cap G = 0$, $E \cap G = 0$. Hence $Q \cong E$. Set $M_0 = 0$. There exists a positive integer $m \le n$ such that $Q \cap \sum_{i=0}^{m-1} M_i = 0$ and $Q \cap H \ne 0$, where $H \equiv \sum_{i=0}^{m} M_i$. Since Q is uniform, Q is an essential extension of $Q \cap H$. Let $K = 0$ for $m = n$ and $K = \sum_{i=m+1}^{n} M_i$ for $m < n$. Let $u : F \to H$ be the projection with kernel K. Since Q is uniform and $Q \cap H \ne 0$, $Q \cap K = 0$. Let $v : H \to M_m$ be the projection with kernel $\sum_{i=0}^{m-1} M_i$. Hence $E \cong Q \cong (vuh)(E) \subseteq M_m$. ▷

3.43 For a ring A, the following conditions are equivalent.

(1) Each right A-module is isomorphic to a submodule of a direct sum of distributive (resp. uniserial, t-distributive) modules.

(2) Each injective right A-module is a direct sum of distributive (resp. uniserial, t-distributive) modules.

(3) A is right Noetherian and each indecomposable injective right A-module is distributive (resp. uniserial, t-distributive).

(4) A is right Noetherian and each uniform right A-module is distributive (resp. uniserial, t-distributive).

(5) Each factor ring X of A is right Noetherian, and each uniform right X-module is distributive (resp. uniserial, t-distributive).

◁ (3)$\Longleftrightarrow$(4) follows from the fact that a module M is uniform $\Longleftrightarrow$ the injective hull of M is indecomposable. (4)$\Longleftrightarrow$(5) is directly verified. (3)$\Longrightarrow$(2) follows from the fact that any injective right module over a right Noetherian ring is a direct sum of indecomposable modules. $(2)\Longrightarrow(1)$ follows from the fact that every module is isomorphic to a submodule of an injective module.

$(1)\Longrightarrow(4)$ By 3.42, each uniform right A-module is distributive (resp. uniserial, t-distributive). By 2.19(4), A is right Noetherian. $\triangleright$

3.44 Let B be a right ideal of a ring A, M_A be a cogenerator, and let $\ell_M(B) \equiv \{m \in M \mid mB = 0\}$.
Then $B = r(\ell_M(B))$.

3.45 Let all simple right modules over a ring A be isomorphic to a simple module T (e.g., this is the case if $A/J(A)$ is a simple Artinian ring), and let the injective hull M of T be distributive.
(1) All ideals of A are linearly ordered with respect to inclusion.
(2) If A is right invariant, then A is right uniserial.

$\triangleleft$ Since (2) follows from (1), it is sufficient to prove (1). Let B and C be two ideals of A. Since M is a distributive module over the local ring A, 2.9 shows us that M is uniserial. Therefore, either $r(\ell_M(B)) \subseteq r(\ell_M(C))$ or $r(\ell_M(C)) \subseteq r(\ell_M(B))$. For example, let $r(\ell_M(B)) \subseteq r(\ell_M(C))$. By 2.38, M_A is a cogenerator. By 3.44, $B = r(\ell_M(B)) \subseteq r(\ell_M(C)) = C$. $\triangleright$

3.46 Let A be a right invariant ring. Then
A is right distributive $\Longleftrightarrow$ each factor ring of A has no ideals which are direct sums of two isomorphic minimal right ideals.

$\triangleleft$ 3.46 follows from 2.32. $\triangleright$

3.47 Let A be a Noetherian right invariant ring A.
(1) If A is not right distributive and all proper factor rings of A are right distributive, then A is a local Artinian ring, and the injective hull E of the simple module $(A/J(A))_A$ is not distributive.
(2) If the injective hull of any simple right A-module is distributive, then A is a finite direct product of Artinian right uniserial rings and invariant hereditary Noetherian domains.

$\triangleleft$ (1) By 3.46 and by assumption, A has an ideal $N = F_A \oplus G_A$, where F and G are isomorphic minimal right ideals. The right annihilator M of N is a maximal right ideal. Since all direct products of right distributive rings are right distributive and A is not right distributive, A is an indecomposable right invariant ring. Therefore M is an essential right ideal. Hence $N \subseteq M$ and $NM = 0$, whence $N^2 = 0$ and $F^2 = G^2 = 0$. Since the ideal F is nilpotent, all idempotents of A/F

are lifted to idempotents of the indecomposable right invariant ring A. Therefore A/F is a Noetherian indecomposable right distributive ring which has the nonzero nilpotent ideal N/F. By 3.35(1), A/F is an Artinian right uniserial ring. Hence A is a local Artinian ring. Since A is not right uniserial, 3.45(2) shows us that E is not distributive.

(2) Assume the contrary. By 3.35(1), A is not right distributive. Since the ring A is Noetherian, A has a Noetherian right invariant factor ring B such that B is not right distributive, and all proper factor rings of B are right distributive. Since any essential extension of each simple A-module is distributive, B possesses the same property. Therefore, the injective hull of any right B-module is distributive; this is a contradiction to (1) applied to B. ▷

3.48 Let Q be the left ring of quotients of a ring A with respect to a left denominator set T, and let $f : A \to Q$ be the canonical homomorphism.

(1) Every finitely generated submodule of $Q_{f(A)}$ is isomorphic to a finitely generated right ideal of $f(A)$.

(2) If A is a right Bezout ring, then $Q_{f(A)}$ is a Bezout module (in particular, Q is a right Bezout ring).

(3) If A is right distributive, then $Q_{f(A)}$ is a distributive module (in particular, Q is right distributive).

◁ (1) follows from 1.53(3). (2) follows from (1).

(3) M is distributive $\Longleftrightarrow$ all 2-generated submodules of M are distributive. Therefore, (3) follows from (1). ▷

3.49 [236],39.16. A ring A is right hereditary $\Longleftrightarrow$ all submodules of projective right A-modules are projective $\Longleftrightarrow$ all factor modules of injective right A-modules are injective.

3.50 [60],20.1. A ring A is right Noetherian $\Longleftrightarrow$ all direct sums of injective right A-modules are injective.

3.51 [192]. *For an invariant ring A, the following conditions are equivalent.*

(1) Each right A-module is isomorphic to a submodule of a semidistributive module.

(2) All injective right A-modules are semidistributive.

(3) A is a finite direct product of uniserial Artinian rings and hereditary Noetherian domains.

◁ Since A is invariant, 3.43 and 3.35(1) show us that conditions (1) and (2) are equivalent to condition (3^*): A is an invariant Noetherian ring and any indecomposable injective right A-module E is distributive.

$(3^*) \Longrightarrow (3)$ follows from 3.47(2).

$(3) \Longrightarrow (3^*)$ We may assume that A is either a uniserial Artinian ring or an invariant hereditary Noetherian domain. It can directly be verified that every uniform module over a uniserial Artinian (invariant) ring is uniserial. Thus, we may assume that A is an invariant hereditary Noetherian domain with a division ring of quotients Q. By 3.48(3) and 3.35(1), the module Q_A is distributive. Let M be a nonzero cyclic submodule of E. There exists an epimorphism $f : A_A \to M$. Since E is injective, f is extended to a nonzero homomorphism $g : Q_A \to E$. It can be verified that Q_A is injective. By 3.49, every homomorphic image of any injective module over a hereditary ring is injective. Therefore $g(Q)$ is a nonzero injective submodule of the indecomposable module E. Hence $g(Q) = E$. Therefore E is distributive. $\triangleright$

3.52 Let M_A be a nonzero distributive module over a right invariant ring A.

(1) Assume that for any nonzero cyclic submodule N of M, each factor ring of the ring $A/r(N)$ is a finite direct product of right uniform rings.

Then M is semiuniform.

(2) If F is a locally Noetherian submodule of M and each right Noetherian factor ring of A is left Noetherian, then F is semiuniform.

(3) If F is a locally Artinian submodule of M, then F is semiuniform.

$\triangleleft$ (1) By 2.46(4), it is sufficient to prove that any nonzero cyclic submodule N of M is semiuniform. Since N_A is a cyclic module over the right invariant ring A, we have $N \cong (A/r(N))_A$. Our assertion follows now from the assumption.

(2) Let N be a nonzero indecomposable cyclic submodule of F, and let $B \equiv A/r(N)$. Since $A/r(N)$ is a right distributive right Noetherian ring, it follows from the assumption that $A/r(N)$ is a Noetherian right distributive ring. By 3.35(1), $A/r(N)$ is a finite direct product of right uniform rings. We apply now (1) to F.

(3) follows from 2.46(2) and from the fact that right distributive right Artinian rings are finite direct products of right uniserial rings by 3.18. $\triangleright$

3.53 Let M be a right module over a right invariant ring A.

(1) If M is a locally Artinian module, then

M is distributive $\Longleftrightarrow$ M is the unique direct sum of uniform distributive modules M_i (where $i \in I$), and $A = r(m_i) + r(m_j)$ for all distinct subscripts $i, i \in I$ and all $m_i \in M_i$, $m_j \in M_j$.

(2) M is a locally Artinian distributive uniform module $\Longleftrightarrow M$ is a uniserial Artinian module, and there exists a maximal ideal B of A such that the annihilator of any nonzero element of M is a power of B.

◁ (1) By 3.52(3), M is semiuniform. We apply now 2.46(3) and 2.48.

(2) It is sufficient to prove only $\Longrightarrow$. Let B be a maximal ideal of A such that the simple module $(A/B)_A \equiv T$ is an essential submodule of M. It is sufficient to prove that for an arbitrary nonzero finitely generated Artinian submodule F of M, there exists a positive integer n such that $F \cong (A/B^n)_A$. The module F is an essential extension of T and is cyclic by 2.27. Hence $F \cong D_A$, where D is some right Artinian right distributive indecomposable factor ring of A. Our assertion follows now from the fact that D is a right uniserial right Artinian ring by 3.18. Therefore, each proper right ideal of D is a finite power of the radical of D. ▷

3.54 [184]. *Let M be a right module over an invariant ring A.*

(1) If M is a locally Noetherian module, then

M *is distributive* $\Longleftrightarrow M$ *is the unique direct sum of uniform distributive modules M_i $(i \in I)$, and $A = r(m_i) + r(m_j)$ for all distinct subscripts $i, j \in I$ and all $m_i \in M_i$, $m_j \in M_j$.*

(2) M is a locally Noetherian distributive uniform module $\Longleftrightarrow$ either M is a uniserial Artinian module and there exists a maximal ideal B of A such that the annihilator of any nonzero element of M is a power of B, or $A/r(M)$ is an invariant hereditary Noetherian domain with a division ring of quotients Q and M is isomorphic to a submodule of Q_A.

◁ (1) By 3.52(2) and 3.52(3), M is semiuniform. We apply now 2.46(3) and 2.48.

(2)

$\Longleftarrow$ follows from 3.35(1) and 3.48(3).

$\Longrightarrow$ Let M be a distributive uniform module, and let N be a nonzero cyclic submodule of M. There exists an indecomposable invariant distributive Noetherian factor ring D of A, such that $N \cong D_A$. By 3.35(1), the following two assertions hold.

(∗) If N contains a simple submodule T, then N is a uniserial module with a finite composition series, and all simple subfactors of N are isomorphic to T.

(∗∗) If $\mathrm{Soc}(N) = 0$, then $A/r(N)$ is an invariant hereditary Noetherian domain, and all nonzero cyclic submodules of N are isomorphic to N.

Assume that M contains a simple submodule T. Let $F = \sum_{i=1}^{n} F_i$ be a finitely generated submodule of M, where $F_1, \ldots, F_n$ are nonzero cyclic submodules of the uniform module M. Then $T \subseteq F_1 \cap \ldots \cap F_n$, and by $(*)$ applied to $F_1, \ldots, F_n$, these modules are Artinian. Therefore M is locally Artinian, and we apply 3.53(2). It is sufficient to consider now the case $\mathrm{Soc}(M) = 0$. Let $0 \neq f, g \in M$. Since M is uniform, there exists a nonzero element $d \in fA \cap gA$. The assertion $(**)$ implies that $fA \cong dA \cong gA$. In addition, the ring A is invariant. Therefore $r(fA) = r(f) = r(g) = r(gA)$. Hence $r(M) = r(f)$. By $(**)$, $A/r(M)$ is an invariant hereditary Noetherian domain with a division ring of quotients Q, and M is a uniform torsion-free module over $A/r(M)$. Therefore M is isomorphic to a submodule of Q_A. $\triangleright$

3.55 [213]. *Let M_A be a finite direct sum of n distributive modules $M_1, \ldots, M_n$, and let P be the prime radical of $\mathrm{End}(M)$.*

(1) P contains all right or left nil-ideals of the ring $\mathrm{End}(M)$.

(2) If M is nonsingular, then $P^n = 0$.

In addition, if $\mathrm{End}(M)$ is also a prime ring, then all the rings $\mathrm{End}(M_i)$ are domains.

(3) If $\mathrm{End}(M)$ is a right or left Rickartian ring, then $P^n = 0$.

$\triangleleft$ By 2.63, all M_i are nilpotently invariant modules. Therefore, (1) and (2) follow from 3.39.

(3) By 2.50(2), all $\mathrm{End}(M_i)$ are normal rings. In addition, $\mathrm{End}(M_i)$ are right or left Rickartian rings. By 3.36, all $\mathrm{End}(M_i)$ are reduced rings. By 3.38(4), $P^n = 0$. $\triangleright$

3.4   Serial modules and rings

3.56 Let A be a semiperfect ring, $P = \oplus_{i=1,\ldots,m} e_i A$ and $Q = \oplus_{j=1,\ldots,n} e_j A$ be finitely generated projective right A-modules, where e_i and e_j are indecomposable idempotents, and let $f : P \to Q$ be a homomorphism. Denote by A_{ij} the Abelian group $e_i A e_j$. Since every homomorphism from $e_i A$ to $e_j A$ is given by left multiplication on an element of A_{ij}, f can be written as left multiplication on the $m \times n$ matrix (f_{ij}), where $f_{ij} \in A_{ij}$.

From the case $P = Q = A_A$, we obtain that every element a of the ring A can be written in the form $a = \sum_{i,j} a_{ij}$, where $a_{ij} = e_i a e_j \in A_{ij}$. Hence a can be considered as the matrix $M(a) = (a_{ij})$, and the multiplication of elements of A with respect to this representation is the matrix mutiplication (i.e., the equality $M(ab) = M(a) \cdot M(b)$ holds for all $a, b \in A$).

For a semiperfect ring A, the following assertions hold.

(1) [43] Let P_1, P_2, and P_3 be any three finitely generated projective right A-modules, and let Q_1, Q_2, and Q_3 be any three finitely generated projective left A-modules. Denote by $P^*{}_i$ and $Q^*{}_i$ the left A-modules $\operatorname{Hom}(P_i, A_A)$ and the right A-modules $\operatorname{Hom}(Q_i, {}_A A)$, respectively. For any $f \in \operatorname{Hom}(P_i, P_j)$ and $g \in \operatorname{Hom}(Q_i, Q_j)$ denote by f^* and g^* homomorphisms $P^*{}_j \to P^*{}_i$ and $Q^*{}_j \to Q^*{}_i$ induced by f and g, respectively.

Then $P^*{}_i$ are finitely generated projective left A-modules, $(f_{23}f_{12})^* = f_{12}{}^* f_{23}{}^*$ for any $f_{12} \in \operatorname{Hom}(P_1, P_2)$ and $f_{23} \in \operatorname{Hom}(P_2, P_3)$, $Q^*{}_i$ are finitely generated projective right A-modules, and $(g_{23}g_{12})^* = g_{12}{}^* g_{23}{}^*$ for any $g_{12} \in \operatorname{Hom}(Q_1, Q_2)$ and $g_{23} \in \operatorname{Hom}(Q_2, Q_3)$.

(2) Let the ring A be (right and left) serial, $P = \oplus_{i=1,\ldots,n} e_i A$ and $Q = \oplus_{j=1,\ldots,m} e_j A$ be finitely generated projective right A-modules, where e_i and e_j are local idempotents, and $f = (f_{ij}) : P \to Q$ be homomorphisms with $f_{ij} \in A_{ij}$.

(i) There are automorphisms $\alpha : P \to P$ and $\beta : Q \to Q$ such that α and β are given by matrices which are products of elementary matrices, and $\beta f \alpha : P \to Q$ is given by the matrix $G = (g_{ij})$ such that G has at most one nonzero element in every row and every column.

(ii) For every element a of A, there are invertible elements u and v of A such that all rows and columns of the matrix $M(uav)$ contain at most one nonzero element, and matrices $M(u)$ and $M(v)$ are products of elementary matrices.

In addition, for some positive integer m, the matrix $M((uav)^m)$ is diagonal.

(3) [43] The following conditions are equivalent.

(i) A is right serial.

(ii) For any three local projective right modules P, P_1, and P_2 and for any two homomorphisms $f_1 : P_1 \to P$ and $f_2 : P_2 \to P$, either there exists $x \in \operatorname{Hom}(P_1, P_2)$ such that $f_1 = f_2 x$ or there exists $y \in \operatorname{Hom}(P_2, P_1)$ such that $f_2 = f_1 y$.

(iii) For any three local projective left modules Q, Q_1, and Q_2 and for any two homomorphisms $f_1 : Q \to Q_1$ and $f_2 : Q \to Q_2$, either there exists $x \in \operatorname{Hom}(Q_2, Q_1)$ such that $f_1 = x f_2$ or there exists $y \in \operatorname{Hom}(Q_1, Q_2)$ such that $f_2 = y f_1$.

(4) [43] If all finitely presented indecomposable right A-modules have indecomposable projective hulls, then A is left serial.

(5) [66] Let M be a right A-module. Then

M is distributive $\iff$ Me_i is a right distributive $e_i A e_i$-module for any e_i $\iff$ Me_i is a right uniserial $e_i A e_i$-module for any e_i.

(6) [184] A is right semidistributive $\iff$ $e_j A e_i$ is a right uniserial

$e_i A e_i$-module for any e_i and any e_j.

◁ (1) can directly be verified.

(2)

(i) Without loss of generality, we may assume that $m = n$ (i.e., G is a square matrix). By the induction on n, we prove that the diagonal form can be obtained by elementary transformations of rows and columns. The case $n = 1$ is trivial. Assume that for all numbers $< n$, the assertion holds. Let us prove the assertion for n. Without loss of generality, we may assume that we have chosen the matrix element f_{11} such that the left ideal Af_{11} is the largest element of the set $\{Af_{i1}\}$. By elementary transformations of rows, we can annihilate all elements below f_{11}, and then, by induction, we can transform the complemented minor of f_{11} to the diagonal form without changing the first column. Therefore, our matrix has the form

$$\begin{pmatrix} f_{11} & f_{12} & \cdots & f_{1n} \\ 0 & f_{22} & \cdots & 0 \\ \vdots & \vdots & \ddots & \vdots \\ 0 & 0 & \cdots & f_{nn} \end{pmatrix}$$

If $f_{1i} \in f_{11}A$ for some $i > 1$, then we can annihilate f_{1i} by elementary transformation of columns, and then we can apply the induction hypothesis to the complemented minor of f_{ii} (if $i > m$ then the i-th column is zero). Analogously, the induction hypothesis can be applied if $f_{1i} \in Af_{ii}$ for some $i > 1$. Otherwise, $f_{11} \in f_{1i}A$, and $f_{ii} \in Af_{1i}$ holds for every $1 < i \leq n$. Choose f_{1j} such that the right ideal $f_{1j}A$ is the largest element of the set $\{f_{1i}A\}$. By assumption, $j > 1$. Then all elements of the first row and j-th column (excluding f_{1j}) can be annihilated by elementary transformations, and we can apply the induction hypothesis to the complemented minor of f_{1j}.

(ii) The first assertion follows from (i). The second assertion follows from the fact that for any positive integer n, there exists a positive integer m such that $\pi^m \equiv 1$ for an arbitrary permutation π on the set $\{1, \ldots, n\}$.

(3)

(ii)$\Longleftrightarrow$(iii) follows from (1).

(i)$\Longrightarrow$(ii) Since A is right serial, P is uniserial. Therefore, either $f_1(P_1) \subseteq f_2(P_2)$ or $f_2(P_2) \subseteq f_1(P_1)$. Since P_1 and P_2 are projective modules, either there exists $x \in \mathrm{Hom}(P_1, P_2)$ such that $f_1 = f_2 x$ or there exists $y \in \mathrm{Hom}(P_2, P_1)$ such that $f_2 = f_1 y$.

(ii)$\Longrightarrow$(i) Assume that A is not right serial. Then some local projective module P has incomparable submodules M_1 and M_2. Let

$m_1 \in M_1 \setminus M_2$, $m_2 \in M_2 \setminus M_1$. There exist local idempotents e_1 and e_2 of A such that $m_1 e_1 \in M_1 \setminus M_2$ and $m_2 e_2 \in M_2 \setminus M_1$. Denote by f_i homomorphisms $e_i A \to P$ such that $f_i(e_i) = m_i e_i$. Set $P_i \equiv e_i A$. Then $f_1(P_1)$ and $f_2(P)$ are not comparable. Hence both equations $f_1 = f_2 x$ and $f_2 = f_1 y$ have no solutions; we have a contradiction.

(4) Assume the contrary. By (3), there exist three local projective right modules P, P_1, P_2 and two homomorphisms $f_1 : P_1 \to P$, $f_2 : P_2 \to P$ such that both equations $f_1 = f_2 x$ and $f_2 = f_1 y$ have no solutions. Set $Q \equiv P_1 \oplus P_2$. Denote by f the homomorphism $P \to Q$ with the matrix $\begin{pmatrix} f_1 \\ f_2 \end{pmatrix}$. Since f_i are not isomorphisms, $f(P) \subseteq J(Q)$. Since $f_i \neq 0$, $\mathrm{Ker}(f) \subseteq J(P)$. Set $M \equiv Q/f(P)$. Then Q is a projective hull of the finitely presented right module M. Since Q is not indecomposable, M is not indecomposable. Therefore, there exist automorphisms u and v of P and Q, respectively, such that either $vfu = \begin{pmatrix} g_1 \\ 0 \end{pmatrix}$, or $vfu = \begin{pmatrix} 0 \\ g_2 \end{pmatrix}$, where $g_i \in \mathrm{Hom}(P, P_i)$. Assume that v has the matrix

$$\begin{pmatrix} v_{11} & v_{12} \\ v_{21} & v_{22} \end{pmatrix}, \quad v_{ij} \in \mathrm{Hom}(P_j, P_i).$$

Then either $v_{11}f_1 + v_{12}f_2 = 0$ or $v_{21}f_1 + v_{22}f_2 = 0$. For example, let $v_{21}f_1 + v_{22}f_2 = 0$. Then either v_{21} or v_{22} is an isomorphism, whence at least one of the equations $f_1 = f_2 x, f_2 = f_1 y$ has a solution; we have a contradiction.

(5) The first equivalence follows from 2.41(11). Since all the rings $e_i A e_i$ are local, the second equivalence follows from 2.9.

(6) follows from (5). ▷

3.57 Let A be a right serial ring. Then

A is a right semiuniform semiperfect ring, and each cyclic right A-module M is generated by uniform submodules.

In addition, A is right nonsingular $\iff$ A is right semihereditary

◁ Since each uniserial module is uniform, A is right semiuniform. The identity element of the right serial ring A is a finite sum of local orthogonal idempotents. Hence A is semiperfect. There exists an epimorphism $h : A_A = \oplus_{i=1}^n A_i \to M$, where the A_i are uniserial modules. Hence M is a sum of uniserial modules $h(A_i)$ which are, in particular, uniform.

Let us prove the second assertion. $\impliedby$ holds for any right semihereditary ring.

$\Longrightarrow$ Let $1 = e_1 + \ldots + e_n$, where e_i are nonzero orthogonal idempotents such that all $e_i A$ are uniserial right A-modules. Consider some $e_i \equiv e$. By 1.34(5), it is sufficient to prove that eA is a semihereditary A-module. Let B be a finitely generated submodule of eA. Since eA is uniserial, $B = ebA$ for some element $eb \in eA$. Then $\sum_{j=1}^{n} ebe_j A = ebA$ is a submodule of the uniserial module eA. Hence there exists a subscript j such that $ebe_j A = \sum_{k=1}^{n} ebe_k A = ebA$. By 3.37(2), $ebe_j A = B$ is a projective module. $\triangleright$

3.58 Let M be a distributive right module over a right serial ring A.

(1) M is a finite direct sum of uniform modules, and M is a completely finite-dimensional Bezout module.

(2) If all cyclic submodules of M are serial, then M is a finite direct sum of uniserial modules.

Therefore, the class of all semidistributive right A-modules coincides with the class of all serial right A-modules.

$\triangleleft$ (1) follows from 2.46(4), 3.57, and 2.29.

(2) By (1), we may assume that M is a uniform distributive Bezout module. It is sufficient to prove that any finitely generated submodule N of M is uniserial. Since M is a uniform Bezout module, N is a uniform cyclic module. By assumption, N is serial. Hence N is uniserial whence M is uniserial. $\triangleright$

3.59 (Drozd-Warfield's Theorem) [43], [232]. *For a ring A, the following conditions are equivalent.*

(1) A is a serial ring.

(2) All finitely presented right A-modules are serial.

(3) All finitely presented left A-modules are serial.

(4) Every finitely presented right A-module is a direct sum of uniserial modules of the form $e_j A / e_j a_{ji} e_i A$, where e_j are local idempotents of A.

(5) Every finitely presented left A-module is a direct sum of uniserial modules of the form $Ae_j / Ae_i a_{ji} e_j$, where e_j are local idempotents of A.

(6) A is a semiperfect ring, all finitely presented indecomposable right A-modules are local, and all finitely presented indecomposable left A-modules are local.

$\triangleleft$ It is sufficient to prove the equivalence of conditions (1), (2), (4), and (6).

(4)$\Longrightarrow$(2) and (2)$\Longrightarrow$(6) are directly verified.

$(6)\Longrightarrow(1)$ It can be directly verified that projective hulls of all finitely presented indecomposable (right or left) A-modules are indecomposable. (1) follows now from 3.56(4).

$(1)\Longrightarrow(4)$ Let M_A be a finitely presented A-module, P and Q be two finitely generated projective right A-modules such that $M \cong Q/f(P)$, $f : P \to Q$ be a homomorphism, and let $P = \oplus_{i=1}^{s} e_i A$, $Q = \oplus_{j=1}^{t} e_j A$ for local idempotents e_i and e_j. By 3.56(2(i)), there are automorphisms $g : P \to P$ and $h : Q \to Q$ such that $hfg(e_i A) \subseteq e_j A$ and subscripts j are different for distinct subscripts i. Then $M \cong Q/f(P) = Q/fg(P) \cong Q/hfg(P)$, and the last module is a direct sum of modules $e_j A/hfg(e_i A)$ for suitable subscripts i. Since hfg is a homomorphism from $e_i A$ into $e_j A$ which is left multiplication on some element $a_{ji} \in e_j A e_i$, $e_j A/hfg(e_i A) = e_j A/a_{ji} A$, and all modules $e_j A/_{ji} a A$ are uniserial. $\triangleright$

3.60 (Ivanov's Theorem) [88]. *For a ring A, the following conditions are equivalent.*

(1) All finitely generated right A-modules are serial.

(2) A is a serial ring, and all uniform right A-modules are uniserial.

(3) All 2-generated right A-modules are serial.

(4) A is a right semiuniform ring, and all indecomposable injective right A-modules are uniserial.

(5) A is a right serial ring, and for every isomorphism $f : M \to N$ between arbitrary submodules M and N of any primitive cyclic A-modules xA and yA, respectively, either f or f^{-1} can be extended to a homomorphism between xA and yA.

$\triangleleft$ An element m of M_A will be called *primitive* if $m = me$ for a indecomposable idempotent $e \in A$. A generating set of a module will be called *minimal* if no element can be deleted.

$(1)\Longrightarrow(2)$ By 3.59, A is a serial ring. By (1), all indecomposable finitely generated right A-modules are uniserial. Since all submodules of uniform modules are indecomposable, all finitely generated submodules of uniform right A-modules are uniserial. Hence all uniform right A-modules are uniserial.

$(2)\Longrightarrow(3)$ is directly verified.

$(3)\Longrightarrow(4)$ It is verified directly that A is right serial. Let M be any indecomposable injective right A-module. Then M is uniform. If M is not uniserial, M contains a 2-generated submodule which is a direct sum of at least two uniserial submodules; this is a contradiction to the uniformity of M.

$(4)\Longrightarrow(1)$ It is sufficient to prove that for any positive integer n, every finitely generated right A-module with a minimal set of n primitive generators is a direct sum of n uniserial submodules. We use the induction on the number of primitive generators of finitely generated right A-modules. Let $\{m_1, \ldots, m_{n+1}\}$ be a minimal set of primitive generators for a finitely generated module M_A. Assume that every right A-module with a minimal set of n primitive generators is a direct sum of n uniserial submodules. Set $N \equiv m_1 A + \ldots + m_n A$. Without loss of generality, we may assume that $N = \oplus_{i=1}^n m_i A$. Set $U \equiv N \bigcap x_{n+1} A$. Then U contains no elements $\sum_{i=1}^n m_i a_i$ such that $a_i = 1$ for some i, since otherwise $\{m_j \mid j \neq i, 1 \leq j \leq n+1\}$ would generated M; this is a contradiction to the minimality of $\{m_1, \ldots, m_{n+1}\}$.

By the rule $f_i(\sum_{j=1}^n m_j a_j) = m_i a_i$, the homomorphisms $f_i : U \to m_i A$ are well defined $(1 \leq i \leq n)$. Set $U_i \equiv f_i(U)$. Since U is uniserial, we may assume that $\mathrm{Ker}(f_1) \subseteq \mathrm{Ker}(f_2) \subseteq \ldots \subseteq \mathrm{Ker}(f_n)$. By the rule $g_i(f_1(u)) = f_i(u)$, the homomorphisms $g_i : U_1 \to U_i$ are well defined $(1 \leq i \leq n)$. Let E_i be the injective hull of U_i $(1 \leq i \leq n)$. Homomorphisms g_i can be extended to homomorphisms $\overline{g}_i : m_1 A \to E_i$. Since E_i are uniserial, either $\overline{g}_i(m_1 A) \subseteq m_i A$ or $m_i A \subseteq \overline{g}_i(m_1 A)$. Therefore, there exist elements $b_i \in A$ such that homomorphisms $\overline{g}_i$ induce homomorphisms $h_i : m_1 b_i A \to m_i A$, where either $x_1 b_i = x_1$ or $h_i(x_1) b_i = m_i$. Since $m_1 A$ is a uniserial module, one of the modules $m_1 b_i A$ is contained in the others. Denote the element b_i corresponding to this module by b. Then $m_1 b A$ is contained in all modules $m_1 b_i A$ and coincides with some of these modules. Set $y \equiv m_1 b + h_2(m_1 b) + \ldots + h_n(m_1 b)$. If $u \in U$, then $u = f_1(u) + \ldots + f_n(u)$, whence $u = x_1 b a + h_2(x_1 b a) + \ldots + f_n(x_1 b a) = ya$ for some $a \in A$. Therefore $U \subseteq yA$, and yA is uniserial, since $yA \cong x_1 b A$.

Since all indecomposable injective right A-modules are uniserial, 1_U can be extended to either $\alpha : m_{n+1} A \to yA$ or $\beta : yA \to m_{n+1} A$. If 1_U can be extended to α, then $yA + m_{n+1} A = yA \oplus (m_{n+1} - \alpha(m_{n+1} A))A$. If 1_U can be extended to β, then $yA + m_{n+1} A = (y - \beta(y))A \oplus m_{n+1} A$. Since one of the elements $x_1 b, f_2(x_1 b), \ldots, f_n(x_1 b)$ is an m_i $(1 \leq i \leq n)$, yA is a direct summand of M. Therefore M is serial.

$(4) \Longrightarrow (5)$ Since all indecomposable injective right A-modules are uniserial, all uniform right A-modules are uniserial. Hence all semiuniform right A-modules are serial. Therefore A is a right serial ring. Let $E(yA)$ be the injective hull of yA, and let $g : xA \to E(yA)$ be an extension of f. Since $E(yA)$ is uniserial, either $g(x)A \subseteq yA$ or $yA \subseteq g(x)A$. In the first case, the assertion is proved. Assume that $yA \subseteq g(x)A$. Then there exists an element b in A such that $g(xb) = y$. By the rule $h(ya) = xba$, the extension g of f^{-1} is well defined.

$(5) \Longrightarrow (4)$ Let M be an indecomposable injective right A-module. Assume that M is not uniserial. Then there exist two elements x and

y in M such that $xA \cap yA \equiv U$ is a proper submodule of both xA and yA. Without loss of generality, we may assume that the identity map on U can be extended to $f : xA \to yA$. Set $z \equiv x - f(x)$. Then $xA + yA = zA + yA$. If $za_1 = ya_2 \in zA \cap yA$, then $xa_1 = f(xa_1) + ya_2$, whence $xa_1 \in U$. Since $(f - 1)(U) = 0$, $xa_1 = f(x)a_1$, whence $za_1 = ya_2 = 0$. Hence $zA \cap yA = 0$. Therefore $xA + yA = zA \oplus yA$. Hence M is not uniform; this is a contradiction. $\triangleright$

3.61 For a ring A, the following conditions are equivalent.
(1) A is a serial right Noetherian ring.
(2) A is a right Noetherian ring, and each finitely generated right A-module is serial.
(3) A is right semiuniform, and each injective right A-module is serial.

$\triangleleft$ By 2.19(3), we may assume that A is right Noetherian, and each injective right A-module is a direct sum of indecomposable injective modules. Therefore, each finitely generated right A-module is finitely presented, $(1) \Longleftrightarrow (2)$ follows from 3.59, and $(2) \Longleftrightarrow (3)$ follows from 3.60. $\triangleright$

3.62 [236, 54.3]. Let A be a right Artinian ring such that there exists a finite upper bound for the lengths of finitely generated indecomposable right A-modules.

Then A is a left Artinian ring, and there exist the common finite upper bound for the composition lengths of finitely generated indecomposable left or right A-modules.

In addition, every (right or left) A-module is a direct sum of indecomposable finitely generated modules.

3.63 If every right module over a ring A is a finite direct sum of cyclic right modules, then A is a (right and left) Artinian ring, and there exists a finite upper bound for the lengths of finitely generated indecomposable left A-modules.

$\triangleleft$ 3.63 follows from 2.19(7) and 3.62. $\triangleright$

3.64 [167]. *For a ring A, the following conditions are equivalent.*
(1) All right A-modules are semidistributive.
(2) A is (right and left) Artinian, and each right A-module is a direct sum of distributive completely cyclic modules with a composition series.
(3) A is a (right and left) Artinian ring with basis idempotent e, and for any right A-module M, the right eAe-module $M e_{eAe}$ is a direct sum of completely cyclic modules with a composition series.

◁ 3.64 follows from 2.22, 2.29, 2.39, 2.40(4), and 3.63. ▷

3.65 All right modules over a ring A are semidistributive, and the ring $A/J(A)$ is normal $\Longleftrightarrow$ A is a (right and left) Artinian ring, and each right A-module is a direct sum of completely cyclic modules with a composition series.

◁ If A is a semiperfect ring, then A is selfbasis $\Longleftrightarrow$ the ring $A/J(A)$ is normal. Hence our assertion follows from 3.64. ▷

3.66 (Nakayama-Skornyakov's Theorem) [137], [157]. *All right modules over a ring A are serial $\Longleftrightarrow$ all left A-modules are serial $\Longleftrightarrow$ A is a serial Artinian ring.*

3.67 Any semiprime right Rickartian right finite-dimensional ring A is right semiuniform.

◁ The right finite-dimensional ring A has an essential right ideal $B = \oplus_{i=1}^{n} B_i$, where the B_i are nonzero uniform right ideals. Since A is a semiprime right nonsingular right finite-dimensional ring, 3.7 shows us that the essential right ideal B contains a regular element b of A. Since $bA \cong A_A$, it is sufficient to prove that bA_A is semiuniform. It is sufficient to prove that if a principal right ideal N of A is contained in B and has Goldie dimension $m \leq n$, then N is semiuniform. For $m = 1$, our assertion holds. Let $h_i : B \to B_i$ be the natural projections. Since all $h_i(N)$ are projective modules, $N \bigcap(\oplus_{j \neq i} B_j)$ is a direct summand of N_A for all i, and the corresponding complement is isomorphic to $h_i(N)$. Our assertion can be proved now by the induction on m. ▷

3.68 Let A be a right order in a semisimple Artinian ring Q.
(1) If Q_A is a Noetherian module, then $A = Q$ is a semisimple Artinian ring.
(2) If A is indecomposable and the module Q_A has a nonzero Noetherian direct summand F, then $A = Q$ is a simple Artinian ring.

◁ (1) Let b be a regular element of A. It is sufficient to prove the inclusion $b^{-1} \in A$. Since $\sum_{i=1}^{\infty} (b^{-i}A)$ is a Noetherian module, there exist $a_1, \ldots, a_n \in A$ such that $b^{-n-1} = \sum_{i=1}^{n} b^{-i}a_i$. Hence $b^{-1} = b^n \sum_{i=1}^{n} b^{-i}a_i = \sum_{i=1}^{n}(b^{n-i}a_i) \in A$.
(2) We may assume that the module F_A is indecomposable. There exists a nonzero central idempotent e of Q such that eQ is a simple Artinian ring containing F. The injective module eQ_A is the unique (up to isomorphism) direct sum of nonzero indecomposable isomorphic modules. Hence eQ_A is a finite direct sum of copies of the

Noetherian module F_A. Therefore eQ_A is Noetherian. Let $e = ab^{-1}$, where $0 \neq a, b \in A$. Since the module $\sum_{i=1}^{\infty}(b^{-i}eA)$ is Noetherian, there exist $a_1, \ldots, a_n \in A$ such that $b^{-n-1}e = \sum_{i=1}^{n} b^{-i}ea_i$. Taking into account the fact that e is a central idempotent, we obtain $e = b^{n+1}\sum_{i=1}^{n} b^{-i}ea_i = \sum_{i=1}^{n} eb(b^{n-i}a_i) = \sum_{i=1}^{n} ab^{n-i}a_i \in A$. Hence e is a nonzero central idempotent of the indecomposable ring A. Therefore $e = 1$, and the module $Q_A = eQ_A$ is Noetherian. By (1), $A = Q$ is an indecomposable semisimple Artinian ring. Therefore A is a simple Artinian ring. $\triangleright$

3.69 Let A be a ring and let Q_A be the injective hull of A_A.

(1) If each indecomposable factor module of Q is injective, then each indecomposable injective A-module M_A is a homomorphic image of Q.

(2) If each factor module of Q is injective, then each homomorphic image of any indecomposable injective A-module M_A is injective.

$\triangleleft$ (1) Let $\overline{M}$ be a nonzero cyclic submodule of M. There exists an epimorphism $\overline{h} : A_A \to \overline{M}$. Since M is injective, the nonzero epimorphism $\overline{h}$ can be extended to a nonzero homomorphism $h : Q \to M$. The submodule $h(Q)$ of the uniform module M is indecomposable. By assumption, $h(Q)$ is injective. Hence $h(Q)$ is a nonzero direct summand of the indecomposable module M. Therefore $h(Q) = M$.

(2) follows from (1). $\triangleright$

3.70 For a ring A, the following conditions are equivalent.

(1) A is a right hereditary right Noetherian ring.

(2) All factor modules of all injective right A-modules, and all direct sums of injective right A-modules are injective.

(3) Each homomorphic image of every direct sum of injective right A-modules is injective.

(4) A is right Noetherian, and all factor modules of any indecomposable injective right A-module are injective.

(5) A is right Noetherian, and all factor modules of the injective hull of the module A_A are injective.

$\triangleleft$ (1)$\Longleftrightarrow$(2) follows from 3.49 and 3.50. (2)$\Longleftrightarrow$(3) is obvious. (1)$\Longleftrightarrow$(4)$\Longleftrightarrow$(5) follows from [236, 39.16]. $\triangleright$

3.71 Let A be a right Noetherian ring, and let all uniform right A-modules be uniserial.

(1) If N is a proper submodule of a uniform module L_A, then N is a cyclic uniserial Noetherian module.

(2) If M_A is an indecomposable injective module, then either all factor modules of M are injective, or M is a cyclic uniserial Noetherian module.

(3) If the injective hull Q_A of A_A has no nonzero cyclic uniserial Noetherian direct summands, then A is right hereditary.

◁ (1) Since $N \neq L$, there exists $x \in L \setminus N$. The cyclic module xA over a right Noetherian ring A is Noetherian. By assumption, L is uniserial. Therefore N is a submodule of the Noetherian module xA. Hence N is a uniserial and Noetherian module. Therefore N is cyclic.

(2) By assumption, the uniform module M is uniserial. If M is Noetherian, then our assertion is proved. Assume that M is not Noetherian. Let M/F be a nonzero factor module of M, and let E be the injective hull of M/F. Since M/F is uniserial, E is uniform. By assumption, E is uniserial. It is sufficient to prove that $E = M/F$. Assume the contrary. By (1), M/F and F are Noetherian. Therefore M is Noetherian; this is a contradiction.

(3) Let M_A be an indecomposable injective module. By 3.70, it is sufficient to prove that all factor modules of M are injective. By (2), it is sufficient to prove that M is not cyclic. Assume the contrary. There exists an epimorphism $\overline{h} : A_A \to M$. Since M is injective, the epimorphism $\overline{h}$ is extended to an epimorphism $h : Q \to M$. Since Q is the injective hull of the Noetherian module A_A, we have $Q = \oplus_{i=1}^{n} Q_i$, where all Q_i are indecomposable injective modules. Therefore, all Q_i are uniserial. Hence $h(Q_i) \neq 0$ for some i. Since $h(Q)$ is Noetherian, $h(Q_i)$ is Noetherian. Therefore, $Q_i \bigcap \mathrm{Ker}(h) \equiv G$ is a proper submodule of Q_i. By (1), G is Noetherian. Since $h(Q_i) \cong Q_i/G$, Q_i is a Noetherian uniserial module. Therefore Q_i is cyclic; this is a contradiction to the assumption. ▷

3.72 Let B be a projective right module over a right nonsingular ring A.

(1) If B is an essential extension of a finitely generated module C_A, then B is finitely generated.

(2) If B is finite-dimensional, then B is finitely generated.

◁ (1) There exists a free module M_A such that $M = B \oplus E = \oplus_{i \in I} G_i$ with all G_i free cyclic. Note that M is nonsingular, since A is right nonsingular. Since C is finitely generated, there exists a finite set $W = \{1, \ldots, n\} \subseteq I$ such that $C \subseteq \oplus_{i=1}^{n} G_i \equiv H$. Let $N \equiv \oplus_{i \in I \setminus W} G_i$, and let $u : M \to N$ be the natural projection. Since $C \subseteq \mathrm{Ker}(u)$ and B is an essential extension of C, we have $u(B) \subseteq \mathrm{Sing}(M) = 0$, whence $u(B) = 0$. Hence $B \subseteq H$ and $B \oplus E = M$. Therefore $H = B \oplus H \bigcap E$.

Hence B is a direct summand of a finitely generated module H, whence B is finitely generated.

(2) By 1.17, each finite-dimensional module is an essential extension of a finitely generated module. Our assertion follows now from (1). ▷

3.73 Let A be a right nonsingular right finite-dimensional ring. Then each projective right ideal of A is finitely generated. In particular, if A is right hereditary, then A is right Noetherian.

◁ 3.73 follows from 3.72(2). ▷

3.74 Let Q be the maximal right ring of quotients of a right hereditary ring A.

(1) If $Q_A = \oplus_{i \in I} Q_i$, then for each indecomposable nonzero injective module M_A, there exists a subscript $i \in I$ such that M is a homomorphic image of Q_i.

(2) If the module Q_A is semidistributive (serial), then each indecomposable injective right A-module M_A is distributive (uniserial).

(3) Q_A is a finite direct sum of uniserial modules $\Longleftrightarrow$ each injective right A-module is serial.

◁ By [60, 19.35], Q_A is the injective hull of A_A. By 3.49, each factor module of the injective module Q_A is injective.

(1) By 3.69(1), there exists an epimorphism $h : Q_A \to M$. Therefore, there exists a subscript $i \in I$ such that $h(Q_i) \neq 0$. By 3.49, $h(Q_i)$ is a nonzero injective submodule of the indecomposable module M. Therefore, $h(Q_i) = M$.

(2) follows from (1) and from the fact that every homomorphic image of a distributive (uniserial) module is distributive (uniserial).

(3) It is sufficient to prove only $\Longrightarrow$. By (2), it is sufficient to prove that all injective right A-modules are direct sums of indecomposable modules. Since Q_A is a finite direct sum of uniform modules, Q_A is finite-dimensional. By 3.73, the right finite-dimensional hereditary ring A is right Noetherian. By 2.19(3), all injective right A-modules are direct sums of indecomposable modules. ▷

3.75 [192]. *Let A be a semiprime ring, and let Q be the maximal right ring of quotients of A. Then the following conditions are equivalent.*

(1) Each injective right A-module is serial.

(2) A is right hereditary, and Q_A is a finite direct sum of uniserial right A-modules.

(3) A is a serial right Noetherian ring.

◁ $(1)\Longrightarrow(2)$ By 3.43, A is right Noetherian, and each uniform right A-module is uniserial. Without loss of generality, we may assume that A is an indecomposable right Noetherian semiprime ring which is not simple Artinian ring. By 3.68(2), Q_A has no nonzero Noetherian direct summands. By 3.71(3), A is right hereditary. By [60, 19.35], Q_A is an injective finite-dimensional module. Hence Q_A is a finite direct sum of injective uniform modules. Since each uniform right A-module is uniserial, Q_A is a finite direct sum of uniserial modules.

$(2)\Longrightarrow(3)$ By 3.74(3), each injective right A-module is serial. By 3.43, A is right Noetherian and each uniform right A-module is uniserial. The right hereditary right Noetherian semiprime ring A is right semiuniform by 3.67. By 3.61, A is serial.

$(3)\Longrightarrow(1)$ follows from 3.61. ▷

3.76 If N is a nil-ideal of a ring A and the ring A/N is semiperfect, then A is a semiperfect ring.

◁ Since N is a nil-ideal, $N \subseteq J(A)$, $J(A/N) = J(A)/N$, and all idempotents of A/N are lifted to idempotents of A. ▷

3.77 Let each injective right module over a ring A be serial, and let N be the prime radical of A.

(1) A is a semiperfect right Noetherian ring, and A/N is a serial semiprime ring.

(2) If A is the indecomposable normal ring, then A is a local right Noetherian ring, and A/N is an invariant uniserial Noetherian domain.

(3) If A is right uniform, then A is a uniserial Noetherian ring.

◁ (1) By 3.43, A is right Noetherian. By 3.43, A/N is semiprime, and each injective right A-module is serial. By 3.75, A/N is serial. Hence A/N is semiperfect. In addition, N is a nil-ideal. By 3.76, A is semiperfect.

(2) By (1), A/N is a serial Noetherian ring. The indecomposable normal ring A has no nontrivial idempotents. In addition, all idempotents of A/N are lifted to idempotents of A. Therefore, the serial ring A/N has no nontrivial idempotents. Hence A/N is uniserial. By 3.28, the uniserial Noetherian semiprime ring A/N is an invariant domain.

(3) By 3.61, A is a serial right uniform right Noetherian ring. Therefore A is a uniserial right Noetherian ring. By 3.35(3), A is Noetherian. ▷

3.78 For a ring A, the following conditions are equivalent.

(1) A is right invariant, and each injective right A-module is a serial.

(2) A is left invariant, and each injective left A-module is serial.

(3) A is a finite direct product of uniserial Noetherian rings.

◁ It is sufficient to prove $(1)\Longleftrightarrow(3)$.

$(3)\Longrightarrow(1)$ follows from 3.61 and from the fact that A is right invariant by 2.72.

$(1)\Longrightarrow(3)$ By 3.43, A is right Noetherian. Also, A is right invariant. In particular, A is normal. Therefore, we may assume that A is an indecomposable normal ring. By 3.77(2), A is local. By 3.45(2), A is right uniserial. By 3.77(3), A is a uniserial Noetherian ring. ▷

3.79 Let Q be the right ring of quotients of a ring A with respect to a right denominator set T, $f : A \to Q$ be the canonical homomorphism, and let M_Q be a right Q-module.

(1) If the module M_A defined by f is distributive (uniserial), then M_Q is distributive (uniserial).

(2) If M_Q is uniform, then M_A is uniform.

(3) If all uniform right A-modules are distributive (uniserial), then all uniform right Q-modules are distributive (uniserial).

(4) If all injective right A-modules are semidistributive (serial), then all injective right Q-modules are semidistributive (serial).

◁ (1) follows from the fact that each submodule of M_Q is a submodule of M_A.

(2) Let $0 \neq m_1, m_2 \in M$. Since M_Q is uniform, there exist $q_1, q_2 \in Q$ such that $0 \neq m_1 q_1 = m_2 q_2$. By 1.53(2), there exist $a_1, a_2 \in A$ and $t \in T$ such that $q_i = f(a_i)f(t)^{-1}$. Hence $0 \neq m_1 f(a_1)f(t)^{-1} = m_2 f(a_2)f(t)^{-1}$. Therefore $0 \neq m_1 f(a_1) = m_2 f(a_2) \in m_1 f(A) \bigcap m_2 f(A)$. Hence M_A is uniform.

(3) follows from (1) and (2).

(4) By 3.43, A is right Noetherian, and all uniform right A-modules are distributive (uniserial). By 1.55(3), Q is right Noetherian. By (3), all uniform right A-modules are distributive (uniserial). By 3.43, all uniform right Q-modules are distributive (uniserial). ▷

3.80 If A is a right invariant ring and all injective right A-modules are semidistributive, then A is a right distributive right Noetherian ring.

◁ Let M be a maximal right ideal of the right invariant ring A. Then M is a completely prime ideal. By 3.43, A is right Noetherian. By 1.62, the right ring of quotients A_M exists and is a right invariant right Noetherian ring. By 1.53(14), A_M is local. By 3.79(4), all uniform right A_M-modules are distributive. By 3.45(2), A_M is right uniserial. By 3.22, A is right distributive. ▷

3.81 For a ring A, the following conditions are equivalent.

(1) A is a domain, and all injective right A-modules and all injective left A-modules are semidistributive.

(2) A is a left finite-dimensional domain, and all injective right A-modules are semidistributive.

(3) A is an invariant Noetherian hereditary domain.

◁ (3)$\Longrightarrow$(1) follows from 3.51. (1)$\Longrightarrow$(2) follows from the fact that A is a left Noetherian domain by 3.43.

(2)$\Longrightarrow$(3) By 3.43, A is a right Noetherian domain. Hence A is right uniform. By 3.43, A is a right distributive domain. Since A is a right distributive right Noetherian left finite-dimensional domain, it follows from 3.34 that A is an invariant Noetherian hereditary domain. ▷

3.5 Semidistributive rings

3.82 Let A be a semilocal ring, and let A_A be an essential extension of a finite sum of distributive right ideals.

Then A is right finite-dimensional.

In addition, if A is right hereditary, then A is right Noetherian.

◁ 3.82 follows from 2.13(2) and 3.73. ▷

3.83 Let A be a semiperfect ring, and let all 2-generated right ideals of A be projective. Then

A is right semiuniform $\Longleftrightarrow$ A is right serial.

◁ It is sufficient to prove only $\Longrightarrow$. Let M be a uniform direct summand of A_A. It is sufficient to prove that M_A is uniserial. Assume the contrary. There exist incomparable cyclic submodules F and G of M. Hence $F + G$ is a 2-generated nonlocal module. Since $F + G$ is an indecomposable finitely generated projective module over the semiperfect ring A, the module $F + G$ is local by 2.42; this is a contradiction. ▷

3.84 [215]. *Any semiperfect semiprime right semihereditary right finite-dimensional ring is a right serial ring.*

◁ 3.84 follows from 3.83 and 3.67. ▷

3.85 [187], [215]. *For a semiperfect semiprime right hereditary ring A, the following conditions are equivalent.*

(1) *A is right Noetherian.*

(2) *A is right serial.*

(3) *A is right semidistributive.*

◁ $(1)\Longrightarrow(2)$ follows from 3.84. $(2)\Longrightarrow(3)$ is obvious. $(3)\Longrightarrow(1)$ follows from 3.82. ▷

3.86 Let e be a nonzero idempotent of a prime ring A. Then
eAe is a domain $\Longleftrightarrow$ any nonzero endomorphism of the right ideal eA is a monomorphism $\Longleftrightarrow$ any nonzero endomorphism of the left ideal Ae is a monomorphism.

◁ It is sufficient to prove the first equivalence.
$\Longleftarrow$ is directly verified.
$\Longrightarrow$ We may assume that $eAe = \mathrm{End}(eA)$. Therefore, it is sufficient to prove the following assertion: if $0 \neq a = eae \in eAe$ and $0 \neq b = eb \in eA$, then $eaeb \neq 0$.

Assume that $eaeb = 0$. Then $(eae)(ebAe) = 0$. Since eAe is a domain and $eae \neq 0$, we have $ebAe = 0$. Hence $AebAeA = 0$. Since A is prime, either $eb = 0$ or $e = 0$; this is a contradiction. ▷

3.87 Let e be a nonzero idempotent of a ring A.
(1) If N is an ideal of the ring eAe, then $N = e(ANA)e$.
(2) If A/F is a factor ring with the maximum condition on right annihilators, then eAe/eFe is a ring with the maximum condition on right annihilators.
(3) If a module M_A has Krull dimension, then the module $(Me)_{eAe}$ has Krull dimension.
(4) If the module eA_A is distributive (resp. is semidistributive, is Noetherian, has Krull dimension), then the ring eAe is right distributive (resp. is right semidistributive, is right Noetherian, has right Krull dimension).

◁ (1) is directly verified. (2) follows from 1.5(2) and 1.37(4). (3) follows from 1.5(10). (4) follows from (3) and 1.5(11). ▷

3.88 Let A be a ring, and let $1 = \sum_{i=1}^{n} e_i$, where e_i are orthogonal idempotents. The ring A is called a *piecewise domain (with respect to* $\{e_1, \ldots, e_n\}$*)* if the following three equivalent conditions hold.
(i)   For every e_i and e_k, each nonzero homomorphism $e_iA \to e_kA$ is a monomorphism.
(ii)   For every e_i, each nonzero homomorphism $e_iA \to A_A$ is a monomorphism.
(iii) $ab \neq 0$ for any nonzero elements $a \in e_jAe_i$, $b \in e_iAe_k$ and for all e_i, e_j, e_k.
((i)$\Longleftrightarrow$(ii)$\Longleftrightarrow$(iii) is directly verified.)
Since condition (iii) is symmetric, conditions (i) and (ii) can be replaced by their left-side analogs.

A proper ideal B of A is called *piecewise integral with respect to* $\{e_1, \ldots, e_n\}$ if the ring A/B is a piecewise domain with respect to natural images of $\{e_1, \ldots, e_n\}$ in A/B.

A proper ideal B of A is *nonsingularly prime* (resp. *nonsingularly semiprime*) if A/B is a prime right nonsingular (resp. semiprime right nonsingular) ring.

(1) A is a piecewise domain with respect to $\{e_1, \ldots, e_n\} \iff ab \neq 0$ for all nonzero $a \in Ae_i$, $b \in e_i A$ and for each e_i.

(2) If A is a piecewise domain with respect to $\{e_1, \ldots, e_n\}$, then A is nonsingular.

$\triangleleft$ (1) is directly verified.

(2) Let $a \in \mathrm{Sing}(A_A) \equiv S$. Then $ae_i \in S$ for all i. Therefore, for all i, there exists a nonzero $b_i \in r(ae_i) \bigcap e_i A$. Since A is a piecewise domain and $e_i b_i = b_i$, we obtain $ae_i = 0$ for all i. Therefore $a = 0$ and $S = 0$. Analogously, $\mathrm{Sing}(_A A) = 0$. $\triangleright$

3.89 Let $1 = e_1 + \ldots + e_n$ be a decomposition of the identity element of a ring A into a sum of nonzero orthogonal idempotents, P be the prime radical of A, N be a right ideal of A, and let M_A be a right A-module.

(1) Let A be right nonsingular, and let all right ideals $e_i A$ be uniform.

Then A is a piecewise domain with respect to $\{e_1, \ldots, e_n\}$, and A is left nonsingular.

(2) Let F be a prime ideal of A. Then

F is piecewise integral with respect to $\{e_1, \ldots, e_n\} \iff$ the ideal $e_i F e_i$ of $e_i A e_i$ is completely prime for any $e_i \in A \setminus F$.

(3) Let F be a prime ideal of A, and let all right ideals $(e_i A + F)/F$ of A/F be uniform. Then

F is a nonsingularly prime ideal $\iff F$ is piecewise integral with respect to $\{e_1, \ldots, e_n\} \iff$ the ideal $e_i F e_i$ of $e_i A e_i$ is completely prime for any $e_i \in A \setminus F \iff$ the ideal $e_i F e_i$ of $e_i A e_i$ is completely prime for one or more $e_i \in A \setminus F$.

$\triangleleft$ (1) Since any nonzero module homomorphism into a nonsingular module cannot have essential kernel, A is a piecewise domain with respect to $\{e_1, \ldots, e_n\}$. By 3.88(1), A is left nonsingular.

(2) By 1.5(2), we may assume that $F = 0$ and A is prime.

$\Longrightarrow$ is directly verified.

$\Longleftarrow$ Assume that there exist e_i and nonzero $a = ae_i \in Ae_i$, $b = e_i b \in e_i A$ such that $ae_i b = 0$. For each $d \in A$, the rule $f_d(e_i x) = e_i dae_i x$ defines an endomorphism f_d of $e_i A$. Since $ae_i b = 0$, all f_d are not monomorphisms. By 3.86, all f_d are zero endomorphisms. Therefore $e_i Aae_i A = 0$. Since A is prime, either $e_i A = 0$ or $aA = ae_i A = 0$; this is a contradiction. By 3.88(1), A is a piecewise domain.

(3) By 1.5(2), we may assume that $F = 0$, and A is prime. The first equivalence follows from (1) and 3.88(2). The second equivalence follows from (2). Let us prove the third equivalence.

$\Longrightarrow$ is directly verified.

$\Longleftarrow$ By (2) and 3.86, it is sufficient to prove that for any j, each endomorphism f of $e_j A$ which has the nonzero kernel is the zero endomorphism. Assume that $f(e_j A) \equiv N \neq 0$. Since A is prime, $Ne_i A \neq 0$. Therefore, there exists a nonzero homomorphism $h : e_i A \to N$. Since A is prime, $e_i Ah(e_i A) \neq 0$. Therefore, there exists a nonzero homomorphism $g : h(e_i A) \to e_i A$. Hence gh is a nonzero endomorphism of $e_i A$. Since the module $e_i A$ is projective and $h(e_i A) \subseteq N \subseteq f(e_j A)$, there exists a homomorphism $t : e_i A \to e_j A$ such that $h = ft$. Since $h \neq 0$, we have $t \neq 0$. Since $e_j A$ is uniform, $M \equiv \mathrm{Ker}(f) \bigcap t(e_i A) \neq 0$. Therefore, $t^{-1}(M) \neq 0$ and $(gh)(t^{-1}(M)) = (gft)(t^{-1}(M)) = 0$. By 3.86, $gh \equiv 0$; this is a contradiction. $\triangleright$

3.90 [213]. *Let the identity element of a ring A be a sum of nonzero orthogonal idempotents $e_1, \ldots, e_n$, and let all right A-modules $e_i A$ be distributive.*

(1) The prime radical P of A contains all right nil-ideals and all left nil-ideals of A.

(2) If A is right nonsingular, then $P^n = 0$, and all the rings $e_i Ae_i$ are reduced.

(3) If A is a prime right nonsingular ring, then all the rings $e_i Ae_i$ are domains.

(4) If A is prime, then all modules $e_i A_A$ are uniform, and therefore A is right semiuniform, and the right Goldie dimension of A is equal to n.

(5) If F is a nonsingularly prime ideal of A, then for any $e_i \in A \backslash F$, the factor ring $e_i Ae_i / e_i Fe_i$ is a domain.

(6) If A is prime and all the rings $e_i Ae_i$ are division rings, then all $e_i A$ are minimal right ideals of A, and therefore A is isomorphic to the ring of $n \times n$-matrices over a division ring.

(7) If the ring A is simple, then A is isomorphic to the ring of $n \times n$-matrices over a division ring.

(8) If A is a regular prime ring, then A is isomorphic to the ring of $n \times n$-matrices over a division ring.

(9) If B is an ideal of A and $h : A \to A/B$ is the natural epimorphism, then the factor ring A/B is a direct sum of right distributive ideals $h(e_i A)$.

◁ (1), (2), and (3) By 2.63, all modules $e_i A_A$ are nilpotently invariant. Therefore, we can use 3.40.

(4) follows from 3.12(2). (5) follows from (3), 1.5(2) and 2.63.

(6) Let $N = e_i N$ be a nonzero submodule of $e_i A$, and let $A_{ij} \equiv e_i A e_j$. It is sufficient to prove that $N e_j = A_{ij}$ for any j. Since A is prime, $N e_j \neq 0$. By 2.32, nonzero distributive vector spaces over division rings are one-dimensional. Therefore, it follows from 1.5(11) that A_{ij} is a one-dimensional vector space over the division ring $e_j A e_j$. Hence $0 \neq N e_j = A_{ij}$.

(7) Fix one of the rings $e_i A e_i \equiv R$. By (6), it is sufficient to prove that R is a division ring. By 3.87(4) and 3.87(10), R is a simple right distributive ring. By 2.53(4), each maximal right ideal M of R is an ideal. Therefore R is a division ring.

(8) follows from (4) and from the fact that right finite-dimensional regular rings are Artinian. (9) follows from the fact that inclusions $e_i B \subseteq B$ imply the equality $B = \oplus_{i=1}^n (B \bigcap e_i A)$. ▷

3.91 If F and G are incomparable completely prime submodules of a distributive module M_A, then $F + G = M$.

◁ (1) is directly verified.

(2) Let $h_F : M \to M/F$, $h_G : M \to M/G$ be natural epimorphisms, $f \in F \setminus G$, and let $g \in G \setminus F$. By 2.4, there exist $a, b, c, d \in A$ such that $1 = a + b$, $fa = gc \in G$, and $gb = fd \in F$. Hence $a \in r(h_G(f)) = r(h_G(M))$, whence $Ma \subseteq G$. Analogously, $Mb \subseteq F$. Hence $M = Ma + Mb \subseteq G + F \subseteq M$. ▷

3.92 [213]. *Let A be a right semidistributive ring which is a direct sum of n right distributive ideals $e_1 A, \ldots, e_n A$, where e_i are orthogonal idempotents in A.*

(1) If M is a maximal ideal of A, then A/M is isomorphic to the ring of $m \times m$-matrices over a division ring, where $m \leq n$.

(2) If B is the intersection of some set of maximal ideals of A, then the nilpotence index of any nilpotent element of the factor ring A/B does not exceed n.

(3) Let F be a prime ideal of A. Then

F is a nonsingularly prime ideal $\Longleftrightarrow$ F is piecewise integral with respect to $\{e_1, \ldots, e_n\}$ $\Longleftrightarrow$ for any $e_i \in A \setminus F$, the ideal $e_i F e_i$ of $e_i A e_i$ is completely prime $\Longleftrightarrow$ the ideal $e_i F e_i$ of $e_i A e_i$ is completely prime for one or more $e_i \in A \setminus F$.

(4) Let all left ideals Ae_i be distributive, and let F be a prime ideal of A. Then

F is nonsingularly prime $\Longleftrightarrow$ A/F is left nonsingular.

(5) *If F and G are two nonsingularly prime ideals of A, then either $F + G = A$ or the ideals F and G are comparable.*

(6) *Every proper nonsingularly prime ideal F of A contains precisely one minimal nonsingularly prime ideal.*

◁ (1) follows from 3.90(7),(9).

(2) Since the nilpotence index of nilpotent elements of the ring of $m \times m$-matrices over a division ring does not exceed m, our assertion follows from (1) and from the fact that all subdirect products of rings whose index of nilpotent elements does not exceed m have the same property.

(3) follows from 3.89(3) and 3.12(2). (4) follows from (3).

(5) Assume that $F + G \neq A$. There exists $e_i \equiv e \in A \setminus (F + G)$. By 3.87(4), the ring eAe is right distributive. By (3), eFe and eGe are completely prime ideals of eAe. Since $e \in A \setminus (F + G)$, we have $eAe \neq eFe + eGe$. Since $eFe + eGe \neq eAe$, we may assume that the completely prime ideals eFe and eGe of the right distributive ring eAe are comparable by 3.91. Therefore, we may assume that $eFe \subseteq eGe$. In addition, the ideal G is prime, $e \in A \setminus G$, and $eFe \subseteq G$. Therefore $F \subseteq G$.

(6) By (5) and Zorn's Lemma, it is sufficient to prove that if $G = \bigcap_{j \in J} G_j$ is the intersection of a descending chain of nonsingularly prime ideals G_j which are contained in F, then the ideal G is nonsingularly prime. Since all ideals G_j are prime, G is a prime ideal. Let $e_i \equiv e \in A \setminus F$. By (3), all eG_je are completely prime ideals of eAe. Hence $eGe = \bigcap_{j \in J} eG_je$ is a completely prime ideal. By (3), the ideal G is nonsingularly prime. ▷

3.93 [213]. *Let A be a regular right semidistributive ring which is a direct sum of n right distributive ideals.*

(1) *If M is a prime ideal of A, then the factor ring A/M is isomorphic to the ring of $m \times m$-matrices over a division ring, where $m \leq n$.*

In particular, the ideal M is maximal.

(2) *The nilpotence index of any nilpotent element of every factor ring of A does not exceed n.*

◁ (1) follows from 3.90(8),(9). (2) follows from 3.92(2) and from the fact that any ideal of a regular ring is the intersection of prime ideals. ▷

3.94 [213]. *Let the identity element of a ring A be the sum of orthogonal idempotents $e_1, \ldots, e_n$, and let any ring e_iAe_i be right distributive.*

(1) *Let F and G be two incomparable prime ideals of A, and let rings $e_i A e_i / e_i F e_i$, $e_i A e_i / e_i G e_i$ be either domains or zero rings for all i.*

Then $F + G = A$.

(2) *If F and G are two incomparable nonsingularly prime ideals of A, then $F + G = A$.*

(3) *The prime radical P of A contains all right nil-ideals and all left nil-ideals of A.*

◁ (1) Assume the contrary. Since $1 \in A \setminus (F + G)$, there exists an idempotent $e_i \in A \setminus (F + G)$. It follows from the assumption and from 1.5(17) that $e_i F e_i$ and $e_i G e_i$ are incomparable completely prime ideals of the right distributive ring $e_i A e_i$ for all i. Also, $e_i F e_i + e_i G e_i \neq e_i A e_i$; this is a contradiction to 3.91.

(2) Fix an idempotent e_i. By (1), it is sufficient to prove that either $e_i A e_i = e_i F e_i$ or $e_i A e_i / e_i F e_i$ is a (nonzero) domain. Assume that $e_i A e_i / e_i F e_i \neq 0$. Then $e_i \in A \setminus F$. By 3.90(5), $e_i A e_i / e_i F e_i$ is a domain.

(3) Fix an idempotent e_i. Let P_i be the prime radical of $e_i A e_i \equiv R_i$. By 2.64(10), all right or left nil-ideals of the right distributive ring R_i are contained in P_i. By 2.41(4), P contains all right or left nil-ideals of A. ▷

3.95 [213]. *Let the identity element of a ring A be a sum of nonzero orthogonal idempotents $e_1, \ldots, e_n$ such that the sum of any two incomparable prime ideals of any ring $e_i A e_i$ is equal to $e_i A e_i$. (This is the case if the ideal lattice of each ring $e_i A e_i$ is a chain.)*

(1) $A = P + Q$ *for any two incomparable prime ideals P and Q of the ring A.*

(2) *If A is semiprime and A contains no infinite direct sums of nonzero ideals, then A is a finite direct product of prime rings.*

(3) *Let A be a semiprime ring which contains no infinite direct sums of nonzero ideals. Then*

A is a piecewise domain with respect to $\{e_1, \ldots, e_n\} \iff e_i A e_i$ is a domain for any e_i.

(4) *Let F be a semiprime ideal of A such that the factor ring A/F contains no infinite direct sums of nonzero ideals. Then*

F is piecewise integral with respect to $\{e_1, \ldots, e_n\} \iff e_i F e_i$ is a completely prime ideal of $e_i A e_i$ for any $e_i \in A \setminus F$.

(5) *Assume that F is a semiprime ideal of A, the factor ring A/F contains no infinite direct sums of nonzero ideals, and all right ideals $(e_i A + F)/F$ of A/F are uniform. Then*

F is a nonsingularly semiprime ideal $\iff$ F is piecewise integral with respect to $\{e_1, \ldots, e_n\}$ $\iff$ the ideal $e_i F e_i$ of $e_i A e_i$ is completely prime for any $e_i \in A \setminus F$.

◁ (1) Assume the contrary. Since $1 \in A \setminus (P + Q)$, there exists an idempotent $e_i \in A \setminus (P+Q)$. By assumption and by 1.5(17), the ideals $e_i P e_i$, $e_i Q e_i$ $e_i A e_i$ are incomparable; this is a contradiction.

(2) follows from (1) and 1.24. (3) follows from (2) and 3.89(2). (4) follows from (3) and 3.87(1). (5) follows from (4) and 3.89(3). ▷

3.96 If A is a semiprime right semidistributive ring which contains no infinite direct sums of nonzero ideals, then A is right finite-dimensional.

◁ By 1.22, A is a finite subdirect product of prime rings $A_1, \ldots, A_n$. By 3.90(4), all A_i are right finite-dimensional. Since any finite subdirect product of right finite-dimensional rings is right finite-dimensional, A is right finite-dimensional. ▷

3.97 [213]. *For a semiprime right semidistributive ring A, the following conditions are equivalent.*

(1) A is a ring with the maximum condition on right annihilators.

(2) A is a ring with the maximum condition on left annihilators.

(3) A is a right nonsingular ring which contains no infinite direct sums of nonzero ideals.

(4) A is a right Goldie ring.

(5) A is a finite direct product of prime right Goldie rings $A_1, \ldots, A_n$ which are direct sums of uniform distributive right ideals.

◁ (5)$\Longrightarrow$(4), (4)$\Longrightarrow$(1), and (4)$\Longrightarrow$(2) are directly verified.

(1)$\Longrightarrow$(3) and (2)$\Longrightarrow$(3) By 1.22, A contains no infinite direct sums of nonzero ideals. By 1.37(3), A is right nonsingular.

(3)$\Longrightarrow$(4) follows from 3.96(1) and 3.7.

(4)$\Longrightarrow$(5) By 3.7, A is right nonsingular. By 1.22, there exist distinct minimal prime ideals $P_1, \ldots, P_n$ of A such that $P_1 \cap \ldots \cap P_n = 0$, and any intersection of less than n ideals P_i is not equal to zero. Hence $B_i \equiv \bigcap_{j \neq i} P_j \neq 0$ and $B_i \cap P_i = 0$ for $i = 1, \ldots, n$. By 1.20(9), all P_i are nonsingularly prime ideals. By 3.92(5), $P_i + P_j = A$ for all $i \neq j$. Let $A_i \equiv A/P_i$. By 1.23(2), A is isomorphic to the direct product of prime right Goldie rings $A_1, \ldots, A_n$. By 3.90(4), each right semidistributive prime ring A_i is a direct sum of uniform distributive right ideals. ▷

3.98 [213]. *If A is a semiprime indecomposable right semidistributive ring with the maximum condition on right annihilators, then A is a prime right Goldie ring, A is a finite direct sum of distributive uniform right ideals $e_1 A, \ldots, e_m A$ such that $e_i{}^2 = e_i$, and all the rings $e_i A_i e_i$ are right distributive domains.*

◁ By 3.97, A is a prime right nonsingular right Goldie ring. By 3.90(3), $1 = \sum_{i=1}^n e_i$, where e_i are nonzero orthogonal idempotents of A, all $e_i A$ are right distributive ideals, and all $e_i A_i e_i$ are domains. By 3.12(2), all right ideals $e_i A$ are uniform. By 1.5(11), all the rings $e_i A_i e_i$ are right distributive. ▷

3.99 [213]. *If A is a semilocal semiprime right semidistributive right nonsingular ring, then A is a finite direct product of prime right semiuniform right Goldie rings.*

◁ 3.99 follows from 3.97 and 3.82. ▷

3.100 [213]. *Assume that A is a semiprime right hereditary right semidistributive ring, and either A contains no infinite direct sums of nonzero ideals or A is semilocal.*
Then A is a finite direct product of prime right Noetherian rings.

◁ The right Rickartian ring A is right nonsingular. By 3.97 and 3.99, A is a direct product of prime right finite-dimensional rings. The right hereditary right finite-dimensional ring A is right Noetherian by 3.73. ▷

3.101 [213]. *Let A be a semiperfect right semidistributive ring, and let $1 = \sum_{i=1}^n e_i$ be a decomposition of the identity element of A into a sum of local orthogonal idempotents.*
(1) The sum of any two incomparable prime ideals of A is equal to A.
(2) If A is semiprime, then A is a finite direct product of prime rings $A_1, \ldots, A_n$, and each A_i is a direct sum of distributive uniform right ideals.
(3) Let F be a semiprime ideal of A. Then
F is a nonsingularly semiprime ideal $\iff$ F is piecewise integral with respect to $\{e_1, \ldots, e_n\}$ $\iff$ the ideal $e_i F e_i$ of $e_i A e_i$ is completely prime for any $e_i \in A \setminus F$ $\iff$ A/F is a right Goldie ring.

◁ By 3.82, A is right finite-dimensional. Since A is a semiperfect right semidistributive ring, 3.56(6) shows us that the identity element of A is the sum of nonzero orthogonal idempotents $e_1, \ldots, e_n$ such that all the rings $e_i A e_i$ are right uniserial.

(1) follows from 3.95(1).

(2) By 1.22, there exist distinct minimal prime ideals $P_1, \ldots, P_n$ of A such that $P_1 \cap \ldots \cap P_n = 0$. By (1), $P_i + P_j = A$ for $i \neq j$. By 1.23(2), A is a finite direct product of prime rings $A/P_i \equiv A_i$. By 3.90(4), each prime right semidistributive ring A_i is a finite direct sum of uniform distributive right ideals.

(3) The ring A satisfies the conditions of 3.95. By 3.82, each factor ring of A is right finite-dimensional. Therefore, (3) follows from 3.95 and 3.97. $\triangleright$

3.102 [213]. *Let A be a semiperfect right semidistributive ring, $1 = \sum_{i=1}^{n} e_i$ be a decomposition of the identity element of A into a sum of local orthogonal idempotents, and let N be the intersection of all nonsingularly semiprime ideals of A.*

(1) N is the least nonsingularly semiprime ideal of A.

(2) N is the least semiprime ideal of A which is piecewise integral with respect to $\{e_1, \ldots, e_n\}$.

(3) N is the least ideal of A with the property that A/N is a semiprime right Goldie ring.

$\triangleleft$ Let $\mathcal{E}$ be the set of all nonsingularly semiprime ideals of A. By 3.101(3), $\mathcal{E}$ coincides with the set of all semiprime ideals which are piecewise integral with respect to $\{e_1, \ldots, e_n\}$. In addition, 3.101(3) shows us that $\mathcal{E}$ coincides with the set of all ideals X of A such that the factor ring A/X is a semiprime right Goldie ring. Therefore, it is sufficient to prove that N is a nonsingularly semiprime ideal. Since N is the intersection of some set of semiprime ideals of A, N is a semiprime ideal. Let $e_i \in A \setminus N$. By 3.101(3), it is sufficient to prove that the ideal $e_i N e_i$ of $e_i A e_i$ is completely prime. Let $\mathcal{E} = \{F_j\}_{j \in J}$. Then $e_i N e_i = e_i (\bigcap_{j \in J} F_j) e_i = \bigcap_{j \in J} (e_i F_j e_i)$. By 3.101(3), all ideals $e_i F_j e_i$ of $e_i A e_i$ are completely prime. By 3.56(5), $e_i A e_i$ is right uniserial. Therefore $e_i N e_i$ is the intersection of a chain of completely prime ideals $e_i F_j e_i$ of $e_i A e_i$. Hence the ideal $e_i N e_i$ of $e_i A e_i$ is completely prime. $\triangleright$

3.103 Let the identity element of a ring A be the sum of nonzero orthogonal idempotents $e_1, \ldots, e_n$ such that all right ideals $e_i A$ are uniform.

(1) $\mathrm{Sing}(A_A)$ coincides with the set of all elements $a \in A$ such that for any e_i, there exists a nonzero element $b \in e_i A$ such that $ab = 0$.

(2) Let $i \in \{1, \ldots, n\}$, $a \in A e_i$. Then
$a \in \mathrm{Sing}(A_A) \iff$ there exists a nonzero element $b \in e_i A$ such that $ab = 0$.

(3) A is right nonsingular $\iff$ for any idempotent e_i, we have $ab \neq 0$ for each nonzero $a = a e_i \in A e_i$ and for any nonzero $b = e_i b \in e_i A$.

(4) A is right nonsingular $\Longleftrightarrow$ for all idempotents e_i, e_j, and e_k, we have $ab \neq 0$ for each nonzero $a = e_j a e_i \in e_j A e_i$ and for any nonzero $b = e_i b e_k \in e_i A e_k$.

(5) Assume that A is a ring with the maximum condition on right annihilators, $a \in A e_i$, and there exists a nonzero element $b \in e_i A$ such that $ab = 0$.

Then the ideal $\mathrm{Id}(a)$ of A generated by a is nilpotent.

◁ (1) follows from 2.41(7), from the equalities $a = a e_i$, $b = e_i b$, and from the fact that all nonzero submodules of a uniform module are essential. (2) follows from (1). (3) follows from (1) and 2.41(8). (4) follows from (3) and from the fact that if $a e_i \neq 0, e_i b \neq 0$, then $e_j a e_i \neq 0, e_i b e_k \neq 0$ for some e_j, e_k.

(5) By (2), $\mathrm{Id}(a) \subseteq \mathrm{Sing}(A_A) \equiv Z$. By 1.37(1), the ideal Z is nilpotent. ▷

3.104 Let the identity element of a ring A be the sum of nonzero orthogonal idempotents $e_1, \ldots, e_n$ such that all right ideals $e_i A$ and all left ideals $A e_i$ are uniform. Then the following conditions are equivalent.

(1) A is right nonsingular.

(2) A is left nonsingular.

(3) For all idempotents e_i, e_j, e_k, we have $ab \neq 0$ for each nonzero $a = e_j a e_i \in e_j A e_i$ and for any nonzero $b = e_i b e_k \in e_i A e_k$.

◁ 3.104 follows from 3.103(3),(4), and from the symmetry of condition (3). ▷

3.105 [111] If a ring A is (right and left) Noetherian and its right Krull dimension does not exceed 1, then $\bigcap_{i=1}^{\infty}(J(A))^i = 0$.

In 3.106, we use the fact that for any nonzero idempotent e of a right (left) Noetherian ring A, the ring eAe is right (left) Noetherian (see 3.87(4)).

3.106 Let the identity element of a ring A be a sum of nonzero orthogonal idempotents $e_1, \ldots, e_n$.

(1) If M is a right A-module such that each right $e_i A e_i$-module $M e_i$ has Krull dimension α_i for all $i = 1, \ldots, n$, then M_A has the Krull dimension which is equal to $\max\{\alpha_i\}$.

(2) If the right $e_i A e_i$-module $e_j A e_i$ has Krull dimension α_{ij} for all i and j, then the module A_A has the Krull dimension which is equal to $\max\{\alpha_{ij}\}$.

(3) If the ring A is right Noetherian and the right Krull dimension of the right Noetherian ring $e_i A e_i$ is equal to γ_i for $i = 1, \ldots, n$, then $\mathrm{Kdim}(A_A) = \max\{\gamma_i\}$.

(4) If the ring A is Noetherian and the right Krull dimension of the Noetherian ring $e_i A e_i$ does not exceed 1 for $i = 1, \ldots, n$, then $\mathrm{Kdim}(A_A) \le 1$ and $\bigcap_{i=1}^{\infty}(J(A))^i = 0$.

(5) If the ring A is Noetherian and the Noetherian ring $e_i A e_i$ is right distributive for $i = 1, \ldots, n$, then $\mathrm{Kdim}(A_A) \le 1$ and $\bigcap_{i=1}^{\infty}(J(A))^i = 0$.

◁ (1) follows from the definition of Krull dimension.

(2) Let $\beta_j \equiv \max \alpha_{ij}$ $(1 \le i \le n)$. By (1), the module $e_j A_A$ has Krull dimension which is equal to β_j. Since $A_A = \oplus_{j=1}^{n} e_j A$, the Krull dimension of A_A exists and is equal to $\max\{\beta_j\} = \max\{\alpha_{ij}\}$.

(3) Fix $i, j \in \{1, \ldots, n\}$. Since the module $e_j A_A$ is Noetherian, the right $e_i A e_i$-module $e_j A e_i$ is Noetherian and has the Krull dimension α_{ij}. Since $e_j A e_i$ is a finitely generated right $e_i A e_i$-module and $e_i A e_i$ has the right Krull dimension γ_i, $\alpha_{ij} \le \gamma_i \le \mathrm{Kdim}(A_A)$. By (2), $\mathrm{Kdim}(A_A) = \max\{\alpha_{ij}\}$. Hence $\mathrm{Kdim}(A_A) = \max\{\gamma_i\}$.

(4) Fix $i \in \{1, \ldots, n\}$. Since the ring $e_i A e_i$ is Noetherian and its right Krull dimension does not exceed 1, (3) shows us that $\mathrm{Kdim}(A_A) \le 1$. By 3.105, $\bigcap_{i=1}^{\infty}(J(A))^i = 0$.

(5) It follows from 3.35(1) and 3.31 that the right Krull dimension of the Noetherian ring $e_i A e_i$ does not exceed 1 for $i = 1, \ldots, n$. By (4), $\bigcap_{i=1}^{\infty}(J(A))^i = 0$. ▷

3.107 *If A is a Noetherian right semidistributive ring, then* $\mathrm{Kdim}(A_A) \le 1$ *and* $\bigcap_{i=1}^{\infty}(J(A))^i = 0$.

◁ By 3.87(4), the identity element of A is a sum of nonzero orthogonal idempotents $e_1, \ldots, e_n$ such that all the rings $e_i A e_i$ are right distributive. We can apply now 3.106. ▷

3.6 Distributively generated modules

3.108 A module M is *distributively generated* if M is a sum of distributive submodules.

A module is *arithmetical* if the lattice of its fully invariant submodules is distributive.

(1) Any semidistributive module M is arithmetical.

(2) The lattice of two-sided ideals of a ring A is distributive $\iff$ the module A_A is arithmetical $\iff$ the module ${}_A A$ is arithmetical.

(3) Finite direct products of arithmetical rings are arithmetical.

(4) If all factor rings of a ring A are semiprime, then A is arithmetical and $B = B^2$ for any ideal B of A.

(5) If A is a regular ring, then A is arithmetical and $B = B^2$ for any right or left ideal B of A.

(6) If any two nonzero ideals of a ring A have the nonzero intersection and each proper factor ring of A is arithmetical, then A is arithmetical.

(7) If A is a prime ring, and all proper factor rings of A are right semidistributive, then A is arithmetical.

(8) Each hereditary Noetherian semiprime ring is arithmetical.

$\lhd$ (1) follows from the fact that if N is a fully invariant submodule of M and $M = \oplus_{i=1}^{n} M_i$, then $N = \oplus_{i=1}^{n}(N \bigcap M_i)$. (2) and (3) are directly verified.

(4) Since the ring A/B^2 is semiprime, $B = B^2$. Let F, G, and H be three ideals of A, and let $M \equiv F \bigcap (G + H)$. Then $M = M^2 \subseteq F(G + H) = FG + FH \subseteq F \bigcap G + F \bigcap H \subseteq M$.

(5) If B is a right or left ideal, then $b \in bAb$ for any $b \in B$, whence $B^2 = B$. Therefore, all factor rings of A are semiprime. Hence A is arithmetical.

(6) Let F, G, and H be three ideals of A. We have to prove the equality (*): $F \bigcap (G + H) = F \bigcap G + F \bigcap H$. Without loss of generality, we may assume that F, G, and H are nonzero ideals. Let $B \equiv F \bigcap G \bigcap H$, and let $h : A \to A/B$ be the natural epimorphism. By assumption, $D \neq 0$ and $h(A)$ is arithmetical. Therefore, $h(F) \bigcap (h(G) + h(H)) = h(F) \bigcap h(G) + h(F) \bigcap h(H)$. This yields (*).

(7) follows from (1), (6), and from the fact that any two nonzero ideals of a prime ring have nonzero intersection.

(8) Since any hereditary Noetherian semiprime ring is a finite direct product of hereditary Noetherian prime rings [60, 20.30], we have that (8) follows from (7) and from the fact that each proper factor ring of a hereditary Noetherian prime ring is serial [60, 25.5.1]. $\rhd$

3.109 (1) If N is an invariant module which is a direct summand of an arithmetical module M, then N is a distributive module.

(2) If M is a semi-Noetherian or semi-Artinian module, then

M is semidistributive $\Longleftrightarrow$ M is a direct sum of invariant modules and M is arithmetical.

(3) If A is a semiperfect right semi-Noetherian ring, then

A is right semidistributive $\Longleftrightarrow$ A is arithmetical and eA_A is an invariant module for each local idempotent e of A.

$\lhd$ (1) Let $e = e^2 : M \to N$ be a projection. For each fully invariant submodule Y of M, we have $Y = e(Y) \oplus (1 - e)(Y)$. Hence

$e(Y + Z) = e(Y) + e(Z)$ and $e(Y \bigcap Z) = e(Y) \bigcap e(Z)$ for any two fully invariant submodules Y and Z of M. Set $E \equiv \mathrm{End}(M)$. Then $\mathrm{End}(N) = eEe$. Also, $X = e(Ee(X))$ for each $X \in \mathrm{Lat}(N)$, where $(Ee(X))$ is a fully invariant submodule of M. Let $F, G, H \in \mathrm{Lat}(N)$. Since M is arithmetical, $Ee(F) \bigcap (Ee(G) + Ee(H)) = Ee(F) \bigcap Ee(G) + Ee(F) \bigcap Ee(H)$. Hence $F \bigcap (G+H) = eEe(F) \bigcap (eEe(G) + eEe(H)) = e(Ee(F) \bigcap (Ee(G) + Ee(H))) = e(Ee(F) \bigcap Ee(G) + Ee(F) \bigcap Ee(H)) = e(Ee(F) \bigcap Ee(G) + Ee(F) \bigcap Ee(H)) = (F) \bigcap G + F \bigcap H$.

(2) follows from (1) and 2.71(2). (3) follows from (2). $\triangleright$

3.110 (1) Assume that F is a fully invariant submodule of a module M over a ring A, G_1 and G_2 are two submodules of M such that $G_1 \bigcap G_2 = 0$, and each idempotent endomorphism of every 2-generated submodule of M is extended to an endomorphism of M.

Then $F \bigcap (G_1 + G_2) = F \bigcap G_1 + F \bigcap G_2$, and for any $g_1 \in G_1$ and $g_2 \in G_2$, there exist endomorphisms d_1, d_2 of M such that $d_1(g_1) = d_2(g_2) = 0$, $d_1(g_2) = g_2$, and $d_2(g_1) = g_1$.

(2) If each idempotent endomorphism of any 2-generated right ideal of an arbitrary factor ring B of a ring A can be extended to an endomorphism of B_B, then A is an arithmetical ring.

$\triangleleft$ (1) Let $H \equiv g_1 A \oplus g_2 A$, and let $h_i : H \to g_i A$ be the natural projections. Idempotent endomorphisms h_i can be extended to endomorphisms $d_i \in \mathrm{End}(M)$. Then $d_1(g_1) = d_2(g_2) = 0, d_1(g_2) = g_2$, and $d_2(g_1) = g_1$. If $f = g_1 + g_2 \in F \bigcap (G_1 + G_2)$, then $g_1 = d_2(g_1 + g_2) = d_2(f)$ and $g_2 = d_1(f)$. Hence $d_i(f) \in F$, since F is fully invariant in M. Therefore $f = d_2(f) + d_1(f) \in F \bigcap G_1 + F \bigcap G_2$ and $F \bigcap (G_1 + G_2) \subseteq F \bigcap G_1 + F \bigcap G_2 \subseteq F \bigcap (G_1 + G_2)$.

(2) Let F, G, and H be three ideals of A such that $F \subseteq G + H$, $h : A \to A/(G \bigcap H)$ be the natural epimorphism, and let $M \equiv F \bigcap G + F \bigcap H$. By (1) applied to $h(A)$, $h(F) = h(F) \bigcap h(G) + h(F) \bigcap h(H)$. Using the modular law, we obtain the equalities $F + G \bigcap H = M + G \bigcap H$ and $F = F \bigcap (M + G \bigcap H = M + F \bigcap G \bigcap H = M$. $\triangleright$

3.111 (1) Each homomorphic image of a distributively generated module is distributively generated.

(2) If a module M is a sum of distributively generated submodules, then M is a distributively generated module.

(3) Every distributively generated module is a homomorphic image of a semidistributive module.

(4) For each finitely generated distributively generated module M, there exist distributive submodules $M_1, \ldots, M_n$ of M such that $M = \sum_{i=1}^{n} M_i$, and M is a homomorphic image of the module $\oplus_{i=1}^{n} M_i$.

(5) Every right module over a right distributively generated ring is a distributively generated module.

(6) Every finitely generated right module over a right distributively generated ring A is a homomorphic image of a finite direct sum of distributive right ideals of the ring A.

◁ 3.111 can directly be verified. ▷

3.112 Let M be a direct summand of a module $\oplus_{i=1}^{n} M_i$.

(1) If all modules $M_1, \ldots, M_n$ are nilpotently invariant, then the prime radical of the ring $\text{End}(M)$ contains all right nil-ideals and all left nil-ideals of the ring $\text{End}(M)$.

(2) If all modules $M_1, \ldots, M_n$ are distributive, then the prime radical of the ring $\text{End}(M)$ contains all right nil-ideals and all left nil-ideals of the ring $\text{End}(M)$.

◁ (1) Let $X \equiv \oplus_{i=1}^{n} M_i$, $R \equiv \text{End}(X)$, and let P be the prime radical of the ring R. By 3.39, P contains all right or left nil-ideals of R. There exists an idempotent $e \in R$ such that $M = e(X)$ and $\text{End}(M) = eRe$. By 3.37(3), the prime radical of the ring eRe contains all right or left nil-ideals of eRe.

(2) follows from (1) and from the fact that each distributive module is nilpotently invariant by 2.63. ▷

3.113 Let M be a projective module which is a finite sum of its submodules $M_1, \ldots, M_n$, and let P be the prime radical of the ring $\text{End}(N)$.

(1) M is isomorphic to a direct summand of the module $\oplus_{i=1}^{n} M_i$.

(2) If all modules $M_1, \ldots, M_n$ are nilpotently invariant, then P contains all right nil-ideals and all left nil-ideals of the ring $\text{End}(N)$.

(3) If all modules $M_1, \ldots, M_n$ are distributive, then P contains all right nil-ideals and all left nil-ideals of the ring $\text{End}(N)$.

◁ (1) follows from existence of an epimorphism from the module $\oplus_{i=1}^{n} M_i$ onto the projective module N. (2) and (3) follow from (1) and 3.112. ▷

3.114 *Let M be a finitely generated projective right module over a ring A, and let P be the prime radical of the ring $\text{End}(M)$.*

(1) If M is a distributively generated module, then P contains all right nil-ideals and all left nil-ideals of $\text{End}(M)$.

(2) If A is a right distributively generated ring, then P contains all right nil-ideals and all left nil-ideals of $\text{End}(M)$.

◁ (1) follows from 3.111(4) and 3.113(3). (2) follows from (1) and 3.111(5). ▷

Chapter 4

Flat modules and invariant rings

4.1 Characterizations of flat modules

4.1 Let A be a ring, E_A and $_AM$ be right and left A-modules, respectively, and let $E \times M$ be the cartesian product of these modules. The *tensor product* $E \otimes_A M$ is the Abelian group F/H, where F is the free Abelian group with basis indexed by $E \times M$, and H is the subgroup of F generated by all elements of the form $(x + u, y) - (x, y) - (u, y)$, $(x, y+v) - (x, y) - (x, v)$, and $(xa, y) - (x, ay)$, where $x, u \in E$, $y, v \in M$, and $a \in A$. (We write $E \otimes M$ instead of $E \otimes_A M$ if there is no doubt about A.) The image of (x, y) under the natural map $E \times M \to E \otimes M$ is denoted by $x \otimes y$. If $_C E_A$ and $_A M_B$ are bimodules, then the group $E \otimes_A M$ can be naturally regarded as a (C, B)-bimodule ($c(\sum e_i m_i) = \sum c e_i m_i$, and $(\sum e_i m_i)b = \sum e_i m_i b$ for all $c \in C, b \in B$). In particular, $E \otimes_A M$ is an $(\mathrm{End}(E), \mathrm{End}(M))$-bimodule, and $E \otimes_A A$ is a right A-module. A map t from $E \times M$ into an Abelian group G is called *balanced* if $t(x + u, y) = t(x, y) + t(u, y)$, $t(x, y + v) = t(x, y) + t(x, v)$, and $t(xa, y) = t(x, ay)$ for all $x, u \in E$, $y, v \in M$, and $a \in A$. A module E_A is *flat* if for each monomorphism of left A-modules $u : M_1 \to M_2$, the group homomorphism $E \otimes M_1 \to E \otimes M_2$ is a monomorphism.

(1) For each $x \in E \otimes M$, there exists a finite subscript set J such that $x = \sum_{i \in J} x_i \otimes y_i$.

For any $x, u \in E$, $y, v \in M$, and $a \in A$, we have the equalities $(x+u) \otimes y = x \otimes y + u \otimes y$, $x \otimes (y+v) = x \otimes y + x \otimes v$, and $xa \otimes y = x \otimes ay$.

(2) For any module homomorphisms $f : E_A \to X_A$ and $g : {}_A M \to {}_A N$, the rule $(f \otimes g)(\sum x_i \otimes y_i) = \sum f(x_i) \otimes g(y_i)$ defines the group

homomorphism $f \otimes g : E \otimes M \to X \otimes N$.

(3) The natural map $t : E \times M \to E \otimes M$ is balanced, and for every balanced map u from $E \times M$ into an Abelian group G, there exists the unique group homomorphism $f : E \otimes M \to G$ such that $u = ft$.

(4) For each left ideal B of A, the rule $f_B(\sum e_j \otimes b_j) = \sum e_j b_j$ defines the canonical group epimorphism $h_B : E \otimes_A B \to EB$.

In addition, if C is a left ideal of A such that $B \subseteq C$, and $u : B \to C$ is the natural embedding, then $h_B = h_C(1_E \otimes u)$.

(5) The canonical group epimorphism $h : E \otimes_A A \to E$ is an isomorphism from the right A-module $E \otimes A$ onto the module E_A.

(6) If P and Q are two submodules of $_AM$, then the intersection in $E \otimes (P+Q)$ of the canonical images of $E \otimes P$ and $E \otimes Q$ coincides with the canonical image of $E \otimes (P \bigcap Q)$.

(7) Let $_AM = \sum_{i \in I} Am_i$, and let $\{e_i\}_{i \in I}$ be a set of elements of E such that $e_i = 0$ for all but finitely many i. Then

$\sum_{i \in I} e_i \otimes m_i = 0 \iff$ there exist some finite sets $\{x_j \in E\}_{j \in J}$ and $\{a_{ji} \in A\}_{j \in J, i \in I}$ ($a_{ji} = 0$ for all but finitely many i, j) such that $\sum_{i \in I} a_{ji} m_i = 0$ for all $j \in J$, and $e_i = \sum_{j \in J} x_j a_{ji}$ for all $i \in I$.

(8) Let $_AM_B$ be a bimodule over rings A and B, and let E_A be a module.

If E_A and M_B are two free (resp. projective, flat) modules, then the module $(E \otimes_A M)_B$ is free (resp. projective, flat).

(9) If $E_A = \oplus_{j \in J} E_j$, then $(\oplus_{j \in J} E_j) \otimes M$ is naturally isomorphic (as an Abelian group) to $\oplus_{j \in J}(E_j \otimes M)$.

(10) If E_A is a free module with basis $\{e_j\}_{j \in J}$, then E is isomorphic (as an Abelian group) to $\otimes M \cong M^{(J)}$, where $M^{(J)}$ is a direct sum of J pairwise isomorphic copies of M.

(11) E_A is flat $\iff$ for each left ideal B of A, the canonical group epimorphism $E \otimes B \to EB$ is an isomorphism.

(12) Let $0 \longrightarrow E_1 \overset{f}{\longrightarrow} E_2 \overset{g}{\longrightarrow} E_3 \longrightarrow 0$ be an exact sequence of right A-modules.

Then $E_1 \otimes M \overset{f \otimes 1}{\longrightarrow} E_2 \otimes M \overset{g \otimes 1}{\longrightarrow} E_3 \otimes M \longrightarrow 0$ is an exact sequence of Abelian groups.

In particular, $g \otimes 1$ is an epimorphism, and $E_3 \otimes M \cong (E_2 \otimes M)/(f \otimes 1)(E_1 \otimes M)$.

◁ The proof of 4.1 is left to the reader. ▷

4.2 A module E_A is *Hattori torsion-free* if for any $e \in E$ and $a \in A$ such that $ea = 0$, there exist $e_1, \ldots, e_n \in E$ and $a_1, \ldots, a_n \in A$ such that $e = \sum_{i=1}^{n} e_i a_i$ and $a_i a = 0$ for all i.

E_A is Hattori torsion-free $\iff$ given any $a \in A$, the natural group epimorphism $E \otimes Aa \to Ea$ is an isomorphism.

$\triangleleft \implies$ Assume that $\sum_i e_i b_i a = 0$, where $e_i \in E$ and $b_i \in A$. Set $e \equiv \sum_i e_i b_i \in E$. Then $ea = 0$. Since E is Hattori torsion-free, $e = \sum_j x_j a_j$, where $x_j \in E$, $a_j \in A$, and $a_j a = 0$ for all j. Then $\sum_i e_i \otimes b_i a = e \otimes a = \sum_j x_j a_j \otimes a = \sum_j x_j \otimes a_j a = \sum_j x_j \otimes 0 = 0$.

$\Longleftarrow$ Assume that $ea = 0$, where $e \in E$ and $a \in A$. By assumption, $e \otimes a = 0$. By 4.1(7), there exist $x_j \in E$ and $a_j \in A$ such that $e = \sum_j x_j a_j$ and $a_j a = 0$ for all j. Then E is Hattori torsion-free. $\triangleright$

4.3 (1) All direct sums and direct summands of flat modules are flat.

(2) All projective modules are flat.

$\triangleleft$ (1) can be verified using 4.1(9).

(2) By 4.1(5), all free cyclic module are flat. Every projective module is isomorphic to a direct summand of a direct sum of cyclic free modules. The assertion follows now from 4.3(1). $\triangleright$

4.4 *For a right module E over a ring A, the following conditions are equivalent.*

(1) *E is flat.*

(2) *For each finitely generated left ideal B of A, the canonical group epimorphism $E \otimes B \to EB$ is an isomorphism.*

(3) *Given any $x_1, \ldots, x_n \in E$ and $b_1, \ldots, b_n \in A$ such that $\sum_{i=1}^n x_i b_i = 0$, there exist $u_1, \ldots, u_m \in E$ and $a_{ij} \in A$ ($i = 1, \ldots, n; j = 1, \ldots, m$) such that $x_i = \sum_{j=1}^m u_j a_{ij}$ and $\sum_{i=1}^n a_{ij} b_i = 0$ for all i, j.*

$\triangleleft$ (1)$\implies$(2) follows from 4.1(11).

(2)$\implies$(3) Set $B \equiv \sum_{i=1}^n A b_i \subseteq A$. Since $\sum_{i=1}^n x_i b_i = 0$, it follows from 4.1(11) that $\sum_{i=1}^n x_i \otimes b_i = 0$. Apply now 4.1(7).

(3)$\implies$(1) Let B be a right ideal of A, $h : E \otimes B \to EB$ be the canonical group epimorphism, and let $\sum_{i=1}^n x_i \otimes b_i \in \mathrm{Ker}(h)$. Then $\sum_{i=1}^n x_i b_i = 0$, and there exist $u_1, \ldots, u_m \in E, a_{ij} \in A, (i = 1, \ldots, n, j = 1, \ldots, m)$, such that $x_i = \sum_{j=1}^m u_j a_{ij}$ and $\sum_{i=1}^n a_{ij} b_i = 0$. Hence $\sum_{i=1}^n x_i \otimes b_i = \sum_{i=1}^n \sum_{j=1}^m u_i a_{ij} \otimes b_i = \sum_{j=1}^m \sum_{i=1}^n u_i \otimes a_{ij} b_i = \sum_{i=1}^n u_i \otimes 0 = 0$. Therefore h is an isomorphism. By 4.1(11), E is flat. $\triangleright$

4.5 *For a right module E over a ring A, the following conditions are equivalent.*

(1) *E is flat.*

(2) *E is Hattori torsion-free, and $EB \bigcap EC = E(B \bigcap C)$ for all left ideals B, C of A.*

(3) *E is Hattori torsion-free, and $EB \bigcap EC = E(B \bigcap C)$ for any two finitely generated left ideals B and C of A.*

◁ (1)$\Longrightarrow$(2) follows from 4.1(6), 4.2, and 4.1(11). (2)$\Longrightarrow$(3) is obvious.

(3)$\Longrightarrow$(1) Let M be a finitely generated left ideal of A. By 4.4, it is sufficient to prove that the canonical group epimorphism f_M : $E \otimes M \to EM$ is an isomorphism. We proceed by the induction on k. In the case $k = 1$, the assertion follows from 4.2. Assume now that M is k-generated. Obviously, $M = B + C$, where B is a $(k-1)$-generated left ideal, and C is a principal left ideal. Let $f_B : E \otimes B \to EB$, $f_C : E \otimes C \to EC$, $w : E \otimes A \to EA$, $h_B : E \otimes B \to E \otimes M$, $h_C : E \otimes C \to E \otimes M$, and $g : E \otimes M \to E \otimes A$ be the canonical homomorphisms of Abelian groups. Let $x = \sum e_i \otimes (b_i - c_i) = \sum e_i \otimes b_i - \sum e_i \otimes c_i \in \mathrm{Ker}(f_M)$, $b_i \in B$, $c_i \in C$, $z \equiv \sum e_i c_i \in EB$, and let y be an element of $E \otimes B$ such that $\sum e_i \otimes b_i = h_B(y)$. Since $0 = f_M(x) = \sum e_i(b_i - c_i)$, $z = \sum e_i c_i \in EB \bigcap EC$. By assumption, $EB \bigcap EC = E(B \bigcap C)$, whence $z = \sum u_j d_j$ for some $u_j \in E$ and $d_j \in B \bigcap C$. Let v be an element of $E \otimes B$ such that $\sum u_j \otimes d_j = h_B(v)$. Then $0 = f_M(\sum u_j \otimes d_j - \sum e_i \otimes b_i) = wg(h_B(v - y)) = f_B(v - y)$, whence $v = y$, since f_B is a monomorphism by induction. Then $\sum e_i \otimes b_i = h_B(y) = h_B(v) = \sum u_j \otimes d_j$. Analogously, $\sum e_i \otimes c_i = \sum u_j \otimes d_j$. Then $x = \sum e_i \otimes b_i - \sum e_i \otimes c_i = 0$. Hence f is an isomorphism. ▷

4.6 *For a submodule Q_A of a flat module P_A, the following conditions are equivalent.*

(1) $(P/Q)_A$ is a flat module.

(2) For each finitely generated left ideal B of A, the natural group epimorphism $PB/QB \to (P/Q)B$ is an isomorphism.

(3) For each left ideal B of A, the natural group epimorphism $PB/QB \to (P/Q)B$ is an isomorphism.

(4) $Q \bigcap PB = QB$ for any finitely generated left ideal B of A.

(5) $Q \bigcap PB = QB$ for any left ideal B of A.

◁ (3)$\Longleftrightarrow$(5) follows from the fact that $(P/Q)B = (PB + Q)/Q \cong PB/(Q \bigcap PB) = PB/QB$. (5)$\Longrightarrow$(4) and (3)$\Longrightarrow$(2) are obvious.

(4)$\Longrightarrow$(5) and (2)$\Longrightarrow$(3) Let X_A be an arbitrary module, B be any left ideal of A, and let x be any element in XB. Then there exists a finitely generated ideal $\overline{B}$ of A such that $\overline{B} \subseteq B$ and $x \in X\overline{B}$.

(1)$\Longleftrightarrow$(2) By 4.1(12), the sequence $Q \otimes B \longrightarrow P \otimes B \longrightarrow (P/Q) \otimes M \longrightarrow 0$ is exact. Since P is flat, 4.4 shows us that the canonical epimorphism $P \otimes B \to PB$ is an isomorphism which maps from $Q \otimes B$ onto QB. Therefore $(P/Q)B \cong PB/QB$. By 4.4, P/Q is flat $\Longleftrightarrow$ the canonical epimorphism $(P/Q) \otimes B \to (P/Q)B$ is an isomorphism for any finitely generated left ideal B of A. Hence (P/Q) is flat $\Longleftrightarrow$ a natural epimorphism $PB/QB \to (P/Q)B$ is an isomorphism for any finitely generated left ideal B. ▷

4.7 (1) If Q is a right ideal of a ring A, then
the module $(A/Q)_A$ is flat $\Longleftrightarrow Q \cap B = QB$ for any finitely generated left ideal B of A.

(2) If a is an element of a ring A, then
the module $(A/aA)_A$ is flat $\Longleftrightarrow aA$ is a direct summand of A_A.

◁ Since A_A is flat, (1) follows from 4.6 ($P = A_A$).

(2) It is sufficient only to prove $\Longrightarrow$. By (1), $aA \cap Aa = aAa$, whence $a \in aAa$. Hence aA is a direct summand of A_A. ▷

4.8 A is a regular ring $\Longleftrightarrow N \cap MB = NB$ for every submodule N of any right A-module M and for any left ideal B of A.

◁ $\Longleftarrow$ Let $a \in A$, $M \equiv A_A$, $N \equiv aA$, and let $B \equiv Aa$. Since $a \in N \cap MB$, we have $a \in NB = aAa$ by (2). Then $a = aba$ for some $b \in A$, whence A is regular.

$\Longrightarrow$ Let $f = \sum_{i=1}^{n} m_i b_i \in N \cap MB$, where $m_i \in M$, $b_i \in B$, and $1 \le i \le n$. The left ideal $\sum_{i=1}^{n} Ab_i$ of A is finitely generated, whence this left ideal is generated by some idempotent e. Then each $b_i = a_i e$, where $a_i \in A$. Let $m \equiv \sum_{i=1}^{n} m_i a_i$. Then $f = me \in N, f = fe \in NB$, whence $N \cap MB \subseteq NB \subseteq N \cap MB$. ▷

4.9 *For a ring A, the following conditions are equivalent.*
(1) A is regular.
(2) Given any $a \in A$, the module $(A/aA)_A$ is flat.
(3) All right A-modules and all left A-modules are flat.

◁ (1)$\Longrightarrow$(3) follows from 4.6 and 4.8. (3)$\Longrightarrow$(2) is obvious. (2)$\Longrightarrow$(1) follows from 4.7(2). ▷

4.10 *For a submodule Q of a free module P_A, the following conditions are equivalent.*
(1) The module $(P/Q)_A$ is flat.
(2) Given any $q \in Q$, there exists a homomorphism $h : P \to Q$ such that $h(q) = q$.
(3) For each finite set $q_1, \ldots, q_n \in Q$, there exists a homomorphism $h : P \to Q$ such that $h(q_i) = q_i$ for all i.

◁ (3)$\Longrightarrow$(2) is obvious.

(2)$\Longrightarrow$(1) Let B be a finitely generated right ideal of A, and let $q = \sum_{i=1}^{n} p_i b_i \in Q \cap PB$, where $p_i \in P$ and $b_i \in B$. By assumption, there exists a homomorphism $h : P \to Q$ such that $q = h(q)$. Set $q_i \equiv h(p_i) \in Q$. Then $q = h(q) = \sum_{i=1}^{n} q_i b_i \in QB$, whence $Q \cap PB = QB$. It follows now from 4.6 that P/Q is flat.

$(1) \Longrightarrow (2)$ Let $u = \sum_{i=1}^{n} x_i b_i$, where $b_i \in A$, and $\{x_1, \ldots, x_n\}$ is a subset of a basis $\{x_j\}_{j \in J}$ of the free module P_A. Let $B = \sum_{i=1}^{n} A b_i$. Since $q \in Q \cap PB$ and (by 4.6) $Q \cap PB = QB$, there exist $q_1, \ldots, q_n \in Q$ such that $q = \sum_{i=1}^{n} q_i b_i$. Define the homomorphism $h : P \to P$ by $h(x_i) = q_i$ for all $i = 1, \ldots, n$ and by $h(x_j) = 0$ for all $j \in J \setminus \{1, \ldots, n\}$. Then $h(P) \subseteq Q$ and $h(q) = q$.

$(2) \Longrightarrow (3)$ We use the induction on n. In the case $n = 1$, the assertion holds by (2). Assume now that $n > 1$. Let $f : P \to Q$ be a homomorphism such that $f(q_n) = q_n$. By induction, there exists a homomorphism $t : P \to Q$ such that $t(q_i - f(q_i)) = q_i - f(q_i)$ for all $i = 1, \ldots, n - 1$. Hence $(1_P - t)(1_P - f)(q_n) = (1_P - t)(q_n - q_n) = 0$. For each $i < n$, we obtain $(1_P - t)(1_P - f)(q_i) = (1_P - t)(q_i - f(q_i)) = 0$. Let $h \equiv 1_P - (1_P - t)(1_P - f) \in \mathrm{End}(P)$. Then h has the required properties. $\triangleright$

4.11 (1) Every finitely presented flat module M is projective.

(2) If m is an element of a ring A, then

the module mA is flat $\Longleftrightarrow$ given any $n \in A$ such that $mn = 0$, there exist $a, b \in A$ such that $a + b = 1$, $ma = 0$, and $bn = 0$.

$\triangleleft$ (1) Let $M \cong F/Q$, where F is a finitely generated free module, and Q is generated by $q_1, \ldots, q_n$. By 4.10, there exists a homomorphism $h : F \to Q$ such that $h(q_i) = q_i$ for all i. Hence $h(q) = q$ for all $q \in Q$. Consequently, M is isomorphic to a direct summand of the free module F.

(2) Since $mA \cong A_A / r(m)$ and $\mathrm{Hom}(A_A, r(m))$ can be identified with $r(m)$, 4.10 shows us that mA is flat $\Longleftrightarrow$ for all $n \in r(m)$, there exists $a \in r(m)$ such that $n = an$. $\triangleright$

4.12 A *projective cover* of a module M is any epimorphism $f : P \to M$ such that P is a projective module, and $\mathrm{Ker}(f)$ is superfluous in P. The module P is called a *projective hull* of M. Sometimes M is identified with $P/\mathrm{Ker}(f)$.

If a flat module M has a projective hull P, then $M = P$ is a projective module.

$\triangleleft$ Let Q be a superfluous submodule of P such that $P/Q \cong M$, and let N be a cyclic submodule of Q. It is sufficient to prove that $N = 0$. By 4.10, there exists a homomorphism $f : P \to Q$ such that $(1 - f)(N) = 0$. Since $f(P)$ is superfluous in P and $P = f(P) + (1 - f)(P)$, we have $(1 - f)(P) = P$. Since $1 - f$ is an epimorphism onto the projective module P, the module $\mathrm{Ker}(1 - f)$ is a direct summand of P. If $x \in \mathrm{Ker}(1 - f)$, then $x = f(x) \in Q$. Hence $\mathrm{Ker}(1 - f) \subseteq Q$, and therefore $\mathrm{Ker}(1 - f)$ is superfluous in P. Since $\mathrm{Ker}(1 - f)$ is a direct summand of P, $\mathrm{Ker}(1 - f) = 0$. Therefore $N \subseteq \mathrm{Ker}(1 - f) = 0$. ▷

4.13 If M_A is a nonzero cyclic Hattori torsion-free module over a local ring A, then M is a free cyclic module.

◁ Let $M = mA$, $a \in r(m)$. It is sufficient to prove that $a = 0$. Assume the contrary. Since M is Hattori torsion-free, there exist $m_1, \ldots, m_n \in M$ and $a_1, \ldots, a_n \in A$ such that $m = \sum_{i=1}^{n} m_i a_i$ and $a_i a = 0$ for all i. The left zero-divisors a_i cannot be units. Since A is local, $a_i \in J(A)$ for all i. Therefore $m = \sum_{i=1}^{n} m_i a_i \in M J(A)$, $0 \neq M = M J(A)$; we have a contradiction to Nakayama's Lemma. ▷

4.14 *If M is a right module over a left Bezout ring A, then M is flat $\Longleftrightarrow$ M is Hattori torsion-free.*

◁ $\Longrightarrow$ follows from 4.5.

$\Longleftarrow$ Let B be any finitely generated left ideal of A. By assumption, B is a principal left ideal. By 4.2, the natural group homomorphism $M \otimes B \to MB$ is an isomorphism. By 4.4, M is flat. ▷

4.2 Submodules of flat modules and *pf*-rings

4.15 (1) A module E_A is Hattori torsion-free $\Longleftrightarrow$ every cyclic submodule of E is contained in some Hattori torsion-free submodule of E.

(2) All submodules of a module E are Hattori torsion-free $\Longleftrightarrow$ all cyclic submodules of E are Hattori torsion-free.

◁ (1) follows from the definition of a Hattori torsion-free module. (2) follows from (1). ▷

4.16 *If E_A is a module over a ring A, then E is flat $\Longleftrightarrow$ every finitely generated submodule of E is contained in some flat submodule of E.*

◁ $\Longrightarrow$ is obvious.

$\Longleftarrow$ By 4.15(1), E is Hattori torsion-free. Let B and C be two finitely generated left ideals of A. By 4.5, it is sufficient to prove that $EB \bigcap EC = E(B \bigcap C)$. Assume that $x = \sum_{i=1}^{n} y_i b_i = \sum_{j=1}^{m} z_j c_j \in EB \bigcap EC$, where $y_i, z_j \in E$, $b_i \in B$, and $c_j \in C$. Let N denote the submodule of E generated by all y_i, z_j. Since N is finitely generated, N

is contained, by assumption, in some flat submodule M of E. Hence $x \in MB \cap MC$. By 4.5, $MB \cap MC = M(B \cap C) \subseteq E(B \cap C)$, whence $x \in E(B \cap C)$. Therefore $EB \cap EC = E(B \cap C)$. By 4.5, E is flat. $\triangleright$

4.17 *All submodules of a module E are flat $\Longleftrightarrow$ all finitely generated submodules of E are flat.*

$\triangleleft$ 4.17 follows from 4.16. $\triangleright$

4.18 *For a ring A, the following conditions are equivalent.*
(1) All submodules of flat right A-modules are flat.
(2) All submodules of flat left A-modules are flat.
(3) All right ideals of A are flat.
(4) All left ideals of A are flat.
(5) All finitely generated right ideals of A are flat.
(6) All finitely generated left ideals of A are flat.
(7) For each finitely generated right ideal E and for any finitely generated ideal M of A, the natural group homomorphism $E \otimes M \to EM$ is an isomorphism.

$\triangleleft$ (1)$\Longrightarrow$(3) and (2)$\Longrightarrow$(4) follow from the fact that A_A and $_AA$ are flat modules. (3)$\Longrightarrow$(5) and (4)$\Longrightarrow$(6) are obvious. (6)$\Longrightarrow$(3), (5)$\Longrightarrow$(4), (7)$\Longleftrightarrow$(5), and (7)$\Longleftrightarrow$(6) follow from 4.4 (see condition (2) in 4.4).

(6)$\Longrightarrow$(1) Let Q be a submodule of a flat module E_A, $f : Q \to E$ be the natural embedding, and let B be a finitely generated left ideal of A. Assume that $\sum_{i=1}^{n} q_i b_i = 0$, where $q_1, \ldots, q_n \in Q$ and $b_1, \ldots, b_n \in B$. Since E is flat, 4.4 shows us that the natural group epimorphism $E \otimes B \to EB$ is an isomorphism. Hence $(f \otimes 1)(\sum_{i=1}^{n} q_i \otimes b_i) = 0$. By assumption, $_AB$ is flat, whence $f \otimes 1$ is a monomorphism. Therefore $\sum_{i=1}^{n} q_i b_i = 0$, whence the natural group epimorphism $Q \otimes B \to QB$ is an isomorphism. By 4.4, Q is flat.

(5)$\Longrightarrow$(2) The proof is similar to the proof of (6)$\Longrightarrow$(1). $\triangleright$

4.19 A ring A is a *pf-ring* if the following equivalent conditions hold.
(i)   All principal right ideals of A are flat.
(ii) All principal left ideals of A are flat.
(iii) Given any $m, n \in A$ such that $mn = 0$, there exist $a, b \in A$ such that $a + b = 1$, $ma = 0$, and $bn = 0$.

((i)$\Longleftrightarrow$(iii) follows from 4.11(2). (ii)$\Longleftrightarrow$(iii) follows from symmetry in (3).)
(1) Every right or left Rickartian ring is a *pf*-ring.
(2) If A is a local ring, then

A is a *pf*-ring $\iff$ A is a domain.

(3) Every right (resp. left) ideal of a *pf*-ring A is Hattori torsion-free as a right (resp. left) A-module.

(4) If E is a right ideal of a *pf*-ring A, then

E_A is flat $\iff$ $EB \cap EC = E(B \cap C)$ for any two finitely generated left ideals B and C of A.

(5) If A is a right Bezout right *pf*-ring, then all submodules of flat (right or left) A-modules are flat.

$\lhd$ (1) follows from the fact that all projective modules are flat.

(2) By (1), it is sufficient to prove only $\Longrightarrow$. Assume that A is a *pf*-ring, $m, n \in A$, and $mn = 0$. There exist $a, b \in A$ such that $a + b = 1$, $ma = 0$, and $bn = 0$. Since A is local, either a or b is a unit. Assume that $a \in U(A)$. Then $m = 0$, whence A is a domain.

(3) follows from 4.15(2). (4) follows from (3) and 4.5.

(5) Since A is a right Bezout right *pf*-ring, all finitely generated right ideals of A are flat. The assertion follows now from 4.18. $\rhd$

4.20 *For a reduced ring A, the following conditions are equivalent.*
(1) *A is a pf-ring.*
(2) *For any $m, n \in A$ such that $\mathrm{Id}(m) \cap \mathrm{Id}(n) = 0$, there exist elements $a, b \in A$ such that $a + b = 1$, $\mathrm{Id}(m) \cap \mathrm{Id}(a) = 0$, and $\mathrm{Id}(b) \cap \mathrm{Id}(n) = 0$.*
(3) *For any $m, n \in A$ such that $\mathrm{Id}(m) \cap \mathrm{Id}(n) = 0$, there exist $a, b \in A$ such that $a + b = 1$, $ma = 0$, and $nb = 0$.*
(4) *$mA \oplus nA = (m + n)A$ for any $m, n \in A$ such that $\mathrm{Id}(m) \cap \mathrm{Id}(n) = 0$.*
(5) *For any $m, n \in A$ such that $\mathrm{Id}(m) \cap \mathrm{Id}(n) = 0$, the 2-generated right ideal $mA \oplus nA$ is principal.*
(6) *$(m + n)A = (m + n)A \cap mA \oplus (m + n)A \cap nA$ for any $m, n \in A$ such that $\mathrm{Id}(m) \cap \mathrm{Id}(n) = 0$.*

$\lhd$ (1)$\iff$(2)$\iff$(3) follows from 1.35(2) and 4.19. (4)$\Longrightarrow$(5) and (4)$\Longrightarrow$(6) are obvious.

(3)$\Longrightarrow$(4) It is sufficient to prove the inclusion $m \in (m + n)A, n \in (m + n)A$. By (3), there exist $a, b \in A$ such that $a + b = 1, ma = 0, nb = 0$. Hence $(m + n)b = mb = ma + mb = m$ and $(m + n)a = na = na + nb = n$.

(5)$\Longrightarrow$(3) By assumption, there exist $f, g, u, v \in A$ such that $m = (mf + ng)u$ and $n = (mf + ng)v$. Since $\mathrm{Id}(m) \cap \mathrm{Id}(n) = 0$, we obtain $m = mfu$ and $0 = mfv = n(1 - gv) = nb$, where $b \equiv 1 - gv$. Let $a \equiv 1 - b = gv \in A$. Since $v \in r(mf)$, 1.35(2) shows us that $ugv \in r(mf)$ and $ma = mfua = mfugv = 0$.

$(6)\Longrightarrow(4)$ Set $N \equiv (m + n)A$, $N_1 \equiv mA \oplus nA$. If $h_m : N_1 \to mA$ and $h_n : N_1 \to nA$ are the natural projections, then (6) implies that $N = h_m(N) \oplus h_n(N)$. Therefore $m = h_m(m+n) \in N$, $n = h_n(m+n) \in N$, and $N = N_1$. $\triangleright$

4.21 (1) If ring A is a right distributive ring, then
A is right nonsingular $\Longleftrightarrow$ A is a reduced pf-ring.
(2) If A is a reduced right Bezout ring, then all submodules of flat (right or left) A-modules are flat.
(3) If a ring A is right or left semihereditary, then all submodules of any (right or left) flat A-module are flat.

$\triangleleft$ (1) By 3.2(1), we may assume that A is a reduced ring. Now we can use 4.20.
(2) By 4.20, A is a pf-ring. Now (2) follows from 4.19(5).
(3) follows from 4.18. $\triangleright$

4.22 If Q is a submodule of a flat module P_A, and the factor module P/Q is flat, then Q is flat.

$\triangleleft$ By 4.5, it is sufficient to prove that Q is Hattori torsion-free, and $QB \bigcap QC = Q(B \bigcap C)$ for any finitely generated left ideals B and C of A. By 4.6, the following condition (*) holds: $QX = Q \bigcap PX$ for any left ideal X of A. Since P is flat, 4.5 shows us that P is Hattori torsion-free, and $PB \bigcap PC = P(B \bigcap C)$. This and (*) imply that $QB \bigcap QC = Q \bigcap PB \bigcap PC = Q \bigcap P(B \bigcap C) = Q(B \bigcap C)$. It remains to prove that Q is Hattori torsion-free. Let $qa = 0$, where $q \in Q$ and $a \in A$. Since P is Hattori torsion-free, there exist $p_1, \ldots, p_n \in P$ and $d_1, \ldots, d_n \in A$ such that $q = \sum_{i=1}^{n} p_i d_i$ and $d_i a = 0$ for all i. Let $D = \sum_{i=1}^{n} Ad_i$. Then $q \in Q \bigcap PD$, whence $q \in QD$ by (*). Hence there exist $q_1, \ldots, q_n \in Q$ such that $q = \sum_{i=1}^{n} q_i d_i$. Since $d_i a = 0$ for all i, Q is Hattori torsion-free. $\triangleright$

4.23 Let a_1, a_2, b_1, and b_2 be elements of A such that $a_1 b_1 = a_2 b_2$, and $a_1 A + a_2 A$ is flat as a right A-module.
Then there exist $f_{ij} \in A, 1 \leq i, j \leq 2$ such that $a_1 f_{11} = a_2 f_{21}$, $(1 - f_{11})b_1 = f_{12}b_2$, $a_1 f_{12} = a_2 f_{22}$, and $(1 - f_{22})b_2 = f_{21}b_1$.

$\triangleleft$ Let D_A be the free A-module of rank 2 with a basis $\{d_1, d_2\}$, $d \equiv d_1 b_1 + d_2 b_2$, $B \equiv a_1 A + a_2 A$, and let $g : D_A \to B_A$ be an epimorphism such that $g(d_1 y_1 + d_2 y_2) = a_1 y_1 - a_2 y_2$ for any $y_1, y_2 \in A$. Set $N \equiv \text{Ker}(g)$. Since $d \in N$ and B is flat, 4.10 shows us that there exists a homomorphism $f : D \to N$ such that $f(d) = d$. Let $f_{ij}, 1 \leq i, j \leq 2$ be elements of A such that $f(d_1) = d_1 f_{11} + d_2 f_{21} \in N$, and $f(d_2) = d_1 f_{12} +$

$d_2 f_{22} \in N$. Then $0 = g(f(d_1)) = g(f(d_2))$, $a_1 f_{11} = a_2 f_{21}$, $a_1 f_{12} = a_2 f_{22}$, $d = f(d) = f(d_1 b_1 + d_2 b_2) = d_1(f_{11} b_1 + f_{12} b_2) + d_2(f_{21} b_1 + f_{22} b_2)$, $b_1 = f_{11} b_1 + f_{12} b_2$, $b_2 = f_{21} b_1 + f_{22} b_2$, $(1 - f_{11}) b_1 = f_{12} b_2$, and $(1 - f_{22}) b_2 = f_{21} b_1$.▷

4.24 Let all 2-generated right ideals of a ring A be flat.

(1) Given any $u, v, w, z \in A$ such that $uv = zw$, there exist $f, g, h \in A$ such that $uf = zg$ and $(1 - f)v = hw$.

(2) Given any $u, v, w \in A$ such that $uv = vw$, there exist $f, g, h \in A$ such that $uf = vg$ and $(1 - f)v = hw$.

(3) If the ring A is local, then A is a domain.

In addition, if B and C are two principal right ideals of A and $B \bigcap C \neq 0$, then either $B \subseteq C$ or $C \subseteq B$.

(4) If A is a local right uniform ring, then A is a right uniserial domain.

◁ (1) follows from 4.23. (2) follows from (1) for $z = w$.

(3) By 4.19(2), A is a domain. Let $B = uA$, $C = zA$. Assume that $B \nsubseteq C$. We have to prove that $C \subset B$. By assumption, there exist $u, v \in A$ such that $uv = zw \neq 0$. By (1), there exist $f, g, h \in A$ such that $uf = zg$ and $(1 - f)v = hw$. The element f is not a unit for if f is a unit, $u \in C$, whence $B \subseteq C$; this is a contradiction. Since A is local, $1 - f \in U(A)$. Hence $v = (1 - f)^{-1} hw$ and $zw = uv = u(1 - f)^{-1} hw$. Since A is a domain, $z = u(1 - f)^{-1} h$, whence $C = zA \subseteq uA = B$.

(4) follows from (3). ▷

4.25 *Let M be a finitely generated flat right module over a unitary subring A of a ring B.*

(1) If $M \otimes_A B$ is a projective right B-module, then M_A is a projective module.

(2) If B is a semisimple ring, then M_A is projective.

◁ (1) Choose some exact sequence $0 \to Q \to P \to M \longrightarrow 0$, where P is a finitely generated free module. Since M is flat, the sequence of Abelian groups $0 \to Q \otimes_A B \to P \otimes_A B \to M \otimes_A B \longrightarrow 0$ is exact. Since $M \otimes B$ is a projective B-module, $(Q \otimes B)_B$ is isomorphic to a direct summand of $(M \otimes B)B$. By 4.22, Q_A is flat, since P_A and M_A are flat. Let $\{q_1 \otimes 1, \ldots, q_n \otimes 1\}$ be a set generating $(Q \otimes B)_B$. By 4.10, there exists a homomorphism $f : P \to Q$ such that $f(q_i) = q_i$ for all i. Consider any $q \in Q$. Then $q \otimes 1 = \sum_{i=1}^{n}(q_i \otimes 1) b_i$, where $b_i \in B$. Since Q_A is flat, $Q \cong Q \otimes A \to Q \otimes B$ is a monomorphism. Also, $f(q) \otimes 1 = (f \otimes 1)(q \otimes 1) = \sum (f \otimes 1)(q_i \otimes b_i) = \sum f(q_i) \otimes b_i = \sum q_i \otimes b_i = q \otimes 1$ (as elements of $(Q \otimes B)_B$). Therefore $f(q) = q$ for all $q \in Q$, whence

Q is a direct summand of P. Hence M is projective, since M is a copy of a direct summand of a projective module.

(2) follows from (1) and from the fact that all modules over a semisimple Artinian ring are projective. $\triangleright$

4.26 *If A is a unitary subring of a semisimple ring and all finitely generated right ideals of A are flat, then A is right and left semihereditary.*

$\triangleleft$ By 4.25(2), A is right semihereditary. By 4.18, all left ideals of A are flat. By 4.25(2), A is left semihereditary. $\triangleright$

4.27 Let X and Y be two submodules of a module M.

(1) There exists an epimorphism $f : X \oplus Y \to X + Y$ such that $\mathrm{Ker}(f) \cong X \cap Y$.

(2) If the module $X + Y$ is n-generated and projective, then $X \cap Y$ is isomorphic to a direct summand of $X + Y$, whence the module $X \cap Y$ is n-generated and projective.

$\triangleleft$ (1) The rule $f((x, y)) = x - y$ defines an epimorphism $f : X \oplus Y \to X + Y$. Then $\mathrm{Ker}(f) = \{(x, y) \mid x = y \in X \cap Y\}$. Therefore $\mathrm{Ker}(f) = \{(x, x) \mid x \in X \cap Y\} \cong X \cap Y$.

(2) follows from (1). $\triangleright$

4.28 *Each right Bezout domain A is a right and left semihereditary right uniform domain, and the intersection of any two nonzero principal right ideals of A is a nonzero principal right ideal.*

$\triangleleft$ (1) By 4.21(2), all right ideals of the domain A are flat. By 3.13, A is a right uniform domain. Hence A is a right order in a division ring. By 4.26, A is semihereditary.

(2) Let $0 \neq x, y \in A$. Since A is a right uniform domain, $xA \cap yA \neq 0$. By assumption, $xA + yA$ is a principal right ideal. Therefore $xA + yA$ is a cyclic projective module. By 4.27(2), $xA \cap yA$ is a principal right ideal. $\triangleright$

4.29 Let T be a right denominator set in a ring A, $f : A \to A_T \equiv Q$ be the canonical ring homomorphism, and let N be a right module over A. The right Q-module $N \otimes_A Q$ is called the *module of quotients for N with respect to T* and is denoted by N_T. An A-module homomorphism $g : N \to N_T$ such that $g(n) = n \otimes 1$ is denoted by g_T. If $T = A \setminus M$, where M is a right ideal of A, we write N_M, g_M instead of N_T, g_T.

Let $\sim$ be a relation on the cartesian product $N \times T$ such that $(x, s) \sim (y, t) \iff \exists c, d \in A : xc = yd, sc = td \in T$. It is verified directly that $\sim$ is an equivalence relation on $N \times T$.

Let h be the natural surjective map $N \times T \to (N \times T)/ \sim\, \equiv N(T)$, and let us define a map v from N to $N(T)$ by $v(n) = h(n,1)$. For $N(T)$, define the addition and the multiplication by elements of Q as follows:

$h((x,s)) + h((y,t)) = h((xc + yd, u))$, where $u = sc = td \in T$,

$h((x,c)) \cdot f(b)f(t)^{-1} = h((xc,tu))$ where $b \in A$, $t \in T$, $sc = bu$, and $u \in T$.

One can verify that $N(T)$ is a right Q-module, v is an A-module homomorphism, the map $w : N_T \to N(T)$ defined by $w(n \otimes x) = h((n,1))x$ is a Q-module homomorphism, and $v = g_T w$. Therefore, we will identify N_T with $N(T)$.

4.30 Let Q be the right ring of quotients of a ring A with respect to a right denominator set T, N be a right A-module, and let $f : A \to Q$, $g : N \to N_T$ be the canonical homomorphisms. Given any subset X of N, $\overline{X}$ denotes $g(X)$, and X_T denotes the submodule $g(X)Q$ of $(N_T)_Q$.

(1) $\mathrm{Ker}(g) = \{n \in N \mid \exists t \in T : nt = 0\}$.

(2) Let $0 \neq x = \overline{n}f(t)^{-1} \in N_T$, where $0 \neq n \in N$, $t \in T$.
Then $0 \neq \overline{n} = xf(t) \in \overline{N} \bigcap qf(A)$, and $(N_T)_{f(A)}$ is an essential extension of $\overline{N}_{f(A)}$.

(3) For any $x_1, \ldots, x_k \in N_T$, there exist $t \in T$ and $n_1, \ldots, n_k \in N$ such that $x_i = \overline{n}_i f(t)^{-1}$ for $i = 1, \ldots, k$.

(4) Let Y be a finitely generated submodule of the left module $_{f(A)}(N_T)$.
Then there exists $t \in T$ such that $Yf(t) \subseteq \overline{Y}$.
Therefore $_{f(A)}Y$ is isomorphic to a finitely generated submodule $Yf(t)$ of $_{f(A)}\overline{N}$.

(5) If X is a submodule of N_A, then $X_T = \{\overline{x}f(t)^{-1} | x \in X, t \in T\}$.
Therefore, if $x \in X_T$, then $xf(t) \in \overline{X}$ for some $t \in T$.

(6) $(F + G)_T = F_T + G_T$ and $(F \bigcap G)_T = F_T \bigcap G_T$ for any two submodules F and G of N such that $\mathrm{Ker}(g) \subseteq F \bigcap G$.

(7) If m is a cardinal and P is a nonzero m-generated submodule of a right Q-module N_T, then $g^{-1}(P)$ is a nonzero submodule of N_A, $P = (g^{-1}(P \bigcap g(N)))_T$, and there exists an m-generated submodule H of N such that $P = H_T$.

(8) If $P \in \mathrm{Lat}((N_T)_Q)$, then $P \bigcap g(N)$ is a submodule of $g(N)_{f(A)}$ and $P = (P \bigcap g(N))Q$.

(9) If $P_Q = \oplus_{i=1}^{n} P_i \in \mathrm{Lat}((N_T)_Q)$, then the submodule $P \bigcap g(N)$ of $g(N)_{f(A)}$ is a direct sum of the $P_i \bigcap g(N)$.

(10) $_A Q$ is flat.

(11) $(F/G)_T \cong F_T/G_T$ for any $F \in \mathrm{Lat}(N)$ and $G \in \mathrm{Lat}(F)$.

(12) If $n \in N$, $F \in \mathrm{Lat}(N)$ and $g(n) \in F_T$, then $nt \in F$ for some $t \in T$.

(13) If the module N_A is free (resp. projective, flat), then $(N_T)_Q$ is free (resp. projective, flat).

(14) If m is a cardinal number and all m-generated right ideals of A are free (resp. projective, flat) as A-modules, then all m-generated right ideals of Q are free (resp. projective, flat) as Q-modules.

(15) If $\mathcal{E}$ is the set of all submodules of N_A which contain $\mathrm{Ker}(g)$, then $\mathcal{E}$ is a sublattice of the lattice $\mathrm{Lat}(M)$, and $\mathcal{E}$ is isomorphic to the lattice $\mathrm{Lat}((N_T)_Q)$.

(16) If the module N_A is distributive (resp. Noetherian, Artinian), then $(N_T)_Q$ is a distributive (resp. Noetherian, Artinian) module.

(17) If the module $(N_T)_Q$ is distributive, then $(F \bigcap (G + H))_T = (F \bigcap G + F \bigcap H)_T$ for any submodules F, G, H of N_A.

$\lhd$ (1) follows from 4.29 and 1.48. The proof of (2), (3), (4), (5), and (6) is similar to the proof of 1.53(1),(2),(3),(4),(5) (use 4.29). The proof of (7), (8), and (9) is similar to the proof of 1.53(11),(8),(9).

(10) Let $h : L_A \to M_A$ be a monomorphism of right A-modules, $u : M \to M_T$ be the canonical module homomorphism, $x \in L$, and let $u(h(x)) = 0$. By (1), there exists $t \in T$ such that $h(x)t = 0$. Therefore $xt \in \mathrm{Ker}(h) = 0$, whence $(xA)_T = 0$.

(11) follows from (10) and from the fact that $(X_T)_B \cong (X \otimes_A Q)_Q$ for any module X_A. (12) follows from (11) and (1).

(13) Since Q_Q is free, (13) follows from 4.1(8) and from the fact that $N_T = N \otimes_A Q$.

(14) follows from (13) and 1.53(11). (15) follows from (6) and (7). (16) follows from (15). (17) follows from (6). $\rhd$

4.31 Let N be a right module over a right localizable ring A, and let $g_M : N \to N_M$ be the canonical homomorphisms.

(1) Let $x \in N$, $F \in \mathrm{Lat}(N)$, and let $g_M(x) \in F_M$ for any $M \in \mathrm{max}(A_A)$.

Then $x \in F$.

(2) If $F, G \in \mathrm{Lat}(N)$ and $F_M = G_M$ for all $M \in \mathrm{max}(A_A)$, then $F = G$.

(3) N is a distributive A-module $\iff N_M$ is a distributive A_M-module for all $M \in \mathrm{max}(A_A)$.

(4) N is a distributive A-module $\iff N_M$ is a uniserial A_M-module for all $M \in \mathrm{max}(A_A)$.

$\lhd$ (1) Set $D \equiv (x : F)$. It is sufficient to prove that $D = A$. Assume the contrary. Then $D \subseteq M \in \mathrm{max}(A_A)$, whence $D \bigcap (A \setminus M) = \emptyset$; this is a contradiction to 4.30(5).

(2) follows from (1).

(3)

$\Longrightarrow$ follows from 4.30(16).

$\Longleftarrow$ Let $F, G, H \in \mathrm{Lat}(N_A)$, and let $M \in \max(A_A)$. By 4.30(17), $(F \bigcap (G+H))_M = (F \bigcap G + F \bigcap H)_M$. By (2), $F \bigcap (G+H) = F \bigcap G + F \bigcap H$.

(4) By 1.53(14), A_M is a local ring for all $M \in \max(A_A)$. By 2.8, each distributive module over a local ring is uniserial. The assertion follows now from (3). $\triangleright$

4.32 Let R be a unitary central subring of a ring A, $\max(R)$ be the set of all maximal ideals of R, and let N be a right A-module. For each $M \in \max(R)$, let A_M denote the ring of quotients of A with respect to the central multiplicative subset $R \setminus M$ of A, N_M be the module of quotients of the module N with respect to $R \setminus M$, f_M be the canonical ring homomorphism $R \to R_M$, and let g_M be the canonical A-module homomorphism $N \to N_M$. For each $x \in N$, we write x_M instead of $g_M(x)$. Note that the module N_M is a right A_M-module and R_M-module.

(1) Let $M \in \max(R)$, H be an A_M-submodule of N_M, and let L be the inverse image of H in N.

Then $H = L_M$ and $L \supseteq \mathrm{Ker}(g_M) = \{x \in N \mid \exists t \in R \setminus M : xt = 0\}$.

(2) For each $M \in \max(R)$, R_M is a local ring, $J(R_M) = M_M$, and $R/M \cong R_M/M_M$.

Also, given any $a \in R$, we have $a \in M \Longleftrightarrow a_M \in M_M$.

(3) Let $F \in \mathrm{Lat}(N_R)$. For each $x \in N$ such that $x_M \in F_M$ for all $M \in \max(R)$, we have $x \in F$.

In particular, if $x_M = 0$ for all $M \in \max(R)$, then $x = 0$.

(4) $(F + G)_M = F_M + G_M$ and $(F \bigcap G)_M = F_M \bigcap G_M$ for any $M \in \max(R)$ and any $F, G \in \mathrm{Lat}(N)$.

(5) If $F, G \in \mathrm{Lat}(N_R)$, then
$F = G \Longleftrightarrow F_M = G_M$ for all $M \in \max(R)$.

(6) If $F \in \mathrm{Lat}(N_R)$ and $G \in \mathrm{Lat}(A_R)$, then $(FG)_M = F_M G_M$ for all $M \in \max(R)$.

(7) If G is a right ideal of A, then
G is an ideal of $A \Longleftrightarrow G_M$ is an ideal of A_M for all $M \in \max(R)$.

(8) If $M \in \max(R)$ and S is a right (resp. left) ideal of A_M, then there exists a right (left) ideal B of A such that $B_M = S$.

(9) A is a reduced (resp. semiprime, regular, strongly regular) ring $\Longleftrightarrow$ for all $M \in \max(R)$, A_M is a reduced (resp. semiprime, regular, strongly regular) ring.

(10) If for all $M \in \max(R)$, the ring A_M is right quasi-invariant, then A is right quasi-invariant.

(11) A is right invariant $\Longleftrightarrow$ for all $M \in \max(R)$, A_M is right invariant.

(12) If n is a positive integer, then

n is a unit of $A \Longleftrightarrow$ for all $M \in \max(R)$, n is a unit of A_M.

(13) Let $a, b \in U(R)$, $M \in \max(R)$, and let $h : R \to R/M$ and $v : R_M \to R_M/M_M$ be the natural epimorphisms. Then

the equation $x^2 - h(a)y^2 - h(b)z^2 = 0$ has the unique solution $x = y = z = 0$ in $R/M \Longleftrightarrow$ the equation $x^2 - v(a_M)y^2 - v(b_M)z^2 = 0$ has the unique solution $x = y = z = 0$ in R_M/M_M.

(14) N is a distributive A-module $\Longleftrightarrow$ N_M is a distributive A_M-module for all $M \in \max(R)$.

(15) If $A_M/J(A_M)$ is a simple Artinian ring for all $M \in \max(R)$, then

N_A is distributive $\Longleftrightarrow$ N_M is a uniserial A_M-module for all $M \in \max(A)$.

$\lhd$ (1) – (6) are directly verified. (7) follows from (6) and (5). (8) follows from (1). (9) is verified with the use of (3), (5), (6), and (8).

(10) Let $T \in \max(A)$, $M \in \max(R)$, and let K be the inverse image of the ideal $A_M T_M$ of A_M in A. By (7), it is sufficient to prove that T_M is an ideal of A_M. If $T_M = A_M$, the assertion is proved. Assume that $T_M \neq A_M$. Then T_M is contained in some maximal right ideal of A_M which is an ideal of A_M by assumption. Therefore $(AT)_M = A_M T_M \neq A_M$, whence $K \neq A$ and $T \subseteq K$. Hence $T = K$, whence $T_M = K_M = A_M T_M$ is an ideal of A_M.

(11) follows from (7).

(12) Note that n is invertible in $R \Longleftrightarrow R = nR$, and then apply (5).

(13) Note that kernel of the homomorphism $A \to A_M$ is contained in M. The assertion follows now from (2).

(14)

$\Longrightarrow$ follows from 4.30(16).

$\Longleftarrow$ Let $F, G, H \in \mathrm{Lat}(N_A)$, and let $M \in \max(R)$. By 4.30(17), $(F \cap (G+H))_M = (F \cap G + F \cap H)_M$. By (5), $F \cap (G+H) = F \cap G + F \cap H$.

(15) follows from (14) and 2.7. $\rhd$

4.33 (1) If A is right localizable pf-ring, then A is a reduced ring, and A_M is a local domain for all $M \in \max(A_A)$.

(2) Let A be a right invariant pf-ring which satisfies one of the following conditions.

(i)   Given any $a \in A$ such that $a^2 = 0$, we have $a \in C(A)$.

(ii) Given any $a \in A$, there exists an integer $n = n(a)$ such that $r(a^n) = r(a^{n+1})$.

(iii) A is a ring with the maximum condition on right annihilators.

(iv) A is semiprime.

Then A is a right localizable reduced ring, and A_M is a right uniform local domain for all $M \in \max(A_A)$.

◁ (1) Set $Q \equiv A_M$. By 1.53(14) and 4.30(14), Q is a local pf-ring. By 4.19(2), Q is a domain. It follows now from 1.60(4) that A is reduced.

(2) By 4.15(2), A is right localizable. By 4.33(1), A is reduced. Let $M \in \max(A_A)$, $Q \equiv A_M$, and let H be kernel of the canonical homomorphism $f : A \to Q$. By (1), Q is a local domain. Hence the ideal H is completely prime, whence the right invariant domain A/H is right uniform. By 1.55(11), Q is right uniform. ▷

4.34 Let E be a right ideal of a pf-ring A, and let R be a central subring of A. Then

E_A is a flat module $\Longleftrightarrow E_M B_M \bigcap E_M C_M = E_M(B_M \bigcap C_M)$ for any two finitely generated left ideals B,C of A and for each maximal ideal M of R.

◁ By 4.19(4), E_A is flat $\Longleftrightarrow EB \bigcap EC = E(B \bigcap C)$ for any finitely generated left ideals B, C of A. By 4.32(5) and 4.32(6), $EB \bigcap EC = E(B \bigcap C) \Longleftrightarrow E_M B_M \bigcap E_M C_M = E_M(B_M \bigcap C_M)$ for each maximal ideal M of R. ▷

4.35 *For a ring A, the following conditions are equivalent.*

(1) All submodules of flat (right or left) A-modules are flat.

(2) $EB \bigcap EC = E(B \bigcap C)$ for any finitely generated right ideal E and any two finitely generated left ideals B, C of A, and for any elements $m, n \in A$ such that $mn = 0$, there exist $a, b \in A$ such that $a + b = 1$, $ma = 0$, and $bn = 0$.

(3) A is a pf-ring, and $E_M B_M \bigcap E_M C_M = E_M(B_M \bigcap C_M)$ for any two finitely generated left ideals B and C of A and each maximal ideal M of the centre R of A.

(4) A is a pf-ring, and there exists a unitary central subring S of A such that $E_M B_M \bigcap E_M C_M = E_M(B_M \bigcap C_M)$ for any two finitely generated left ideals B, C of A and each maximal ideal M of S.

◁ By 4.19, we may assume that A is a pf-ring. By 4.18, (1) holds $\Longleftrightarrow$ each finitely generated right ideal E is flat. Therefore, (1)$\Longleftrightarrow$(2) follows from 4.19(4). (2)$\Longrightarrow$(3) and (4)$\Longrightarrow$(2) follow from 4.34. (3)$\Longrightarrow$(4) is obvious. ▷

4.36 Let A be a *pf*-ring, and let R be a unitary central subring of A.

(1) If E is a right ideal of A and E_M is a flat right A_M-module for all $M \in \max(R)$, then E_A is flat.

(2) If for all $M \in \max(R)$, all submodules of flat A_M-modules are flat, then all submodules of flat A-modules are flat.

(3) If for all $M \in \max(R)$, the ring A_M is right or left semihereditary, then all submodules of flat A-modules are flat.

(4) If for all $M \in \max(R)$, the ring A_M is a reduced right or left Bezout ring, then all submodules of flat A-modules are flat.

◁ (1) follows from 4.34 applied to A and from 4.4 applied to A_M. (2) follows from 4.35 applied to A_M and from 4.5 applied to A_M. (3) follows from (2) and 4.18. (4) follows from (2) and 4.19(5). ▷

4.37 *For a ring A, the following conditions are equivalent.*

(1) *A is a pf-ring.*

(2) *All submodules of Hattori torsion-free right A-modules are Hattori torsion-free.*

(3) *All submodules of Hattori torsion-free left A-modules are Hattori torsion-free.*

(4) *All principal right ideals of A are Hattori torsion-free.*

(5) *All principal left ideals of A are Hattori torsion-free.*

◁ By symmetry of (1), it is sufficient to prove the equivalence of conditions (1), (2), and (4).

$(1) \Longrightarrow (2)$ Let M_A be a Hattori torsion-free module, $N \in \mathrm{Lat}(M)$, and let B be a principal left ideal of A. Since A is a *pf*-ring, $_A B$ is flat. Therefore, the natural group homomorphism $f : N \otimes B \to M \otimes B$ is a monomorphism. Since M_A is Hattori torsion-free, 4.2 shows us that the natural group homomorphism $g : M \otimes B \to MB$ is an isomorphism. Let $h : N \otimes B \to NB$ be the natural group epimorphism. Since $h = gf$, h is an isomorphism. By 4.2, N_A is Hattori torsion-free.

$(2) \Longrightarrow (4)$ By 4.5, the free module A_A is Hattori torsion-free.

$(4) \Longrightarrow (1)$ Let $m, n \in A$, and let $mn = 0$. Since mA is a Hattori torsion-free module, there exist $m_1, \ldots, m_k \in mA$ and $x_1, \ldots, x_k \in A$ such that $m = \sum_{i=1}^{k} m_i x_i$ and $x_i n = 0$ for any i. Let $m_i = m y_i$, $y_i \in A$, $i = 1, \ldots, k$, $a \equiv 1 - \sum_{i=1}^{k} y_i x_i \in A$, and let $b \equiv \sum_{i=1}^{k} y_i x_i \in A$. Then $bn = 0$ and $ma = m - mb = m - m \sum_{i=1}^{k} y_i x_i = m - \sum_{i=1}^{k} m_i x_i = 0$. Therefore A is a *pf*-ring. ▷

4.3 Coherent and reduced rings

4.38 (Schanuel's Lemma) *Assume that $h_1 : P_1 \to M_1$ and $h_2 : P_2 \to M_2$ are module epimorphisms with kernels Q_1 and Q_2, respectively, where the modules P_1 and P_2 are projective, and there exists an isomorphism $f : M_1 \to M_2$.*
Then $P_1 \oplus Q_2 \cong P_2 \oplus Q_1$.

◁ Since P_1 is projective and h_2 is an epimorphism, there exists a homomorphism $u : P_1 \to P_2$ such that $h_2 u = f h_1$. Then $u(P_1) + Q_2 = P_2$ and $u^{-1}(Q_2) = Q_1$. By the rule $t(p_1 + q_2) = u(p_1) - q_2$, the homomorphism $t : P_1 \oplus Q_2 \to P_2$ is defined. Since P_2 is projective, t is split, whence $P_1 \oplus Q_2 \cong P_2 \oplus \mathrm{Ker}(t)$. Set $K \equiv \{q_1 + u(q_1) \,|\, q_1 \in Q_1\} \subseteq Q_1 \oplus Q_2$. Then $K \cong Q_1$. It is sufficient to prove that $\mathrm{Ker}(t) = K$. Since $t(q_1 + u(q_1)) = u(q_1) - u(q_1) = 0$, $K \subseteq \mathrm{Ker}(t)$. If $p_1 + q_2 \in \mathrm{Ker}(t)$, then $q_2 = u(p_1)$ and $p_1 \in u^{-1}(Q_2) = Q_1$. Hence $\mathrm{Ker}(t) \subseteq K$. ▷

4.39 A module M is *finitely presented* if $M \cong P/Q$, where P and Q are finitely generated modules, and P is a projective module.
For a finitely presented module M, the following assertions hold.
(1) If $M \cong \overline{P}/\overline{Q}$, where $\overline{P}$ is a finitely generated projective module, then the module $\overline{Q}$ is finitely generated.
(2) If N is a submodule of M, then
N is finitely generated $\Longleftrightarrow$ the module M/N is finitely presented.

◁ (1) follows from Schanuel's Lemma 4.38.
(2) By assumption, there exists an epimorphism $h : P \to M$, where P is a finitely generated projective module, and $\mathrm{Ker}(h) \equiv Q$ is a finitely generated module. Since M is finitely presented, Q is finitely generated. The epimorphism h induces an isomorphism $P/h^{-1}(N) \cong M/N$.
$\Longrightarrow$ Since N is finitely generated, there exists a finitely generated submodule S of P such that $h(S) = N$. Hence $Q + S$ is a finitely generated submodule of P and $M/N \cong P/(Q + S)$. Therefore M/N is finitely presented.
$\Longleftarrow$ Since P is a finitely generated projective module, M/N is finitely presented and $P/h^{-1}(N) \cong M/N$. By (1), $h^{-1}(N)$ is finitely generated. Hence N is a homomorphic image of a finitely generated module $h^{-1}(N)$. Therefore N is finitely generated. ▷

4.40 (1) A direct sum of two finitely presented modules is a finitely presented module.
(2) If N_1 and N_2 are finitely presented submodules of a module M such that $N_1 + N_2$ is a finitely presented module, then $N_1 \bigcap N_2$ is a finitely generated module.

(3) Any finitely generated module M_A over a right Noetherian ring A is a finitely presented Noetherian module.

(4) If $F = \sum_{i=1}^{s} m_i A$ and $G = \sum_{j=1}^{t} n_j A$ are submodules of a distributive module M_A, then $F \bigcap G = \sum_{j=1}^{t} \sum_{i=1}^{s} (m_i A \bigcap n_j A)$.

◁ (1) Let N_1 and N_2 be finitely presented modules, and let $N \equiv N_1 \oplus N_2$. By assumption, for $i = 1, 2$, there exist epimorphisms $h_i : P_i \to N_i$ such that P_i are finitely generated projective modules, and $\mathrm{Ker}(h_i)$ are finitely generated modules. Let $P \equiv P_1 \oplus P_2$, and let $h : P \to N$ be the homomorphism such that $h(x_1 + x_2) = h_1(x_1) + h_2(x_2)$ for $x_i \in N_i$. Then P is a finitely generated projective module, h is an epimorphism, and $\mathrm{Ker}(h) = \mathrm{Ker}(h_1) \oplus \mathrm{Ker}(h_2)$ is a finitely generated module.

(2) By 4.27(1), there exists an epimorphism $h : N_1 \oplus N_2 \to N_1 + N_2$ such that $\mathrm{Ker}(h) \cong N$. By (1), $N_1 \oplus N_2$ is finitely presented. By 4.39(2), N is finitely generated.

(3) There exists a finitely generated free module P such that $M \cong P/Q$, where $Q \in \mathrm{Lat}(P)$. Since finite direct sums of Noetherian modules are Noetherian, P is Noetherian, whence Q is finitely generated. Since $M \cong P/Q$, M is Noetherian. Hence M is finitely presented.

(4) follows from the distributivity of M. ▷

4.41 A module M_A over a ring A is *coherent* if the following equivalent conditions hold.

(1) Each finitely generated submodule of M is finitely presented.

(2) The intersection of any two finitely generated submodules of M is finitely generated, and each cyclic submodule of M is finitely presented.

(3) The intersection of any two finitely generated submodules of M is finitely generated, and $r(m)$ is a finitely generated right ideal of A for any $m \in M$.

◁ (2)$\Longleftrightarrow$(3) follows from 4.39(1) and from the fact that $mA \cong A_A/r(m)$.

(1)$\Longrightarrow$(2) Let N_1 and N_2 be two finitely generated submodules of M, and let $N \equiv N_1 \bigcap N_2 \in \mathrm{Lat}(M)$. By assumption and by 4.40, $N_1 \oplus N_2$ and $N_1 + N_2$ are finitely presented. By 4.27(1), there exists an epimorphism $h : N_1 \oplus N_2 \to N_1 + N_2$ such that $\mathrm{Ker}(h) \cong N$. By 4.40(2), N is finitely generated.

(2)$\Longrightarrow$(1) Let n be a positive integer. Let us prove by the induction on n that any n-generated submodule T of M is finitely presented. For $n = 1$, the assertion holds by assumption. Let $n > 1$, and let any $(n - 1)$-generated submodule of M be finitely presented. Then

$T = N_1 + N_2$, where N_1 and N_2 are finitely presented. By assumption, $N_1 \cap N_2 \equiv N$ is finitely generated. By 4.40(1), $N_1 \oplus N_2$ is finitely presented. By 4.27(1), there exists an epimorphism $h : N_1 \oplus N_2 \to T$ such that $\mathrm{Ker}(h) \cong N$. By 4.39(2), T is finitely presented. $\triangleright$

4.42 (1) Every right module over a right Noetherian ring is coherent.

(2) Every semihereditary module is coherent.

(3) Assume that M_A is a module over a ring A such that the intersection of any two cyclic submodules of M is finitely generated, and M is either a distributive module or a Bezout module. Then

M is a coherent module $\Longleftrightarrow$ the annihilator of any element of M is a finitely generated right ideal of A.

$\triangleleft$ (1) follows from 4.40(3). (2) follows from the fact that each finitely generated projective module is finitely presented. (3) follows from 4.40(4) and 4.41. $\triangleright$

4.43 *For a right module N over a right localizable ring A, the following conditions are equivalent.*

(1) N is distributive.

(2) N_M is a distributive right A_M-module for all $M \in \max(A_A)$.

(3) N_M is a uniserial right A_M-module for all $M \in \max(A_A)$.

(4) For any $m, n \in N$, there exist $a, b \in A$ such that $mA \cap nA = maA + nbA$ and $a + b = 1$.

(5) The set $\overline{\mathrm{Lat}}(N)$ of all finitely generated submodules of N is a distributive sublattice of the lattice $\mathrm{Lat}(N)$. Also, the intersection of any s-generated submodule $F \in \overline{\mathrm{Lat}}(N)$ and any t-generated submodule $G \in \overline{\mathrm{Lat}}(N)$ is a $2st$-generated module for any positive integers s and t (in particular, the intersection of any two cyclic submodules of M is a 2-generated module).

$\triangleleft$ (1)$\Longleftrightarrow$(2) and (2)$\Longleftrightarrow$(3) follow from 4.31(3),(4).

(1)$\Longrightarrow$(4) By 2.4, there exist $a, b \in A$ such that $1 = a+b$ and $F \subseteq G$, where $F \equiv maA+nbA$ and $G \equiv mA \cap nA$. Let $M \in \max(A_A)$, $f : A \to A_M \equiv Q$ be the canonical ring homomorphism, and let $g : N \to N_M$ be the canonical module homomorphism. By 1.53(14), the ring Q is local. Since $f(a) + f(b) = 1_Q$, either $f(a)A = Q$ or $f(b)A = Q$. In both cases $F_M = G_M$. By 4.31(2), $F = G$.

(4)$\Longrightarrow$(5) By 2.4, N is distributive. It follows from the assumption that the intersection of any two cyclic submodules of M is 2-generated. The assertion follows now from 4.40(4).

(5)$\Longrightarrow$(1) follows from the fact that distributivity of a module is equivalent to distributivity of its 2-generated submodules. $\triangleright$

4.44 Let M be a right module over a right localizable ring A.

(1) If M is finitely generated, then

M is distributive $\iff$ the module of quotients M_N is a cyclic uniserial module over the right ring of quotients A_N of A for each maximal right ideal N of A.

(2) If M is a Bezout module, then M is a distributive module, the intersection of any two finitely generated submodules of M is a cyclic module, and the A_M-module N_M is a uniserial module for all $M \in \max(A_A)$.

(3) If M is distributive, then

M is coherent $\iff$ the annihilator of any element $m \in M$ is a finitely generated right ideal of A.

$\triangleleft$ (1) Every finitely generated uniserial module is cyclic. Also, the module of quotients of a finitely generated module is a finitely generated module over the ring of quotients. The assertion follows now from 4.43.

(2) By 1.60(7), A is right quasi-invariant. By 2.35, a Bezout module M_A is distributive. (2) follows now from 4.43.

(3) By 4.43, the intersection of any two finitely generated submodules of M is finitely generated. (3) follows now from 4.42(3). $\triangleright$

4.45 Let Q be the left ring of quotients of a ring A with respect to a left denominator set T, and let $f : A \to Q$ be the canonical homomorphism.

(1) Every finitely generated submodule of $Q_{f(A)}$ is isomorphic to a finitely generated right ideal of $f(A)$.

(2) If A is a right Bezout ring, then $Q_{f(A)}$ is a Bezout module (in particular, Q is a right Bezout ring).

(3) If A is right distributive, then $Q_{f(A)}$ is a distributive module (in particular, Q is right distributive).

$\triangleleft$ (1) follows from 1.53(3). (2) follows from (1). (3) follows from (1) and from the fact that M is distributive $\iff$ all 2-generated submodules of M are distributive. $\triangleright$

4.46 Let A be a right distributive ring, N be either a maximal right ideal or a completely prime ideal of A, $T \equiv A \setminus N$, and let $K(T) \equiv \{a \in A \mid at = 0$ for some $t \in T\}$.

(1) N is a completely prime ideal, T is a right permutable multiplicative set, $K(T)$ is an ideal of A, and the ring $A/K(T)$ is right uniform.

In addition, if either the right singular ideal $S/K(T)$ of $A/K(T)$ is a nil-ideal or $A/K(T)$ is a ring with the maximum condition on right annihilators, then T is a right denominator set in A, and the right ring of quotients A_N is right uniserial.

(2) If T is a left denominator set, then the left ring of quotients $_N A$ is right uniserial.

◁ By 2.54 and 1.43(11), N is a completely prime ideal. Hence the set T is multiplicative.

(1) By 3.6(1), T is right permutable. By 1.57, $K(T)$ is an ideal of A. Let $h : A \to A/K(T)$ be the natural epimorphism. Assume that $h(A)$ is not right uniform. There exist $x, y \in A \setminus K(T)$ such that $h(x)A \cap h(y)A = 0$. By 2.6(1), there exist $a, b \in A$ such that $a + b = 1$ and $0 = xa = yb$. Since $a + b = 1$, either $a \in T$ or $b \in T$. Assume that $a \in T$. Since $h(x)h(a) = 0$, the element $h(a)$ is not left regular in $h(A)$; this is a contradiction to 1.57(1).

If $h(A)$ is a ring with the maximum condition on right annihilators, then 1.37(1) shows us that $\mathrm{Sing}(h(A)_{h(A)})$ is a nilpotent ideal of $h(A)$. Hence we may assume that $\mathrm{Sing}(h(A)_{h(A)})$ is a nil-ideal of $h(A)$. Since $h(A)$ is right uniform, $\mathrm{Sing}(h(A)_{h(A)})$ coincides with the set of all left zero-divisors of $h(A)$. Hence all left zero-divisors of $h(A)$ are nilpotent. By 1.57(4), T is a right denominator set in A. By 1.53(14), A_N is local. By 4.30(16), A_N is right distributive. By 2.8, A_N is right uniserial.

(2) By 1.53(14), $_N A$ is local. By 4.45(3), $_N A$ is right distributive. By 2.8, $_N A$ is right uniserial. ▷

4.47 Let A be a right Prüfer ring.
(1) The intersection of any two finitely generated right ideals of A is a finitely generated right ideal.
(2) A is right coherent $\Longleftrightarrow$ $r(a)$ is a finitely generated right ideal of A for all $a \in A$.

◁ (1) follows from 4.43. (2) follows from 4.44(3). ▷

4.48 *Assume that either all square-zero elements of A are central, or for any element $a \in A$, there exists a positive integer $n = n(a)$ such that $r(a^n) = r(a^{n+1})$, or A is a ring with the maximum condition on right annihilators.*

(1) A is right distributive $\Longleftrightarrow$ A is a right Prüfer ring $\Longleftrightarrow$ all maximal right ideals of A are completely prime ideals, and for any completely prime ideal M of A, the right ring of quotients A_M exists and is right uniserial.

(2) Let A be right distributive.
Then the intersection of any two finitely generated right ideals of A is a finitely generated right ideal.

In addition, A is right coherent $\Longleftrightarrow r(a)$ is a finitely generated right ideal of A for all $a \in A$.

(3) Let A be a right quasi-invariant right Bezout ring.

Then the intersection of any two principal right ideals is a principal right ideal.

In addition, A is right coherent $\Longleftrightarrow r(a)$ is a finitely generated right ideal of A for all $a \in A$.

◁ (1) follows from 3.22 and 3.6(3). (2) follows from (1) and 4.47. (3) follows from (2) and from the fact that A is right distributive by 2.35. ▷

4.49 For a ring A, the following conditions are equivalent.

(1) The prime radical $N(A)$ of A coincides with the set of all nilpotent elements of A.

(2) $A/N(A)$ is a reduced ring.

(3) For every minimal prime ideal H of A, the factor ring A/H is a reduced ring.

(4) Each minimal prime ideal of A is completely prime.

◁ (1)$\Longleftrightarrow$(2) follows from the fact that the prime radical is a nil-ideal.

(2)$\Longrightarrow$(4) Each prime ideal of the factor ring $A/N(A)$ is equal to $P/N(A)$, where P is a prime ideal of A. Therefore, (4) follows from 1.35(7).

(4)$\Longrightarrow$(3) is directly verified.

(3)$\Longrightarrow$(2) The prime radical is the intersection of the minimal prime ideals. Therefore, (2) follows from the fact that a subdirect product of reduced rings is reduced. ▷

4.50 For a right distributive ring A, the following conditions are equivalent.

(1) The prime radical $N(A)$ of A coincides with the set of all nilpotent elements of A.

(2) $A/N(A)$ is right nonsingular.

(3) $A/N(A)$ is reduced.

(4) For every minimal prime ideal P of A, the factor ring A/P is right nonsingular.

(5) Each minimal prime ideal of A is completely prime.

◁ By 3.2(1) if A is right distributive, then A is reduced $\Longleftrightarrow A$ is right nonsingular. The assertion follows now from 4.49. ▷

4.51 Let A be a right distributive ring.

(1) If M and N are two incomparable completely prime ideals of A, then $A = M + N$ and $M \cap N = MN + NM$.

(2) Any two completely prime ideals M and N of A which are contained in the same proper left or right ideal are comparable.

(3) A sum of any two distinct minimal completely prime ideals of A is equal to A.

(4) $A/J(A)$ is a reduced ring.

(5) Let M be a maximal left ideal or maximal right ideal or completely prime ideal of A, and let $\mathcal{E}$ be the set of all completely prime ideals of A which are contained in M.

Then $\mathcal{E}$ is a nonempty linearly ordered set which contains precisely one minimal element.

◁ (1) Let $m \in M \setminus N$, $n \in N \setminus M$. By 2.4, there exist $a, b \in A$ such that $1 = a + b$, $ma \in N$, and $nb \in M$. Hence $a \in N$, $b \in M$, $A = M + N$, and $M \cap N = (M + N)(M \cap N) \subseteq MN + NM$.

(2) and (3) follow from (1). (4) follows from 2.52 and 2.54.

(5) By (2), $\mathcal{E}$ is either empty or nonempty and linearly ordered. In the last case, $\mathcal{E}$ contains precisely one minimal element, since the intersection of a descending chain of completely prime ideals is a completely prime ideal. By 1.43(11), it is sufficient to prove that $\mathcal{E}$ is nonempty if M is a maximal left ideal. Let $h : A \to A/J(A)$ be a natural epimorphism. By (4), $h(A)$ is a reduced ring. By 1.35(8), $h(M)$ contains a completely prime ideal $h(B)$ of $h(A)$, where B is an ideal of A which contains $J(A)$. Hence $B \in \mathcal{E}$. ▷

4.52 *For a ring A, the following conditions are equivalent.*

(1) *A is a semiprime right distributive ring, and for any minimal prime ideal H of A, the ring A/H is right nonsingular.*

(2) *A is a right distributive right nonsingular ring.*

(3) *A is a reduced right distributive ring.*

(4) *A is a right Prüfer right nonsingular ring.*

(5) *A is a reduced right Prüfer ring.*

(6) *For each maximal right ideal M of A, the right ring of quotients A_M exists and is a right uniserial domain.*

(7) *All maximal right ideals of A are completely prime ideals, for any completely prime ideal N, the right ring of quotients A_N exists and is a right uniserial domain, the kernel H of the canonical homomorphism $f : A \to A_N$ is the unique minimal prime ideal H of A contained in N, and H coincides with the set $\{a \in A \mid ta = 0 \text{ for some } t \in A \setminus N\}$.*

◁ (1)$\Longleftrightarrow$(2)$\Longleftrightarrow$(3) follow from 4.50. (3)$\Longleftrightarrow$(5) follows from 4.48(1). (7)$\Longrightarrow$(6) is obvious. (6)$\Longrightarrow$(5) follows from 1.60(4).

$(4)\Longleftrightarrow(5)$ By 3.2(1) if A is right distributive, then A is reduced $\Longleftrightarrow A$ is right nonsingular. By 3.22, all right Prüfer rings are right distributive. Therefore, (5) holds.

$(5)\Longrightarrow(7)$ Set $T \equiv A\setminus N$. By assumption and by 1.52, the right ring of quotients A_N is a right uniserial reduced ring. By 1.35(10), A_N is a domain. Therefore H is a completely prime ideal. By 4.51(4), N contains the unique minimal completely prime ideal E which is contained in H. Hence $H = E$, since $H = \{a \in A \mid at = 0 \text{ for some } t \in T\}$. By 1.35(2), $H = \{a \in A \mid ta = 0 \text{ for some } t \in A \setminus N\}$. $\triangleright$

4.53 Let H be a minimal prime ideal of a right distributive reduced ring A, and let $h : A \to A/H$ be the natural epimorphism.

(1) $h(A)$ is a domain, all elements of H are not regular in A, and the right ring of quotients A_H exists and is a division ring.

(2) For each regular element a of A, the element $h(a)$ is regular in $h(A)$.

(3) If H is a minimal prime ideal of A which is a maximal ideal, then A/H is a division ring which is the right ring of quotients A_H, and the natural epimorphism h coincides with the canonical homomorphism $A \to A_H$.

(4) The set of all regular elements of A coincides with the set of all elements which are not contained in any minimal prime ideal.

$\triangleleft$ (1) can be verified using 4.52. (2) follows from (1). (3) follows from (1), 1.43(11), and 4.52.

(4) By (2), it is sufficient to prove that if an element a is not contained in any minimal prime ideal, then a is regular. Let $b \in A$, and let $ab = 0$. Since all minimal prime ideals of A are completely prime, b is contained in the intersection of all minimal prime ideals. Therefore $b = 0$. $\triangleright$

4.54 Let A be a weakly right invariant ring.

(1) $\mathrm{Id}(M)\mathrm{Id}(N) = \mathrm{Id}(N)\mathrm{Id}(M) = 0$ for any two right ideals M and N such that $M \bigcap N = 0$.

(2) If $m \in A$ and $m^2 = 0$, then $m \in \mathrm{Sing}(A_A)$.

(3) Let A be right nonsingular, M be a right ideal of A, H be a complement to M in A_A, and let $\widehat{M}$ be a closure of M in A_A.

Then $H = r(M) = \ell(M) = r(\widehat{M}) = \ell(\widehat{M})$.

(4) Let A be right nonsingular, M and E be two right ideals of A, and let E_A be an essential extension of M_A.

Then $r(M) = r(E)$.

(5) If A is right nonsingular and H is a closed right ideal of A, then $H = r(E) = \ell(E)$, where E is a complement to H in A_A.

◁ (1) The closures $\widehat{M}$ and $\widehat{N}$ of the right ideals M and N are disjoint ideals. Therefore $MN + NM \subseteq (M) \bigcap (N) \subseteq \widehat{M} \bigcap \widehat{N} = 0$.

(2) If N is a complement to mA in A_A, then $N \oplus mA$ is an essential right ideal, and $m(N \oplus mA) = 0$.

(3) follows from (1). (4) and (5) follow from (3). ▷

4.55 Let A be a right distributive right nonsingular ring.

(1) If $a \in A$ and $r(a)$ is an essential extension of a finitely generated right ideal, then $r(a) = \ell(a) = eA$, where e is a central idempotent.

(2) If $a \in A$ and $r(a)$ is either finitely generated or right finite-dimensional ideal, then $r(a) = \ell(a) = eA$, where e is a central idempotent.

(3) If A is a right order in a ring Q, then Q is a reduced ring, and all idempotents of Q are contained in A.

◁ It follows from 2.46(5), 3.2(1), and 2.50(3) that A is a reduced normal weakly right invariant ring.

(1) Let $r(a) \equiv H$ be an essential extension of a right ideal $T = \sum_{i=1}^{n} t_i A$, E be a complement to H in A_A such that E is an essential extension of aA, and let $E_i \equiv r(t_i)$ for $i = 1, \ldots, n$. By 4.54(5), $H = r(E)$ and $E = r(H)$. By 1.35(2), $E = E_1 \bigcap \ldots \bigcap E_n$. By 2.31, $H + E_i = A$ for all i. Hence $A = H + E_1 \bigcap \ldots \bigcap E_n = H + E$ and $H \bigcap E = 0$. Therefore $H = \ell(a) = eA$, where e is a central idempotent.

(2) follows from (1).

(3) By 1.52, Q is reduced. Let $t = t^2 = ab^{-1} \in Q$, where $a \in A$ and b is a regular element of A. Then $ab^{-1}a = a$ and $a \in r_Q(ab^{-1} - 1)$. By 1.35(2), $r_Q(ab^{-1} - 1)$ is an ideal of the reduced ring Q. Therefore $0 = (ab^{-1} - 1)ba = (a - b)a$. By 1.35(2), $\mathrm{Id}(a - b) \bigcap \mathrm{Id}(a) = 0$. By 2.31, there exists $h \in A$ such that $a(1 - h) = 0$ and $(a - b)h = 0$. Hence $(1 - h)a = h(a - b) = 0$, $hb = (1 - h)a + h(a - b) + hb = a$, and $t = ab^{-1} = h \in A$. ▷

4.56 *For a right distributive ring A, the following conditions are equivalent.*

(1) A is right or left Rickartian.

(2) A is right nonsingular, and $r_A(a)$ is an essential extension of a finitely generated right ideal for all $a \in A$.

(3) A is a right order in a right or left Rickartian ring Q.

(4) A is a right order in a strongly regular ring Q.

(5) A is a right coherent reduced Rickartian ring which is a right order in a strongly regular ring Q and contains all idempotents of Q.

◁ By 2.50(3), A is a normal ring.

$(1)\Longrightarrow(2)$ follows from 1.34(2). $(2)\Longrightarrow(1)$ follows from 4.55(1). $(5)\Longrightarrow(4)$ and $(4)\Longrightarrow(3)$ are obvious.

$(3)\Longrightarrow(1)$ By 1.55(3), Q is right distributive. By 2.50(3), Q is normal. By 3.36, Q is a Rickartian reduced ring. By 4.55(3) and 1.34(4), A is right Rickartian.

$(1)\Longrightarrow(5)$ By 3.36, A is a Rickartian reduced ring, and each element of A is a product of a regular element and a central idempotent. Let T be the set of all regular elements of A, $a \in A$, and let $t \in T$. By 2.4, there exist $d, g, h \in A$ such that $ad = bh$ and $b(1 - d) = ag$. Let $d = mf$, where f is a central idempotent, and m is a regular element. Then $b(1 - f) = b(1 - mf)(1 - f) = ag(1 - f)$ is a regular element of $(1 - f)A$. Therefore $r_{(1-f)A}(g(1 - f)) = 0$ and $d = mf$ is a regular element of fA. If $u = mf + g(1 - f)$, then $r_A(u) = 0$. Hence $\ell_A(u) = 0$ and u is a regular element. Since $au = amf + ag(1 - f) = b(h + 1 - f)$, the set T is right permutable, and A has the classical right ring of quotients Q. By 4.55(3), Q is a reduced ring, and all idempotents of Q belong to A. Let $q = xy^{-1} \in Q$, where $x \in A$ and $y \in T$. There exists a central idempotent $s \in A$ and a regular element $v \in T$ such that $x = sv$. Hence $z = vy^{-1} \in U(Q)$, the idempotent s of the reduced ring Q is central in Q, and $q = sz$. Therefore Q is strongly regular. By 4.48(2), A is right coherent. $\triangleright$

4.57 *For a ring A, the following conditions are equivalent.*

(1) A is right semihereditary.

(2) A is right coherent, and all finitely generated right ideals of A are flat.

(3) The intersection of any two finitely generated right ideals of A is a flat finitely generated right ideal, and the right annihilator of any element of A is a finitely generated right ideal.

$\triangleleft$ $(2)\Longleftrightarrow(3)$ follows from 4.41. $(1)\Longrightarrow(2)$ follows from 4.42(2). $(2)\Longrightarrow(1)$ follows from the fact that each finitely presented flat module is projective by 4.11(1). $\triangleright$

4.58 For a ring A, the following conditions are equivalent.

(1) A is strongly regular.

(2) A is a right distributive right nonsingular ring, and all completely prime ideals of A are maximal ideals.

(3) A is right localizable, and A_M is a division ring for any $M \in \max(A_A)$.

(4) A is right localizable, and A_M is a regular ring for any $M \in \max(A_A)$.

(5) The right annihilator of any subset of A is an ideal of A, and there exists an injective ring endomorphism φ of A such that $\forall a \in A \exists b \in A \ : \ \varphi(a) = \varphi(a)ab$.

$\lhd$ (1)$\Longrightarrow$(2) and (3)$\Longrightarrow$(4) can directly be verified. (2)$\Longrightarrow$(3) follows from 4.53(3). (3)$\Longrightarrow$(1) follows from 4.32(9). (1)$\Longrightarrow$(5) follows from the fact that we may assume that $\varphi \equiv 1_A$.

(4)$\Longrightarrow$(3) Let $M \in \max(A_A)$. By 1.53(14), the ring A_M is local. The local regular ring A_M is a division ring.

(5)$\Longrightarrow$(1) Let $a \in A$. By assumption, $\varphi(a)(1 - ab) = 0$ for some $b \in A$. Set $m \equiv 1 - ab$. Since $r(\varphi(a))$ is an ideal of A by assumption, $\varphi(a)m = 0$ implies that $\varphi(am)am = 0$ (*). By assumption, there exists $g \in A$ such that $\varphi(am) = \varphi(am)amg$. This and (*) imply that $\varphi(am) = 0$. Hence $0 = am = a(1 - ab)$, and we have $a = a^2 b$. Therefore A is strongly regular. $\rhd$

4.4 Invariant rings

4.59 For a module M_A over a right invariant ring A, the following conditions are equivalent.

(1) M is distributive.

(2) $A = (F : G) + (G : F)$ for any two cyclic submodules F, G of M.

(3) $A = (F : G) + (G : F)$ for any two finitely generated submodules F and G of M.

(4) For any two finitely generated submodules F and G of M, there exist $a, b \in A$ such that $1 = a + b$ and $Fa + Gb \subseteq F \bigcap G$.

$\lhd$ (3)$\Longleftrightarrow$(4) and (3)$\Longrightarrow$(2) are directly verified.

(1)$\Longleftrightarrow$(2) By 2.4, $A = (m : nA) + (n : mA)$ for any $m, n \in M$. Since A is right invariant, $(m : nA) = (mA : nA)$.

(2)$\Longrightarrow$(3) Let $F = \sum_{i=1}^{m} f_i A$ and $G = \sum_{j=1}^{n} g_j A$. By (2), $A = (f_i A : g_j A) + (g_j A : f_i A)$ for all i, j. Therefore $A = (f_i A : G) + (g_j A : f_i A)$ for any i, j. By 1.23(1), $A = (f_i A : G) + \bigcap_{j=1}^{n}(g_j A : f_i A) = (f_i A : G) + (G : f_i A)$ for all i. Hence $A = (f_i A : G) + (G : F)$ for all i. By 1.23(1), $A = \bigcap_{i=1}^{m}(f_i A : G) + (G : F) = (F : G) + (G : F)$. $\rhd$

4.60 Let M_A be a module over a right distributive right invariant ring A, and let B and D be two right ideals of A.

Then $MB \bigcap MD = M(B \bigcap D)$.

$\lhd$ Let $x = \sum_{i=1}^{s} m_i b_i = \sum_{j=1}^{t} n_j d_j \in MB \bigcap MD$, where $m_i, n_j \in M$, $b_i \in B$, and $d_j \in D$. Set $F \equiv \sum_{i=1}^{s} b_i A \subseteq A$ and $G \equiv \sum_{j=1}^{t} d_j \subseteq A$. Then F and G are finitely generated right ideals of the right distributive right invariant ring A. By 4.59, there exist $a, b \in A$ such that $1 = a + b$ and $b_i a, d_j b \in F \cap G$ for all i, j. Hence $x = xa + xb = \sum_{i=1}^{s} b_i a + \sum_{j=1}^{t} d_j b \in M(F \cap G) \subseteq M(B \cap D)$. Therefore $MB \cap MD \subseteq M(B \cap D)$. The inverse inclusion is obvious. ▷

4.61 *For a right module M over a distributive invariant ring A, the following conditions are equivalent.*

(1) *M is a flat module.*

(2) *Each cyclic submodule of M is contained in some flat submodule of M.*

(3) *M is a Hattori torsion-free module.*

◁ (1)$\Longrightarrow$(2) is directly verified.

(2)$\Longrightarrow$(3) By 4.5, each flat module is a Hattori torsion-free module. Therefore, each cyclic submodule of M is contained in some Hattori torsion-free submodule of M. By 4.15(1), M is Hattori torsion-free.

(3)$\Longrightarrow$(1) The ring A is invariant. By 4.60, $MB \cap MD = M(B \cap D)$ for any two ideals B and D of A. By 4.5, M is flat. ▷

4.62 *Let M be a module over a distributive invariant ring. Then all submodules of M are flat $\Longleftrightarrow$ all cyclic submodules of M are flat $\Longleftrightarrow$ all cyclic submodules of M are Hattori torsion-free.*

◁ 4.62 follows from 4.61. ▷

4.63 *Let M be a module over a distributive invariant pf-ring. Then M is flat $\Longleftrightarrow$ M is Hattori torsion-free all submodules of M are flat.*

◁ Since A is a pf-ring, 4.37 implies that each submodule of a Hattori torsion-free A-module is Hattori torsion-free. 4.63 follows now from 4.61. ▷

4.64 Let all 2-generated right ideals of A be flat.

(1) If for all $M \in \max(A_A)$, the ring A_M exists and is right uniform, then A is a reduced right distributive ring.

(2) Assume that A is a right invariant ring, and either all square-zero elements of A are central, or for any $a \in A$, there exists a positive integer $n = n(a)$ such that $r(a^n) = r(a^{n+1})$, or A is a ring with the maximum condition on right annihilators, or A is semiprime.

Then A is a right distributive reduced ring.

◁ (1) Let $M \in \max(A_A)$. By 4.30(14), all 2-generated right ideals of A_M are flat. By 1.53(14), A_M is a local ring. By 4.24(4), A_M is a right uniserial domain. By 4.52, A is a reduced right distributive ring.

(2) Let $M \in \max(A_A)$. By 4.33(2), A_M exists and is a right uniform domain. By (1), A is a reduced right distributive ring. ▷

4.65 *Let A be an invariant ring, and let all square-zero elements of A be central. Then the following conditions are equivalent.*
(1) *All submodules of flat (right or left) A-modules are flat.*
(2) *All 2-generated ideals of A are flat as right A-modules.*
(3) *All 2-generated ideals of A are flat as left A-modules.*
(4) *A is a distributive semiprime ring.*
(5) *A is a distributive pf-ring.*

◁ (5)$\Longrightarrow$(1) follows from 4.63. (1)$\Longrightarrow$(2) and (1)$\Longrightarrow$(3) follow from the fact that A_A and $_AA$ are flat modules. (2)$\Longrightarrow$(4) and (3)$\Longrightarrow$(4) follow from 4.64(2).
(4)$\Longrightarrow$(5) An invariant semiprime ring A is reduced. Since A is a reduced right distributive ring, A is a pf-ring by 4.20. ▷

4.66 *Let A be an invariant semiprime ring. Then*
all submodules of flat A-modules are flat $\Longleftrightarrow$ all 2-generated right ideals of A are flat $\Longleftrightarrow$ all 2-generated left ideals of A are flat $\Longleftrightarrow$ A is a distributive ring.

◁ All invariant semiprime rings are reduced. Hence 4.66 follows from 4.65. ▷

4.67 If A is a right invariant right or left Rickartian ring, then A is a Rickartian reduced ring and a right order in a strongly regular ring.

◁ The right invariant ring A is normal and has a classical right ring of quotients Q. By 3.36, A is a Rickartian reduced ring. Each regular element of A is invertible in Q. By 3.36, each element of A is a product of a central idempotent and a regular element. Hence each element of Q is a product of a central idempotent and a unit. Therefore Q is strongly regular. ▷

4.68 If A is a right invariant ring and all 2-generated right ideals of A are projective, then A is a right distributive reduced ring.

◁ By 4.67, A is reduced. By 4.64(2), A is right distributive. ▷

4.69 If for each maximal right ideal M of a ring A, the right ring of quotients A_M exists and is a right distributive right semi-Noetherian ring, then A is a right invariant right distributive ring.

◁ By 3.22, A is right distributive. Let $M \in \max(A_A)$. By 2.71(1), the right distributive right semi-Noetherian ring A_M is right invariant. By 1.60(6), A is right invariant. ▷

4.70 Let A be a semiprime right distributive ring, and let the ring A/P be right Noetherian for each minimal prime ideal P of A.

Then A is a reduced right invariant right Prüfer ring.

◁ By 3.8, the factor ring A/P is a domain for every minimal prime ideal P. By 4.52, A is a reduced right Prüfer ring, the ring A_M is a right uniserial domain for all $M \in \max(A_A)$, and the kernel of the canonical homomorphism $f : A \to A_M$ is a minimal prime ideal $H(M)$. Let $h : A \to A/H(M)$ be the natural epimorphism. By assumption, $A/H(M)$ is right Noetherian. Since $A_M \cong h(A)_{h(M)}$, 4.30(16) implies that A_M is right Noetherian. By 4.69, A is right invariant. ▷

4.71 [183]. *If A is a right hereditary ring, then*
A is right distributive $\Longleftrightarrow$ A is right invariant.

◁ $\Longleftarrow$ follows from 4.68.
$\Longrightarrow$ By 1.34(2), A is right nonsingular. Let $M \in \max(A_A)$. By 4.52, the right ring of quotients A_M exists and is a right uniserial domain. By 4.30(14), A_M is right hereditary. By 3.73, the right hereditary right finite-dimensional ring A_M is right Noetherian. By 2.72, the right distributive right Noetherian ring A_M is right invariant. By 1.60(6), A is right invariant. ▷

4.72 *For an invariant ring A, the following conditions are equivalent.*
   (1) *A is right semihereditary.*
   (2) *A is left semihereditary.*
   (3) *All 2-generated ideals of A are projective right A-modules.*
   (4) *All 2-generated ideals of A are projective left A-modules.*
   (5) *A is a distributive right or left Rickartian ring.*

◁ Since condition (5) is symmetric, it is sufficient to prove $(1)\Longleftrightarrow(3)\Longleftrightarrow(5)$.
   $(1)\Longrightarrow(3)$ is directly verified. $(3)\Longrightarrow(5)$ follows from 4.68.
   $(5)\Longrightarrow(1)$ By 4.56, A is a reduced right coherent ring. By 4.66, all right ideals of A are flat. By 4.57, A is right semihereditary. ▷

4.73 If M is a right module over a ring A, then
   M is distributive $\Longleftrightarrow$ for any $m, n \in M$, there exists a right ideal B of A such that $(m + n)A = mB + nB$.

◁ $\Longrightarrow$ Set $B \equiv (m : (m + n)A)$. Then $(m + n)A \bigcap mA = mB$. If $b \in B$, we have $nb = (m + n)b - mb \in (m + n)A$. Therefore $B = (n : (m + n)A)$ and $(m + n)A \bigcap nA = nB$. Hence $(m + n)A = (m + n)A \bigcap mA + (m + n)A \bigcap nA = mB + nB$.

$\Longleftarrow$ Let $m, n \in M$. By assumption, $(m+n)A = mB+nB$, where B is a right ideal of A. The modular law implies the equalities $mA \bigcap (m+n)A = mA \bigcap (mB + nB) = mB$, $nA \bigcap (m + n)A = nA \bigcap (mB + nB) = nB$, and $(m + n)A = mA \bigcap (m + n)A + nA \bigcap (m + n)A$. Hence M is distributive. ▷

4.74 A module M is a *multiplication* module if for each submodule N, there exists an ideal B of A such that $N = MB$ (this equality is equivalent to the equality $N = M(M : N)$).

If all 2-generated submodules M_A are multiplication modules, then M is distributive.

◁ Let $m, n \in M$. By assumption, there exists an ideal B of A such that $(m + n)A = (mA + nA)B$. Since B is an ideal, $(m + n)A = (mA + nA)B = mB + nB$. By 4.73, M is distributive. ▷

4.75 *For a module M_A over a right invariant ring A, the following conditions are equivalent.*

(1) M is distributive.

(2) All finitely generated submodules of M are multiplication modules.

(3) For each finitely generated submodule N of M and for an arbitrary finitely generated submodule F of N, there exists a finitely generated ideal B of A such that $F = NB$.

(4) All 2-generated submodules of M are multiplication modules.

◁ $(1) \Longrightarrow (2)$ Let n be a positive integer, N be a n-generated submodule of M, and let $T \in \text{Lat}(N)$. We have to prove that there exists an ideal B of A such that $T = NB$. We use the induction on n. It directly can be proved that each cyclic module over a right invariant ring is a multiplication module. Therefore, the assertion holds for $n = 1$. Assume that the assertion holds for $n < k$ and $N = F + G$, where module F has $k - 1$ generators, and G is cyclic. By induction, $F \bigcap G = F(F : G) = G(G : F)$. In addition, there exist two ideals D and E of A such that $T \bigcap F = FD$ and $T \bigcap G = GE$. Also, $A = (F : G) + (G : F)$ by 4.59. Hence $F = F(F : G) + F(G : F) = G(G : F) + F(G : F) = N(G : F)$. Analogously, we can prove the equality $G = N(F : G)$. Denote by B the ideal $(G : F)D + (F : G)E$ of A. Therefore $T = T \bigcap F + T \bigcap G = FD + GE = N(G : F)D + N(F : G)E = NB$, whence B is the required ideal.

$(2)\Longrightarrow(3)$ Let $N = \sum_{i=1}^{m} x_i A$, $F = \sum_{j=1}^{n} f_j$. By (2), there exist elements $b_{ij} \in (N : F)$ $(1 \le i \le m, 1 \le j \le n)$ such that $f_j = \sum_{i=1}^{m} x_i b_{ij}$ for any j. Hence the ideal B of A generated by all b_{ij} is the required finitely generated ideal.

$(3)\Longrightarrow(4)$ and $(2)\Longrightarrow(4)$ are directly verified. $(4)\Longrightarrow(1)$ follows from 4.74. $\triangleright$

4.76 *If M_A is a finitely generated distributive module over a right invariant ring A, then M is invariant.*

$\triangleleft$ Let $N \in \mathrm{Lat}(M)$, $f \in \mathrm{End}(M)$. By 4.75, there exists an ideal B of the ring A such that $N = MB$. Hence $f(N) = f(MB) \subseteq MB = N$. $\triangleright$

4.77 *If M is a Noetherian right module over a right invariant ring A, then*

M is distributive $\Longleftrightarrow$ all submodules of M are multiplication invariant modules.

$\triangleleft$ 4.77 follows from 4.75 and 4.76. $\triangleright$

4.78 *Let A be a right Noetherian ring A. Then*

A is right distributive $\Longleftrightarrow$ all right ideals of A are multiplication invariant right A-modules $\Longleftrightarrow$ all 2-generated right ideals of A are multiplication right A-modules.

$\triangleleft$ By 2.72, all right distributive right Noetherian rings are right invariant. 4.78 follows now from 4.74 and 4.77. $\triangleright$

4.79 *For a module M_A over a right invariant ring A, the following conditions are equivalent.*

(1) M is distributive.

(2) $A = (F : G) + (G : F)$ for any finitely generated $F, G \in \mathrm{Lat}(M)$.

(3) $L \bigcap N = L(L : N)$ for any finitely generated $L \in \mathrm{Lat}(M)$ and for each $N \in \mathrm{Lat}(M)$.

(4) $L(L : N) = N(N : L) = L \bigcap N$ for any finitely generated $L, N \in \mathrm{Lat}(M)$.

(5) $(L : (F + G)) = (L : F) + (L : G)$ for any finitely generated $F, G \in \mathrm{Lat}(M)$ and for each $L \in \mathrm{Lat}(M)$.

(6) $((F \bigcap G) : L) = (F : L) + (G : L)$ for any finitely generated $F, G \in \mathrm{Lat}(M)$ and for each $L \in \mathrm{Lat}(M)$.

$\triangleleft$ $(1)\Longleftrightarrow(2)$ is proved in 4.59. $(1)\Longleftrightarrow(3)\Longleftrightarrow(4)$ follow from 4.75.

$(5)\Longrightarrow(2)$ $A = ((F + G) : (F + G)) = ((F + G) : F) + ((F + G) : G) = (G : F) + (F : G)$.

$(6) \Longrightarrow (2)$ $A = ((F \cap G) : (F \cap G)) = (F : (F \cap G)) + (G : (F \cap G)) = (F : G) + (G : F)$.

$(2) \Longrightarrow (5)$ $(L : F) + (L : G) \subseteq (L : (F + G)) = (L : (F + G))((F : G) + (G : F)) = (L : (F + G))((F + G) : G) + ((F + G) : F)) \subseteq (L : G) + (L : F)$.

$(2) \Longrightarrow (6)$ $(F : L) + (G : L) \subseteq ((F \cap G) : L) = ((F : G) + (G : F))((F \cap G) : L) = ((F : (F \cap G)) + (G : (F \cap G)))((F \cap G) : L) \subseteq (F : L) + (G : L)$. $\triangleright$

4.80 If A is a distributive invariant ring and D is a finitely generated ideal of A containing a regular element d, then D_A and $_A D$ are projective modules.

$\triangleleft$ The invariant ring A has the classical ring of quotients Q. The regular element d is a unit of Q. By 4.75, there exists a finitely generated ideal B of A such that $DB = dA = Ad$. Let $B = \sum_{j=1}^{n} b_j A = \sum_{j=1}^{n} A b_j$, $d = \sum_{j=1}^{n} d_j b_j$ $(d_j \in D)$, and let $y_j \equiv b_j d^{-1} \in Q$, where $1 \leq j \leq n$. Then $Ad = DB$ and $1 = \sum_{j=1}^{n} a_j y_j$. The ideal B is finitely generated, and $Ad = DB \subseteq B$. By 4.75, there exists an ideal C such that $dA = BC$. Hence $Ad = BC = Bd^{-1}DBC = Bd^{-1}Dd$, $A = (Ad)d^{-1} = Bd^{-1}D \supseteq y_j D$, $\forall j$. By 3.32(1), D_A is projective. Analogously, $_A D$ is projective. $\triangleright$

4.81 (1) Let a distributive ring A be an order in a ring Q, and let M be an ideal of A such that $M = Aa_1 + Aa_2 = a_3 A + a_4 A$, where a_1, a_2, a_3, a_4 are regular elements of A.

Then M_A and $_A M$ are projective modules.

(2) Let A be a distributive invariant ring A, and let M be an ideal in A generated by two regular elements a_1, a_2 of A.

Then M_A and $_A M$ are projective modules.

$\triangleleft$ (1) There exist $b_1, b_2, t \in A$ such that t is a regular element, $a_1^{-1} = t^{-1} b_1$, and $a_2^{-1} = t^{-1} b_2$. By 2.4, there exist elements $f_1, f_2, h_1, h_2 \in A$ such that $1 = f_1 + f_2$, $b_1 f_1 = b_2 h_2$, $b_2 f_2 = b_1 h_1$. Then $a_2 a_1^{-1} f_1 = a_2 t^{-1} b_1 f_1 = a_2 t^{-1} b_2 h_2 = a_2 a_2^{-1} h_2 = h_2 \in A$, $a_1 a_2^{-1} f_2 = a_1 t^{-1} b_2 f_2 = a_1 t^{-1} b_1 h_1 = a_1 a_1^{-1} h_1 = h_1 \in A$. Let $C \equiv a_1^{-1} f_1 A + a_2^{-1} f_2 A \subseteq Q$. Then $MC = A f_1 A + A a_1 a_2^{-1} f_2 A + A f_2 A + A a_2 a_1^{-1} f_1 A \subseteq A$. Since $1 = f_1 + f_2 \in MC$, we obtain $A \subseteq MC \subseteq A, A = MC$. Analogously, there exists a submodule D of $_A Q$ such that $A = DM$. It can directly be verified that C and D are subbimodules of the bimodule $_A Q_A$. By 3.32(2), the modules M_A and $_A M$ are projective.

(2) Since A is an invariant ring, A is an order in a ring Q and $M = Aa_1 + Aa_2 = a_1 A + a_2 A$. By (1), M_A and $_A M$ are projective. $\triangleright$

4.82 Let A_1 and A_2 be two ideals of a ring A, and let x_1 and x_2 be elements of A such that $x_1 - x_2 \in A_1 + A_2$.

Then there exists $a \in A$ such that $a - x_1 \in A_1$ and $a - x_2 \in A_2$.

◁ Let $x_1 - x_2 = a_1 - a_2$, where $a_1 \in A_1, a_2 \in A_2$. Set $a \equiv x_1 - a_1 = x_2 - a_2$. Then $a - x_1 = -a_1 \in A_1$ and $a - x_2 = -a_2 \in A_2$. Hence a is the required element. ▷

4.83 For a ring A, the following conditions are equivalent.

(1) A is arithmetical.

(2) For any ideals $A_1, \ldots, A_n$ of A and for any $x_1, \ldots, x_n \in A$ such that $x_i - x_j \in A_i + A_j$ for all i, j, there exists $x \in A$ such that $x - x_i \in A_i$ for $i = 1, \ldots, n$.

(3) For any ideals A_1, A_2, and A_3 of A and for any $d \in (A_1 + A_2) \cap (A_1 + A_3)$, there exists $x \in A_1$ such that $x - d \in A_2 \cap A_3$.

◁ (1)$\Longrightarrow$(2) For $n = 2$, our assertion follows from 4.82. Assume that our assertion holds for $n - 1$. There exists $b \in A$ such that $b - x_i \in A_i$ for $i = 1, \ldots, n-1$. In addition, $x_i - x_n \in A_i + A_n$ for $i = 1, \ldots, n-1$ and $b - x_n = (b - x_i) + (x_i - x_n) \in \bigcap_{i=1}^{n-1}(A_i + A_n)$. Since A is arithmetical, $b - x_n \in A_n + \bigcap_{i=1}^{n-1} A_i$, $b - x_n = a_n - d$, where $a_n \in A_n, d \in \bigcap_{i=1}^{n-1} A_i$. Set $a \equiv b + d$. Then $a - x_n = an \in A_n$. In addition, $a - x_i = b - x_i + d \in A_i$ for $i = 1, \ldots, n - 1$. Hence a is the required element.

(2)$\Longrightarrow$(3) follows from the fact that (2) becomes (3) for $n = 3, x_1 = 0, x_2 = x_3 = d$.

(3)$\Longrightarrow$(1) It is sufficient to prove that if M, B, and C are three ideals of A and $d \in (M + B) \cap (M + C)$, then $d \in M + (B \cap C)$. Let $d = m_1 + b = m_2 + c$, where $m_1, m_2 \in M, b \in B$, and $c \in C$. By (3), there exists $x \in M$ such that $x - d \in B \cap C$. Set $y \equiv d - x \in B \cap C$. Then $d = x + y \in M + B \cap C$. ▷

4.5   Regular and countably injective rings

4.84 A right module M over a ring A is *countably injective* if for any countably generated right ideal B of A, every homomorphism $B_A \to M$ can be extended to a homomorphism $A_A \to M$.

A ring A is a *Baer* ring if the following equivalent conditions hold.

(i) The right annihilator of each subset B of A is generated (as a right ideal of A) by an idempotent.

(ii) The left annihilator of each subset B of A is generated (as a left ideal of A) by an idempotent.

(The equivalence of (i) and (ii) can directly be verified.)

(1) Let all countably generated right ideals of a ring A be projective A-modules, and let $\overline{M}$ be a homomorphic image of a countably injective module M_A.

Then the module $\overline{M}$ is countably injective.

(2) Let M be a nonsingular module, $\{N_i\}_{i \in I}$ be a set of closed submodules N_i of M, and let $N \equiv \bigcap_{i \in I} N_i$.

Then N is a closed submodule of M.

(3) If B is a subset of a right nonsingular ring A, then $r(B)$ is a closed right ideal of A.

(4) If each closed right ideal of a right nonsingular ring A is a direct summand of A_A, then A is a Baer ring.

(5) If A is a direct product of right self-injective rings A_i $(i \in I)$, then the ring A is right self-injective.

$\lhd$ (1) Let $h : M \to \overline{M}$ be an epimorphism, B be a countably generated right ideal of A, and let $\overline{f} \in \mathrm{Hom}(B_A, \overline{M})$. Since B_A is projective, there exists a homomorphism $f : B_A \to M$ such that $\overline{f} = hf$. Since M is countably injective, f can be extended to a homomorphism $g : A_A \to M$. The homomorphism $hg : A_A \to \overline{M}$ is an extension of $\overline{f}$. Therefore $\overline{M}$ is countably injective.

(2) There exists the natural monomorphism from the module M/N into the nonsingular module $\prod_{i \in I} M/N_i$. Since M/N is nonsingular, 1.30(1) shows us that N is closed in M.

(3) Let $b \in B$. Since the module bA_A is nonsingular and $bA \cong A/r(b)$, the module $(A/r(b))_A$ is nonsingular. By 1.30(1), $r(b)$ is a closed right ideal of A. Since $r(B) = \bigcap_{b \in B} r(b)$, (2) shows us that $r(B)$ is a closed right ideal.

(4) follows from (3).

(5) Every right self-injective ring A_i can be considered as an injective right A-module. The direct product B_A of all this injective A-modules is an injective A-module. In addition, $B_A \cong A_A$. $\rhd$

4.85 Let A be a regular ring.

(1) If B is a countably generated right ideal of A, then the module B_A is projective, and there exists a set $\{e_i\}_{i=1}^{\infty}$ of ortogonal idempotents e_i of A such that $B = \oplus_{i=1}^{\infty} e_i A$.

(2) If A is strongly regular, then every countably generated right (left) ideal of A is generated by a countable set of central orthogonal idempotents.

(3) Each homomorphic image of any countably injective right A-module is countably injective.

(4) If A is right countably injective, then every cyclic right A-module is countably injective, and every factor ring of A is right countably injective.

(5) If A is right self-injective, then A is a Baer ring.

(6) Let A be right self-injective, B be an ideal of A, and let e be an idempotent of A such that eA is an injective hull of B_A.

Then the idempotent e is central.

◁ (1) follows from [70, 2.14, 2.15]. (2) follows from (1). (3) follows from (1) and 4.84(1). (4) follows from (3). (5) follows from 4.84(4). (6) is proved in [70, 9.5(a)]. ▷

4.86 Let $\{e_i\}_{i \in I}$ be a set of central orthogonal idempotents of A, and let $B = \oplus_{i \in I} e_i A$ be an ideal of A. Then the following conditions are equivalent.

(1) Each homomorphism $f : B_A \to A_A$ can be extended to an endomorphism g of A_A.

(2) Each homomorphism $_A B \to {_A A}$ can be extended to an endomorphism of $_A A$.

(3) Each endomorphism f of B_A can be extended to an endomorphism g of A_A.

(4) Each endomorphism of $_A B$ can be extended to an endomorphism of $_A A$.

(5) Given any countable set $\{a_i\}_{i=0}^\infty$ of elements of A, there exists $b \in A$ such that $be_i = a_i e_i$ for all i.

◁ Since all idempotents e_i are central, condition (5) is symmetric. Therefore, it is sufficient to prove the equivalence of conditions (1), (3), and (5).

(1)$\Longrightarrow$(3) is directly verified.

(3)$\Longrightarrow$(5) Let f be the endomorphism of B_A defined by the rule: $f(\sum e_i x_i) = \sum a_i e_i x_i$ for $x_i \in A$. Since the central idempotents e_i are orthogonal, f is well defined. By assumption, f can be extended to a homomorphism $g : A_A \to A_A$. Set $b \equiv g(1)$. Then $be_i = a_i e_i$ for all i.

(5)$\Longrightarrow$(1) Let $f : B_A \to A_A$ be a homomorphism, and let $a_i \equiv f(e_i)$. Then $f(e_i) = a_i e_i$ for all i. By assumption, there exists $b \in A$ such that $be_i = a_i e_i$ for all i. The rule $g(x) = bx$ defines the required extension $g : A_A \to A_A$ of the homomorphism f. ▷

4.87 Assume that M_A is a module, and $L(a_{ij}, m_i, M_A)$ is a system of a countable number of linear equations $\{\sum_{j=0}^{t(i)} x_j a_{ij} = m_i\}_{i=0}^\infty$ with coefficients $a_{ij} \in A$ which contains a countable number of unknowns $\{x_j\}_{j=0}^\infty$, which take values in M, and any finite subsystem of this system has a solution. This system of equations is called a *countable finitely solvable system*.

A module M_A is $\aleph_0$-*algebraically compact if any countable finitely solvable system* $L(a_{ij}, m_i, M_A)$ *is solvable in* M.

M_A is an $\aleph_0$-algebraically compact module $\iff$ if $L \equiv L(a_{ij}, m_i, M_A)$ is an arbitrary countable finitely solvable system, then there exists $y_0 \in M$ such that every finite subsystem L^* of L has a particular solution $\overline{y^*} = \{y_0, y_1^*, \ldots\}$, where y_0 depends on the system L only.

◁ We have to prove only implication $\Longleftarrow$, since the inverse implication is obvious. It is necessary to find a solution $\overline{y} = \{y_0, y_1, \ldots\}$ of L. We shall prove by induction the existence of elements $y_n \in M$. The element y_0 exists. Assume that $y_0, \ldots, y_n$ have been found. Consider a system of equations $N \equiv \{\sum_{j=n+1}^{t(i)} x_j a_{ij} = m_i - \sum_{j=0}^{n} y_j a_{ij}\}_{i=0}^{\infty}$. By the induction hypothesis applied to the system N and the unknown x_{n+1}, there exists $y_{n+1} \in M$ such that any finite subsystem N^* of N has a solution $\overline{z^*} \equiv \{y_{n+1}, y_{n+2}^*, \ldots\}$, where y_{n+1} depends on the system N, but y_{n+1} is not changed when we pass to another finite subsystem of N. Let us prove that $\overline{y} \equiv \{y_0, y_1, \ldots\}$ is a desired solution of L. It is sufficient to prove that $\overline{y}$ is a solution of an arbitrary finite subsystem L^* of L. The subsystem L^* contains only a finite number of unknowns $x_0, \ldots, x_n$. It follows from the method of choice of the sequence $\{y_m\}_{m=0}^{\infty}$ that there exists a solution $\overline{y}^*$ of L^* such that elements $y_0, \ldots, y_n$ of M are the first $n+1$ components of this solution. Therefore $\overline{y}$ is a solution of the system L^*. ▷

4.88 *For a strongly regular ring A, the following conditions are equivalent.*

(1) A is right countably injective.

(2) A is left countably injective.

(3) A is a right $\aleph_0$-algebraically compact ring.

(4) A is a left $\aleph_0$-algebraically compact ring.

(5) Each factor ring $\overline{A}$ of A is a countably injective $\aleph_0$-algebraically compact ring, and each cyclic (right or left) $\overline{A}$-module is countably injective.

(6) For each countable set $\{e_i\}_{i=0}^{\infty}$ of central orthogonal idempotents of A and for each countable set $\{a_i\}_{i=0}^{\infty}$ of elements of A, there exists $b \in A$ such that $be_i = a_i e_i$ for all i.

◁ All idempotents of A are central. Since condition (6) is symmetric, it is sufficient to prove the equivalence of conditions (1), (3), (5), and (6).

(5)$\Longrightarrow$(1) is obvious. (1)$\Longleftrightarrow$(6) follows from 4.86 and 4.85(2).

(3)$\Longrightarrow$(6) It is sufficient to prove that the countable system of equations $L \equiv \{xe_i = a_i e_i\}_{i=0}^{\infty}$ is finitely solvable, which is equivalent to the fact that the finite system $L_n \equiv \{xe_i = a_i e_i\}_{i=0}^{n}$ is solvable for any

n. Since idempotents e_i are orthogonal, the element $\sum_{i=0}^{n} a_i e_i$ is a particular solution of L_n.

$(6)\Longrightarrow(3)$ We denote by L a countable finitely solvable system $\{\sum_{j=0}^{t(i)} x_j a_{ij} = m_i\}_{i=0}^{\infty}$ such that $a_{ij} \in A, m_i \in A$, and for any $n \geq 0$, a finite subsystem $L_n \equiv \{\sum_{j=0}^{t(i)} x_j a_{ij} = m_i\}_{i=0}^{n}$ of L has a particular solution $\overline{x(n)} = \{x_j(n)\}_{j=0}^{m(n)}$, where $m(n) = \max_{0 \leq i \leq n}\{t(i)\}$. By 4.87, it is sufficient to find a set of particular solutions $\{\overline{y(n)}_{n=0}^{\infty}\}$ of L_n such that $\overline{y(n)} = \{b, y_1(n), y_2(n), \ldots\}$, where b depends on L only and is not changed with change of n. We fix the number n. Let $H_n \equiv \{\sum_{j=0}^{t(i)} x_j a_{ij} = 0\}_{i=0}^{n}$ be a homogeneous finite system of equations corresponding to the nonhomogeneous system L_n, G_n be the set of all solutions of the homogeneous system H_n, and let T_n be the left ideal of A generated by all the first components of the set G_n. Then G_n is the kernel of an endomorphism h of a free left module $_AF$ of rank $m(n)$. Since the finitely generated free module $_AF$ over the regular ring A is regular [70, 1.11], $h(F)$ is a direct summand of the module F. Therefore, the module $h(F)$ is finitely generated. The module G_n is finitely generated, since G_n is isomorphic to a direct summand of the finitely generated module F. Therefore, the left ideal T_n of the strongly regular ring A is finitely generated as a homomorphic image of G_n. Then $A = T_n \oplus Af_n$, where f_n is a central idempotent. Let us begin to change n. Since $T_n \supseteq T_{n+1}$, there exists a central idempotent e_{n+1} such that $T_n = T_{n+1} \oplus Ae_{n+1}$, $f_n = f_{n+1} + e_{n+1}$. Set $e_0 \equiv f_0$. Then $\{e_i\}_{i=0}^{\infty}$ is a countable set of central orthogonal idempotents and $f_n = \sum_{i=0}^{n} e_i$ for $n \geq 0$. Denote by a_i elements $x_0(i) \in A$. For $i \leq n$, we have $a_i - a_n \in T_i$, since $\overline{x(i)} - \overline{x(n)} \in H_i$. Therefore $a_i f_i = a_n f_i$ for $i \leq n$. By condition (6), there exists $b \in A$ such that $be_i = a_i e_i$ for all i. Then $(b - a_n)f_n = \sum_{i=0}^{n}(b - a_n)e_i = \sum_{i=0}^{n}(b - a_i)e_i = 0$, $(b - a_n)(1 - f_n) = b - a_n$, $b - a_n \in T_n$. Hence for all i, there exists a solution $\overline{z(n)} = \{z_j(n)\}_{j=0}^{m(n)}$ of the homogeneous system H_n such that $b = a_n + z_0(n)$. Let $\overline{y(n)}$ be the sum of the solution $\overline{x(n)}$ of the nonhomogeneous system L_n and the solution $\overline{z(n)}$ of the homogeneous system H_n. Then $\overline{y(n)}$ is a solution of the nonhomogeneous system L_n and $\overline{y(n)} = \{b, y_1(n), y_2(n), \ldots\}$.

$(6)\Longrightarrow(5)$ By 4.86 and 4.85(2), A is countably injective. By 4.85(4), $\overline{A}$ is countably injective. Since $(1)\Longrightarrow(3)$ is proved, $\overline{A}$ is a countably algebraically compact ring. By 4.85(4), each cyclic right $\overline{A}$-module is countably injective. $\triangleright$

4.89 Let $\mathcal{N}$ be the set of all positive integers, A be a ring containing the infinite countable set $\{e_i\}_{i=1}^{\infty}$ of nonzero central orthogonal idempotents e_i, $T \equiv \oplus_{i=1}^{\infty} e_i A$, and let $\overline{A} \equiv A/T$. Assume that for any proper subset M of $\mathcal{N}$, there exists a central idempotent $w_M \in A$ such that $w_M e_i = e_i$ for $i \in M$ and $w_M e_i = 0$ for $i \in \mathcal{N} \setminus M$.

(1) $\overline{A}$ has a countably generated ideal $\overline{B}$ such that the ideal $r_{\overline{A}}(\overline{B})$ is not generated by an idempotent.

(2) $\overline{A}$ is not a Baer ring.

(3) $\overline{A}$ is not a regular right self-injective ring.

◁ (1) For each $a \in A$, denote by $d(a)$ the set $\{i \in \mathcal{N} \mid e_i(a) \neq 0\}$. Represent $\mathcal{N}$ as the union $\mathcal{N} = \bigcup \mathcal{N}_j$ of an infinite number of infinite disjoint sets N_j. By assumption, for any positive integer j, there exists a central idempotent w_j of A such that $e_i w_j = e_i$ for $i \in N_j$ and $e_i w_j = 0$ for $i \in \mathcal{N} \setminus N_j$. Let $h : A \to \overline{A}$ be the natural epimorphism. For each subset $X \subseteq A$, we write $\overline{X}$ instead of $h(X)$. Let B be the ideal of A generated by the countable set $\{w_j\}$ of idempotents, F be the ideal of A which is the full inverse image of the ideal $r_{\overline{A}}(\overline{B})$ of $\overline{A}$. Then $F = \{a \in A \mid |d(a) \cap N_j| < \infty, \forall j \in \mathcal{N}\}$. Assume that $\overline{F} = \overline{f}\overline{A}$, $f \in F$, and $\overline{f} = (\overline{f})^2 \in \overline{F}$. Since $f \in F$, the set $d(f) \cap N_j$ is finite for any positive integer j. Therefore, we can choose a number $n_j \in N_j \setminus (d(f) \cap N_j)$. All numbers n_j are distinct, since the sets N_j are disjoint. Denote by t the element of A such that $e_i(t) = e_i(1)$ if i coincides with one of the numbers n_j and $e_i(t) = 0$ otherwise. For each N_j, the set $d(t) \cap N_j$ contains exactly one element. Therefore $t \in F$. Hence $\overline{f}\overline{t} = \overline{t}$. Therefore, the set $d(ft - t)$ is finite. Since all the numbers n_j are contained in $d(t)$ and are not contained in $d(f)$, the set $d(ft - t)$ contains infinitely many n_j. Therefore $d(ft - t)$ is infinite; we have a contradiction. Hence the ideal $\overline{F}$ of $\overline{A}$ is not generated by an idempotent.

(2) follows from (1). (3) follows from (2) and 4.85(5). ▷

4.90 (1) Let A be a ring, $\{A_i\}_{i=1}^{\infty}$ be an infinite countable set of nonzero rings A_i. Assume that either $A = \prod_{i=1}^{\infty} A_i$ or A is a regular right self-injective ring containing an infinite countable set $\{e_i\}_{i=1}^{\infty}$ of nonzero central orthogonal idempotents e_i.

Then there exists an infinite countable set of nonzero central orthogonal idempotents $\{t_i\}_{i=1}^{\infty}$ of A such that the factor ring $D \equiv A/(\oplus_{i=1}^{\infty} t_i A)$ has a countably generated ideal B with the ideal $r_D(B)$ is not generated by an idempotent, D is not a Baer ring, D is not a regular right self-injective ring.

(2) Let A be a direct product of an infinite countable set of strongly regular rings A_i, and let D be the factor ring $A/(\oplus_{i=1}^{\infty} A_i)$.

Then A and D are strongly regular rings, and D is not right self-injective.

◁ Let $\mathcal{N}$ be the set of all positive integers. By 4.89, it is sufficient to prove that A contains an infinite countable set $\{e_i\}_{i\in\mathcal{N}}$ of nonzero central orthogonal idempotents e_i such that for any proper subset M of $\mathcal{N}$, there exists a central idempotent $w_M \in A$ such that $w_M e_i = e_i$ for $i \in M$, and $w_M e_i = 0$ for $i \in \mathcal{N} \setminus M$.

Let A be a direct product of an infinite countable set $\{A_i\}_{i=1}^{\infty}$ of nonzero rings A_i, and let e_i be the identity elements of A_i. For each proper subset M of $\mathcal{N}$, denote by w_M the central idempotent of A corresponding to the natural projection of A onto $\prod_{i\in M} A_i$. Then $w_M e_i = e_i$ for $i \in M$ and $w_M e_i = 0$ for $i \in \mathcal{N} \setminus M$.

Let A be a regular right self-injective ring containing an infinite countable set $\{e_i\}_{i=1}^{\infty}$ of nonzero central orthogonal idempotents e_i, and let M be any proper subset of $\mathcal{N}$. Since A is right self-injective, there exists an idempotent $w_M \in A$ such that $w_M A$ is an essential extension of $(\oplus_{i\in M} e_i A)$, and $(1-w_M)A$ is an essential extension of $(\oplus_{i\in\mathcal{N}\setminus M} e_i A)$. Hence $w_M e_i = e_i$ for $i \in M$ and $w_M e_i = 0$ for $i \in \mathcal{N} \setminus M$. Since all idempotents e_i are central, $\oplus_{i\in M} e_i A$ is an ideal of A. By 4.85(6), the idempotent w_M is central.

(2) It is obvious that A is strongly regular. Hence D is strongly regular. In particular, A is nonsingular. The assertion follows now from 4.90(1). ▷

4.91 *If A is an invariant arithmetical ring and M is a finitely generated ideal of A, then each endomorphism f of $_A M$ can be extended to an endomorphism of $_A A$.*

◁ Let $M = \sum_{i=1}^{n} Am_i$. Since $_A M$ is a finitely generated distributive module over the invariant ring A, 4.76 shows us that $f(Am_i) \subseteq Am_i$ for all i. Hence there exist $x_1,\ldots,x_n \in A$ such that $f(m_i) = m_i x_i \in m_i A = Am_i$. Then $0 = (Am_i \cap Am_j)(x_i - x_j)$ for all i,j. Since $Am_i = m_i A$ and $Am_j = m_j A$, $x_i - x_j \equiv d_{ij} \in r(m_i A \cap m_j A)$. By 4.59, there exist $a_{ij}, b_{ij} \in A$ such that $1 = a_{ij} + b_{ij}$ and $m_i A a_{ij} + m_j A b_{ij} \subseteq m_i A \cap m_j A$. Hence $a_{ij} d_{ij} \in r(m_i A) \equiv A_i$, $b_{ij} \in r(m_j A) \equiv A_j$, and $x_i - x_j = a_{ij} d_{ij} + b_{ij} d_{ij} \in A_i + A_j$. By 4.83, there exists $x \in A$ such that $x - x_i \in A_i$ for all i. Hence by the rule $g(y) = yx$ (for $y \in A$), the endomorphism $g \in \mathrm{End}(_A A)$ extending f is well defined. ▷

Example 4.92 Let $\mathbf{Q}$ be the field of rational numbers, and let A be the ring of all rational numbers with odd denominators.

Then $\mathbf{Q}$ is a uniserial A-module with Krull dimension 1, and $\mathbf{Q}$ is not an invariant A-module.

Chapter 5

Rings related to commutative

5.1 Bezout rings

5.1 *If A is a right Bezout ring, then*
A is right distributive $\iff$ A is right quasi-invariant.

$\lhd$ 5.1 follows from 2.35 and 2.54. $\rhd$

5.2 A *Dedekind-finite* ring A is any ring such that every right or left invertible element of A is a unit.

Every left or right quasi-invariant ring A is Dedekind-finite.

$\lhd$ Assume that A is left quasi-invariant, $n, m \in A$, $nm = 1$, and n is not left invertible. Then $n \in M \in \max({}_A A)$. Since M is an ideal, $1 = nm \in M$; we have a contradiction. $\rhd$

5.3 Let A be a right Bezout ring, and let A be either left quasi-invariant or left distributive.

Then A is a right distributive right quasi-invariant ring.

$\lhd$ By 5.1, we may assume that A is left quasi-invariant. By 5.1, it is sufficient to prove that any maximal right ideal M of A is an ideal. Assume the contrary. Then $A = AM$ and $1 = \sum_{i=1}^{n} a_i m_i$, where $a_i \in A$ and $m_i \in M$. Since A is a right Bezout ring, there exist $b_1, \ldots, b_n, d_1, \ldots, d_n \in A$ such that $\sum_{i=1}^{n} m_i d_i \equiv m \in M$, $m_i = m b_i$ for all i. Since $m \in M$, $mA \neq A$. By 5.2, $Am \neq A$. Hence m is contained in some maximal left ideal B. By assumption, B is an ideal. Hence $1 = \sum_{i=1}^{n} a_i m b_i \in B$; we have a contradiction. $\rhd$

5.4 *For a Bezout ring A, the following conditions are equivalent.*
(1) *A is right distributive.*
(2) *A is left distributive.*
(3) *A is right quasi-invariant.*
(4) *A is left quasi-invariant.*

◁ (1) $\Longleftrightarrow$ (3) and (2) $\Longleftrightarrow$ (4) follow from 5.1. (4)$\Longrightarrow$(1) and (3)$\Longrightarrow$(2) follow from 5.3. ▷

5.5 Let M be a rectangular $m \times n$-matrix with elements in a ring A, $k = \min(m,n)$. We say that M *admits a diagonal reduction* if there exist two invertible matrices U and V such that $UMV = D$, where D is a diagonal $m \times n$-matrix and $Ad_{ii} \bigcap d_{ii}A \supseteq Ad_{i+1,i+1}$ for $i = 1, \ldots, k-1$.

If every rectangular matrix over A admits a diagonal reduction, A is called an *elementary divisor ring*.

A ring A is *right Hermitian* if the following equivalent conditions hold.
(1) Given any $a, b \in A$, the row (a,b) admits a diagonal reduction.
(2) Given any $a, b \in A$, there exists an invertible 2×2-matrix V (with elements in A) such that $(a,b)V = (d,0)$.

Analogously if each 2×1-matrix, admits a diagonal reduction, then A is a *left Hermitian* ring.

The right and left Hermitian ring is called a *Hermitian* ring.

5.6 Every right Hermitian ring A is a right Bezout ring.

Therefore, any Hermitian ring (in particular, any elementary divisor ring) is a Bezout ring.

◁ Let $a, b \in A$. By assumption, there exists an invertible 2×2-matrix V (with elements in A) such that $(a,b)V = (d,0)$, $(d,0)V^{-1} = (a,b)$. Therefore $aA + bA = dA$. ▷

5.7 If a and b are two elements of a right distributive ring A with $\mathrm{Id}(a) = bA \supseteq Ab$, then $\mathrm{Id}(a) = aA$.

◁ Assume that aA is a proper submodule of a cyclic module bA. There exists a maximal submodule M of bA such that $aA \subseteq M$. Since bA is an ideal of A and M is a fully invariant submodule of the distributive module bA by 2.54, M is an ideal of A. Hence $\mathrm{Id}(a) \subseteq M \neq bA = \mathrm{Id}(a)$; this is a contradiction. ▷

5.8 For a ring A, the following conditions are equivalent.
(1) A is right quasi-invariant.

(2) For any $a_1, \ldots, a_n \in A$, the equality $A = \sum_{i=1}^{n} Aa_i$ implies the equality $A = \sum_{i=1}^{n} a_i A$.

(3) For any $a, b \in A$, the equality $Aa + Ab = A$ implies the equality $aA + bA = A$.

$\triangleleft$ $(1)\Longrightarrow(2)$ Assume that $A \neq \sum_{i=1}^{n} a_i A$. Then the proper right ideal $\sum_{i=1}^{n} a_i A$ is contained in a maximal right ideal M. Since M is an ideal, $A = \sum_{i=1}^{n} Aa_i \subseteq M$; we have a contradiction.

$(2)\Longrightarrow(3)$ is directly verified.

$(3)\Longrightarrow(1)$ Assume that there exists a maximal right ideal M which is not an ideal. There exist $a \in A$ and $x, y \in M$ such that $ax + y = 1$. Hence $Ax + Ay = A$. By assumption, $xA + yA = A$. Therefore $A = M$; this is a contradiction. $\triangleright$

5.9 *For an elementary divisor ring A, the following conditions are equivalent.*

(1) *A is right quasi-invariant.*

(2) *A is left quasi-invariant.*

(3) *A is right distributive.*

(4) *A is left distributive.*

(5) *A is invariant.*

$\triangleleft$ By 5.6, A is a Bezout ring. Therefore, the equivalence of conditions (1), (2), (3), and (4) follows from 5.4. $(5)\Longrightarrow(3)$ and $(5)\Longrightarrow(4)$ are directly verified.

$((3)+(4))\Longrightarrow(5)$ Let $a \in A$. Since the matrix $\begin{pmatrix} a & 0 \\ 0 & a \end{pmatrix}$ admits a diagonal reduction, there exist two invertible 2×2-matrices $P = (p_{ij})$ and $Q = (q_{ij})$ such that

$$\begin{pmatrix} a & 0 \\ 0 & a \end{pmatrix} P = Q \begin{pmatrix} b & 0 \\ 0 & c \end{pmatrix}$$

, where $b, c \in A$ and $Ab \cap bA \supseteq \mathrm{Id}(c)$. It can directly be verified that $\mathrm{Id}(a) = \mathrm{Id}(c)$. By equality (1), we have equalities (*): $ap_{12} = q_{12}c, ap_{22} = q_{22}c$. Since P is an invertible matrix, $Ap_{12} + Ap_{22} = A$. By 5.8, $p_{12}A + p_{22}A = A$. Therefore, there exist $u, v \in A$ such that $p_{12}u + p_{22}v = 1$. Hence equalities (*) imply equalities $a = ap_{12}u + ap_{22}v = q_{12}cu + q_{22}cv \in \mathrm{Id}(c)$. Therefore $\mathrm{Id}(a) \subseteq \mathrm{Id}(c)$. Since $\mathrm{Id}(c) \subseteq Ab \cap bA \subseteq \mathrm{Id}(b) = \mathrm{Id}(a)$, we obtain $\mathrm{Id}(a) = \mathrm{Id}(c)$. Inclusions $\mathrm{Id}(c) \subseteq bA \subseteq \mathrm{Id}(b)$ and $\mathrm{Id}(c) \subseteq Ab \subseteq \mathrm{Id}(b)$ imply equalities $\mathrm{Id}(a) = bA = Ab$. By 5.7, $aA = \mathrm{Id}(a)$ is an ideal of A. Therefore A is right invariant. Analogously, A is left invariant. $\triangleright$

5.10 Let A be a domain A which is algebraic over its centre $C(A)$.

(1) For each nonzero element $a \in A$, there exist $b \in A$ and $t \in C(A)$ such that $ab = ba = t \neq 0$.

(2) Every nonzero right or left ideal of A contains a nonzero ideal generated by a central element.

(3) A is a uniform domain.

(4) If A is right or left primitive, then A is a division ring.

$\lhd$ (1) There exist central elements $t_0, \ldots, t_n$ and nonnegative integer m such that $t_0 \neq 0, t_n \neq 0$ and $a^m \sum_{i=0}^{n} a^i t_i = 0$. We can set $t \equiv -t_0$ and $b \equiv \sum_{i=1}^{n} a^{i-1} t_i$.

(2) follows from (1). (3) and (4) follow from (2). $\rhd$

5.11 Let A be a ring which is integral over its centre.

(1) All factor rings of A are integral over their centres.

(2) If all prime ideals of A are completely prime ideals, then any right or left primitive factor ring R of A is a division ring, and therefore A is quasi-invariant.

$\lhd$ (1) is directly verified.

(2) The ring R is prime. By assumption, R is a domain. By (1) and 5.10(4), R is a division ring. $\rhd$

5.12 *For a right Bezout ring A which is integral over its centre, the following conditions are equivalent.*

(1) *A is right distributive.*

(2) *A is right quasi-invariant.*

(3) *A is left quasi-invariant.*

(4) *A is right invariant.*

(5) *All prime ideals of A are completely prime.*

$\lhd$ (1)$\Longleftrightarrow$(2) foolows from 5.1. (1)$\Longrightarrow$(4) follows from 2.62. (4)$\Longrightarrow$(5) is directly verified. (5)$\Longrightarrow$(3) follows from 5.11(2). (3)$\Longrightarrow$(2) follows from 5.3. $\rhd$

5.13 *Let A be a Bezout ring integral over its centre. Then A is right invariant $\Longleftrightarrow$ A is left invariant.*

$\lhd$ 5.13 follows from 5.4 and 5.12. $\rhd$

5.14 If M is a module over a strongly regular ring A, then M is distributive $\Longleftrightarrow$ M is a Bezout module.

$\triangleleft \Longleftarrow$ follows from 2.35 and from the fact that strongly regular rings are invariant.

$\Longrightarrow$ Let $m, n \in M$, and let $N \equiv mA + nA \in \mathrm{Lat}(M)$. It is sufficient to prove that N is cyclic. By 2.4, there exist two elements a and b in A such that $1 = a + b$, $ma \in nA$, and $nb \in mA$. There exists a central idempotent e of the strongly regular ring A such that $aA = eA$. Then $me \in neA$, $n(1 - e) = (na + nb)(1 - e) = nb(1 - e) \in m(1 - e)A$. Hence $N = Ne \oplus N(1 - e) = meA + neA + m(1 - e)A + n(1 - e)A = (ne + m(1 - e))A$. $\triangleright$

5.15 *Let M_A be a module over a ring A, and let the factor ring $A/J(A)$ be strongly regular. Then*

M is distributive $\Longleftrightarrow$ M is a Bezout module.

$\triangleleft \Longleftarrow$ follows from 2.36 and from the fact that all strongly regular rings are invariant.

$\Longrightarrow$ Let $J \equiv J(A)$, and let N be a 2-generated submodule of M. Since N/NJ is a distributive module over the strongly regular ring A/J, N/NJ is a cyclic A/J-module by 5.14. By 2.25, M is a Bezout module. $\triangleright$

5.16 (1) *A is a right distributive ring, and the factor ring $A/J(A)$ is regular $\Longleftrightarrow$ A is a right Bezout ring, and $A/J(A)$ is strongly regular.*

(2) *A is a right distributive semilocal ring $\Longleftrightarrow$ A is a right Bezout ring, and $A/J(A)$ is a finite direct product of division rings.*

In this situation, all cyclic right A-modules are completely finite-dimensional.

$\triangleleft$ Since strongly regular rings coincide with normal regular rings, (1) follows from 5.15 and 2.50(3). (2) follows from (1) and 2.29. $\triangleright$

5.2 Rings algebraic over their centre

5.17 For a reduced right distributive ring A, the following conditions are equivalent.

(1) A is left distributive.

(2) For each maximal left ideal M, the left ring of quotients ${}_M A$ exists and is a right and left uniserial domain.

(3) For each maximal left ideal M, the left ring of quotients ${}_M A$ exists and is left finite-dimensional.

(4) A is left localizable, and for every minimal prime ideal H of A, the factor ring A/H is left finite-dimensional.

(5) For every minimal prime ideal H of A, the factor ring A/H is a left localizable left finite-dimensional ring.

(6) For every minimal prime ideal H of A, the factor ring A/H is left distributive.

◁ By 1.35(7), the factor ring A/H is a domain for every minimal prime ideal H of A.

$(2)\Longrightarrow(1)$ follows from 4.52. $(1)\Longrightarrow(6)$ is obvious. $(6)\Longrightarrow(5)$ follows from left-side analogs of 3.12(2) and 4.52 applied to A/H.

$(5)\Longrightarrow(4)$ Let M be a maximal left ideal. By 1.35(8), M contains a minimal prime ideal H. Let $h : A \to A/H$ be the natural epimorphism. By 1.35(7), A/H is a domain. By assumption, A/H has the left ring of quotients $_{h(M)}h(A) \equiv Q$. Since $h(A)$ is a domain, the canonical homomorphism $f : h(A) \to Q$ is a monomorphism. By the left-side analog of 4.52 (see condition (7)), $h(M)$ is a completely prime ideal in $h(A)$. Hence M is a completely prime ideal of A. By 4.52, $H = \{a \in A \,|\, ta = 0 \text{ for some } t \in A \setminus M\}$. Therefore Q is the left ring of quotients of A with respect to the set $A \setminus M$, and the homomorphism $g \equiv fh : A \to Q$ is the canonical homomorphism.

$(4)\Longrightarrow(3)$ Let M be a maximal left ideal. By the left-side analog of 4.52 (see condition (7)), M is a completely prime ideal. By assumption and by 4.52, there exist the canonical homomorphisms $f : A \to {}_MA$ and $g : A \to A_M$. Set $H \equiv \mathrm{Ker}(g)$. By 4.52, $H = \{a \in A \,|\, ta = 0$ for some $t \in A \setminus M\}$. Therefore $H = \mathrm{Ker}(f)$. By assumption, A/H is left finite-dimensional. By the left-side analog of 1.55(3), $_MA$ is left finite-dimensional.

$(3)\Longrightarrow(2)$ Let M be a maximal left ideal. By the left-side analog of 1.43(11), M is a completely prime ideal. By 4.52, the ring A_M exists and is a right uniserial domain. By assumption, the ring $_MA$ exists and is left finite-dimensional. By 1.50(3), $_MA \cong A_M$. Therefore $_MA$ is a left finite-dimensional right uniserial domain. By 3.26, $_MA$ is a left uniserial domain. ▷

5.18 Let A be a right quasi-invariant ring, and let for any $x \in A$, there exists a positive integer n such that $x^n A \subseteq Ax$.

(1) Any maximal left ideal M of A is a completely prime ideal, A/N is a left uniform domain, and for any completely prime ideal N of A, the set $A \setminus N$ is a left permutable multiplicative set.

(2) Assume that either all square-zero elements of A are central, or A is a ring with the maximum condition on right annihilators, or for any $a \in A$, there exists a positive integer $n = n(a)$ such that $r(a^n) = r(a^{n+1})$.

Then A is left localizable, and for any completely prime ideal N of A, the left ring of quotients $_NA$ exists.

◁ (1) Assume that M is not an ideal. Then $MA = A$ and $1 = \sum_{i=1}^{n} m_i a_i$, where $m_i \in M$ and $a_i \in A$. By assumption, there exists a positive integer d such that $m_i^d \in Am_i \subseteq M$ for any i. Set $H \equiv \sum_{i=1}^{n} m_i^d A$. Then $H \neq A$. Therefore H is contained in some maximal right ideal B. By assumption, B is an ideal. Hence A/B is a division ring, and the inclusions $m_i^d \in H \subseteq B$ imply the inclusions $m_i \in B$ for $i = 1, \ldots, n$. Therefore $1 = \sum_{i=1}^{n} m_i a_i \in B$; we have a contradiction. Hence M is an ideal, A/M is a division ring, and M is a completely prime ideal. Let $T \equiv A \backslash N$, $t \in T$, and let $a \in A$. By assumption, there exists $b \in A$ such that $t^n a = bt$. Since N is a completely prime ideal, the set T is multiplicative, and $t^n \in T$. Therefore T is left permutable. If $a \in T$, then $b \in T$. Hence A/N is a left uniform domain.

(2) follows from (1) and the left-side analog of 1.51. ▷

5.19 Let A be a right distributive reduced ring.

(1) If for any $a \in A$, there exists a positive integer n such that $a^n A \subseteq Aa$, then A is left distributive.

(2) Assume that for any $a \in A$, there exists a positive integer n such that $h(a^n A) \subseteq h(Aa)$ for an arbitrary minimal prime ideal H of A and for the natural epimorphism $h : A \to A/H$.

Then A is left distributive, and $a^n A \subseteq Aa$ for all $a \in A$.

◁ (1) By 2.54, A is right quasi-invariant. By 5.18(2), A is left localizable. By 1.35(7), for every minimal prime ideal H of A, the ring $A/H \equiv R$ is a domain. By 5.18(1), the domain R is left uniform. By 5.18(2), R is left localizable. By 5.17, A is left distributive.

(2) Let H be an arbitrary minimal prime ideal of A, and let $A/H \equiv R$. By 1.35(7), R is a domain. By (1), the domain R is left distributive. By 5.17, A is left distributive. By the left-side analog of 4.52, for each maximal left ideal M of A, there exists the left ring of quotients $_M A$ which is a left uniserial domain, and kernel H of the canonical homomorphism $f : A \to {_M A}$ is the only minimal prime ideal which is contained in M. Let $a, b \in A$, and let $h : A \to A/H$ be a natural epimorphism. By assumption, there exists a positive integer n such that $h(a^n b) \in h(Aa)$. Hence $f(a^n b) \in f(Aa)$. By the left-side analog of 1.60(1), $a^n b \in Aa$. ▷

5.20 Let A be a reduced ring.

(1) Assume that for any $a \in A$, there exists a positive integer n such that $a^n A = Aa^n$. Then

A is right distributive $\iff$ A is left distributive.

(2) Assume that A is right or left distributive, and for any $a \in A$, there exists a positive integer n such that $h(a^n A) = h(Aa^n)$ for every

minimal prime ideal H of A and for a natural epimorphism $h : A \to A/H$.

Then A is a right and left distributive ring, and $a^n A = Aa^n$.

$\triangleleft$ (1) and (2) follow from 5.19(1) and 5.19(2), respectively. $\triangleright$

5.21 Let A be a right uniserial ring, and let $d_0, \ldots, d_m$ be nonzero elements of A such that $\sum_{i=0}^{m} d_i = 0$.

Then $d_i A = d_j A$ for some $i \neq j$.

$\triangleleft$ We may assume that d_i are indexed as follows: $d_0 A \supseteq d_1 A \supseteq \ldots \supseteq d_m A$. Therefore, there exist $b_1, \ldots, b_m \in A$ such that $d_1 = d_0 b_1, \ldots, d_m = d_0 b_m$. Set $b \equiv \sum_{i=1}^{m} b_i$. Then $d_0(1 + b) = 0$. Assume that $d_0 A$ is properly contained in $d_1 A$. Therefore $b_1, \ldots, b_m \in J(A)$, $b \in J(A)$, and $1 + b \in U(A)$. Hence the equality $d_0(1 + b) = 0$ implies the equality $d_0 = 0$; we have a contradiction. Therefore $d_0 A = d_1 A$. $\triangleright$

5.22 Let A be a right distributive domain, and let $\sum_{i=0}^{n} x^i a_i = 0$, where $0 \neq x \in A$, $a_i \in C(A)$, and $a_n \neq 0$.

Then $x^m A = Ax^m$, where $m \equiv n!$.

$\triangleleft$ Set $F \equiv \max(A_A)$. By 3.23(1), A is a right order in a division ring Q, and for all $M \in F$, there exists a right uniserial subring A_M of the division ring Q such that A_M contains A, A_M is the right ring of quotients of A with respect to M, and the natural embedding $A \to A_M$ is the canonical homomorphism. By 1.60(5) and 3.23(5), for any $y \in A$, the following assertion (*) holds: yA is an ideal of $A \iff yA_M$ is an ideal of A_M for all $M \in F$; Ay is an ideal of $A \iff A_M y$ is an ideal of A_M for any $M \in F$. Let $M \in F$. We have $\sum_{i=0}^{n} x^i a_i = 0$. By 5.21 applied to the right uniserial domain A_M, there exist subscripts i and j such that $0 \leq i < j \leq n$, $a_i \neq 0$, $a_j \neq 0$, and $x^i a_i = x^j a_j$. Let $m \equiv n!$, $d \equiv m/(j - i)$, and let $t \equiv (a_j^{-1} a_i)^d \in C(A_M)$. Then $x^{j-i} A_M = a_j^{-1} a_i A_M$, $x^m A_M = t A_M = A_M t = A_M x^m$. The assertion follows now from (*) applied to $y \equiv x^m$. $\triangleright$

5.23 Let A be a right distributive ring which is algebraic over its centre, B be the ring of quotients of A with respect to the set T of all central regular elements, N be the prime radical of A, and let $A/N \equiv R$.

(1) B is a right distributive right invariant ring integral over its centre, the module B_A is distributive, and A contains all idempotents and nilpotent elements of B.

(2) If A is prime, then A is a distributive domain.

◁ (1) By 4.45(2), B is right distributive, and the module B_A is distributive. By 2.61(4), A contains all idempotents and nilpotent elements of B. Assume that $\sum_{i=0}^{n} x^i a_i = 0$, where $x \in A$, $a_0, \ldots, a_n \in C(A)$, and $a_n \in T$. For each $t \in T$, the equality $\sum_{i=0}^{n}(xt^{-1})^i t^i a_i = 0$ holds, and the inclusions $t^i a_i \in C(A)$, $t^n a_n \in T$ hold. Therefore, the ring B is integral over its centre. By 2.62, B is right invariant.

(2) Since A is prime, B is prime. By (1), B is right invariant. Hence B is a domain. Therefore A is a domain. By 5.22, for any $a \in A$, there exists a positive integer m such that $Aa^m = a^m A$. By 5.20(1), A is left distributive. ▷

5.24 *Let A be a semiprime ring algebraic over its centre. The following conditions are equivalent.*

(1) A is right distributive.

(2) A is left distributive.

(3) A is a distributive reduced ring, for any $x \in A$, there exists a positive integer m such that $x^m A = Ax^m$, and for any minimal prime ideal H of A, the ring A/H is a domain algebraic over its centre.

◁ It is sufficient to prove (1)⟺(3).

(3)⟹(1) is obvious.

(1)⟹(3) Since A is semiprime, B is semiprime. By 5.23(1), B is right invariant. Hence B is reduced. Therefore A is reduced. Let H be a minimal prime ideal of A, $R \equiv A/H$, and let $h : A \to R$ be a natural epimorphism. By 1.35(7), R is a right distributive domain. If a is a regular element of A, then $h(a) \neq 0$ by 4.53(2). Since A is algebraic over its centre, the domain R is algebraic over its centre. By 5.23(2), R is left distributive. The remaining part of the assertion follows from 5.20(2) and 5.22. ▷

5.25 *Let A be a right distributive ring algebraic over its centre, N be the prime radical of A, $R \equiv A/N$, and let B be the ring of quotients of A with respect to the set T of all central regular elements.*

(1) $N = tN = Nt$ for any central regular element $t \in A$, N coincides with the set E of all nilpotent elements of B, and N coincides with the prime radical F of B.

(2) R is a distributive reduced ring algebraic over its centre, and for any $x \in R$, there exists a positive integer m such that $x^m R = Rx^m$.

(3) A has the classical right ring of quotients which is right distributive.

(4) If A is semiprime, then B is an invariant distributive reduced ring.

◁ (1) Set $B/F \equiv D$. A right invariant semiprime ring D is reduced. Hence $F = E$. By 5.23(1), $E \subseteq N$. Therefore $N = E = F$. Hence the subring R of the reduced ring D is reduced. Since $t^{-1} \in B$ and $N = F$ is an ideal of B, we obtain $t^{-1}N \subseteq N$ and $tN \subseteq N$. Therefore $N = tN = Nt$.

(2) Let $h : A \longrightarrow R$ be the natural epimorphism, and let t be an arbitrary element of T. It is verified directly that $h(t)$ is a central regular element of R. Therefore R is algebraic over its centre. The assertion follows now from 5.24.

(3) Since B is right invariant, B has the classical right ring of quotients Q. By 4.30(16), Q is right distributive. Since $B = A_T$, the ring Q is the classical right ring of quotients of A.

(4) Since A is semiprime, B is semiprime. By 5.23(1), B is a right distributive right invariant ring integral over its centre. By 5.24 applied to B, this ring is left distributive. By the left-side analog of 2.62, B is left invariant. The invariant semiprime ring B is reduced. ▷

5.26 Let A be an indecomposable right distributive right Goldie ring algebraic over its centre, and let N be the prime radical of A.

(1) A/N is a right uniform domain, N is a completely prime nilpotent ideal, each element $a \in A \setminus N$ is regular, $aN = N \subset aA$, and there exists a central regular element $t \in aA \setminus N$.

(2) $NM = MN = N$ for any right ideal M of A which properly contains N.

(3) Either N_A is a nonzero module which has no simple factor modules, or $N = J(A)$ and A is a right uniserial right Artinian ring, or A is a right distributive uniform domain, and for any nonzero $a \in A$, there exists a nonzero $b \in A$ such that $ab = ba$ is a nonzero central element of A.

(4) If N is a finitely generated right ideal, then A is either a right uniserial right Artinian ring or a uniform right distributive domain.

◁ (1) By 3.19(1), the following assertion (*) holds: A/N is a right uniform domain, $aN = N \subset aA$ for any $a \in A \setminus N$, and N is a completely prime nilpotent ideal containing all left zero-divisors of A. Let $a \in A \setminus N$. By (*), a is a right regular element. Since A is algebraic over its centre, there exists a polynomial $f(x) \in C(A)[x]$ of a minimal possible degree n with properties that $f(a) \in N$, and the leading coefficient c_n of $f(x)$ is a central regular element. Let t be the constant term of $f(x)$, and let $f(x) = xg(x) + t, g(x) \in C(A)[x]$. It follows from the minimality of n that $g(a) \in A \setminus N$. Since N is a completely prime ideal, $ag(a) \in A \setminus N$. Therefore $t = f(a) - ag(a) \in A \setminus N$. By (*), the element t is right regular. In addition, t is central.

Hence t is regular. We have $f(a) \in N$ and $a \in A \setminus N$. By (*), there exists $m \in N$ such that $f(a) = am$. Then $t = a(m - g(a)) \in aA \setminus N$. Since $t \in aA$ and t is a central regular element, a is left regular.

(2) By (1), there exists a central regular element $t \in M \setminus N$. Using (1) and the fact that t is a central element, we obtain $N = tN = Nt \subseteq NM \subseteq N$.

(3) The following three cases are possible: (*) $N = 0$; (**) N_A is a nonzero module having a simple factor module H; (***) N_A is a nonzero module which has no simple factor modules.

Consider case (*). By (1), A is a right distributive domain. Let $0 \neq a \in A$. By assumption, there exists a nonzero polynomial $f(x) \in C(A)[x]$ such that $f(a) = 0$. Let t be the constant term of $f(x)$. Since A is a domain, $t \neq 0$. In addition, $t \in C(A)$. Hence $t = ab = ba \neq 0$, where $b \in A$. Therefore A is a uniform domain.

Consider case (**). By 2.54, A is right quasi-invariant. Therefore, the annihilator of a simple module H is a maximal right ideal M. Then $NM \neq N$ and $N \subseteq J(A) \subseteq M$. By (2), M cannot properly contain N. Hence $M = J(A) = N$ and A/N is a division ring. By 3.18, A is a right uniserial right Artinian ring, and $N = J(A)$.

It remains for us to consider the case (***). Therefore, the assertion is proved.

(4) follows from (3) and from the fact that each nonzero finitely generated module has a maximal submodule. $\triangleright$

5.27 *Let A be a ring which is algebraic over its centre.*

(1) A is a right distributive right Noetherian ring $\iff$ A is a finite direct product of right uniserial right Artinian rings and invariant hereditary Noetherian domains.

(2) A is a right distributive left Noetherian ring $\iff$ A is a finite direct product of Artinian right uniserial rings and invariant hereditary Noetherian domains.

$\triangleleft$ We may assume that A is an indecomposable ring. Since right Artinian left Noetherian rings are Artinian, we may assume that A is not a right uniserial right Artinian ring. By 5.26(4) and 3.35(2), we may assume that A is a domain. Apply now 5.24 and 3.35(3). $\triangleright$

5.28 Let A be a semiprime right or left distributive ring, and let the ring A/H be algebraic over its centre for any minimal prime ideal H of A.

Then A is a distributive reduced ring, and for any minimal prime ideal H of A, the ring A/H is a domain.

◁ By 5.24, for any minimal prime ideal H, the ring A/H is reduced. By 4.49, A is a reduced ring, and A/N is a domain. By 5.24, A/H is distributive. By 5.17, A is distributive. ▷

5.29 Let R be a unitary central subring of a prime ring A.

Then R is a commutative domain with the classical field of quotients F, the set T of all nonzero elements of R is a central multiplicative subset of A, all elements of T are regular in A, and the ring of quotients A_T exists and is a prime algebra over the field F.

◁ 5.29 can be directly verified. ▷

5.30 Let A be a semiprime right or left distributive ring such that there exists a positive integer n with the following property:

for each $a \in A$ and for an arbitrary minimal prime ideal H of A, there exists a polynomial $f(x) \in C(h(A))[x]$ of the degree $\leq n$ such that $h(a)$ is a root of $f(x)$, where $h : A \to A/H$ is the natural epimorphism.

Then A is a distributive reduced ring, and there exists a positive integer $m = m(n)$ such that $Aa^m = a^m A$ for all $a \in A$.

◁ By 5.29, the factor ring A/H is algebraic over its centre for any minimal prime ideal H. By 5.28, A is a distributive reduced ring, and for any minimal prime ideal H, the factor ring A/H is a domain. Let $a \in A$, $m \equiv n!$, and let $h : A \to A/H$ be the natural epimorphism. It follows from the assumpton and 5.22 that $h(A)h(a)^m = h(a)^m h(A)$. By 5.20(2), $Aa^m = a^m A$. ▷

5.31 Let A be a semiprime ring, and let the set $A \setminus P$ be multiplicative and right permutable for any prime ideal P of A.

(1) A is a reduced ring, any prime ideal of A is a completely prime ideal, and there exists a reduced local right ring of quotients A_P.

(2) If A is integral over its centre, then A is a right localizable reduced ring, and all maximal right or left ideals of A are completely prime ideals.

◁ (1) Let P be a prime ideal. Since $A \setminus P$ is a multiplicative set, P is a completely prime ideal. By 4.49, A is reduced. By 1.52 and 1.53(14), A_P is a reduced local ring.

(2) By 5.11(2), A is quasi-invariant. By 1.43(11), all maximal right or left ideals of A are completely prime ideals. By (1), A is right localizable. ▷

5.32 *Let A be a semiprime ring which is integral over its centre. Then the following conditions are equivalent.*

(1) *All submodules of flat right A-modules are flat, and for any prime ideal P of A, the set $A \setminus P$ is multiplicative and right permutable.*

(2) *All submodules of flat left A-modules are flat, and for any prime ideal P of A, the set $A \setminus P$ is multiplicative and left permutable.*

(3) *A is right distributive.*

(4) *A is left distributive.*

◁ Since $(3) \Longleftrightarrow (4)$ follows from 5.24, it is sufficient to prove $(1) \Longleftrightarrow (3)$.

$(3) \Longrightarrow (1)$ By 5.24, A is distributive. By 2.62, A is invariant. Therefore, for any prime ideal P of A, the set $A \setminus P$ is multiplicative and right permutable. By 4.66, all submodules of flat A-modules are flat.

$(1) \Longrightarrow (3)$ By 5.31(2), A is a reduced right localizable ring. Let $M \in \max(A_A)$, $Q \equiv A_M$, and let N be the kernel of the canonical ring homomorphism $A \to Q$. By 4.33(1), Q is a domain. Therefore N is a completely prime ideal. By assumption, the set $A \setminus N$ is right permutable. Therefore Q is a right uniform domain. By 4.64(1), A is right distributive. ▷

5.3 Rings module-finite over their centre

5.33 Let A be a ring which is a finitely generated module over its central unitary subring R.

(1) Every finitely generated right or left A-module M is a finitely generated R-module.

(2) $AJ(R) \subseteq J(A)$.

(3) Let B be an ideal of A. Then

B is a finitely generated right ideal $\Longleftrightarrow$ B is a finitely generated left ideal $\Longleftrightarrow$ B is a finitely generated R-module.

(4) If R is a right uniserial right Artinian ring, then R is a uniserial Artinian ring.

◁ (1) is directly verified.

(2) Since $J(R)$ is an ideal of the central subring R of A, the set $AJ(R)$ is an ideal of A. By Nakayama's Lemma, it is sufficient to prove that $MAJ(R) \neq M$ for any nonzero finitely generated A-module M. By (1), M is a finitely generated R-module. By Nakayama's Lemma, $M \neq MJ(R) = MAJ(R)$.

(3) follows from (1).

(4) Let N be the nonzero radical of R, F be a field which is the centre of a division ring $Q \equiv R/N$, n be the dimension of the vector space Q over F, and let $T \equiv N/N^2$. Since $T_R \cong Q_R$, the dimension of

the vector space T over F is equal to n. Hence $_R T \cong {}_R Q$. Therefore, the principal right ideal N is also a principal left ideal, whence R is a uniserial Artinian ring. $\triangleright$

5.34 Let A be a prime ring which is a finitely generated module over its central unitary subring R, and let $A = \sum_{i=1}^{n} a_i R$.

(1) R is a commutative domain with the classical field of quotients F, the set T of all nonzero elements of the domain R is a central multiplicative subset of A, all elements of T are regular in A, $A_T = \sum_{i=1}^{n} a_i F$, and the ring of quotients A_T exists and is a prime algebra over the field F.

Hence the dimension of the F-algebra A_T does not exceed n.

(2) A is a nonsingular finite-dimensional (in the sense of Goldie) ring, and A is an order in a simple Artinian ring A_T.

(3) For each $a \in A$, there exists a polynomial $f(x) \in R[x]$ of the degree n such that $f(a) = 0$.

(4) A is algebraic over its centre.

$\triangleleft$ (1) can be verified withy the use of 5.29. (2) follows from (1) and from the fact that prime nonsingular finite-dimensional rings are orders in simple Artinian rings.

(3) Let F_R be a free R-module with basis $x_1, \ldots, x_n$, $h : F_R \to A_R$ be a module epimorphism such that $h(x_i) = a_i$, and let B be the subring of $\mathrm{End}(F_R)$ which consists of the endomorphisms f such that $f(\mathrm{Ker}(h)) \subseteq \mathrm{Ker}(h)$. All endomorphisms of A_R are lifted to endomorphisms of a free module F_R. Therefore $\mathrm{End}(A_R)$ is isomorphic to B. The ring A is isomorphic to a subring of $\mathrm{End}(A_R)$. The ring $\mathrm{End}(F_R)$ is isomorphic to the ring R_n of all $n \times n$-matrices over the commutative domain R. Therefore, there exists a ring monomorphism $A \to R_n$. Let $a \in A$. By Cayley-Hamilton theorem, there exists a polynomial $f(x) \in R[x]$ of the degree n such that $f(a) = 0$.

(4) follows from (3). $\triangleright$

5.35 Let A be a right or left distributive semiprime ring such that for any minimal prime ideal H of A, the ring A/H is a finitely generated module over its centre.

Then A is a distributive reduced ring, and for any minimal prime ideal H of A, the ring A/H is a domain algebraic over its centre.

$\triangleleft$ By 5.34(4), for any minimal prime ideal H of A, the ring A/H is algebraic over its centre. The assertion follows now from 5.28. $\triangleright$

5.36 Assume that A is a semiprime right or left distributive ring, and for each minimal prime ideal H of A, there exists a positive integer n such that the ring A/H is an n-generated module over its centre.

Then A is a distributive reduced ring, and there exists a positive integer $m = m(n)$ such that $Aa^m = a^m A$ for all $a \in A$.

$\triangleleft$ 5.36 follows from 5.30 and 5.34(3). $\triangleright$

5.37 Let R be a unitary central subring of a ring A.

(1) A_M is a semiprime Artinian ring for all $M \in \max(R)$ $\Longleftrightarrow$ A is regular, and A_M is an orthogonally finite ring for all $M \in \max(R)$.

(2) If S is a nil-ideal (a nilpotent ideal) of R, then SA is a nil-ideal (a nilpotent ideal) of A.

(3) If S is a vanishing ideal of R, then the ideal SA of A is right and left vanishing.

$\triangleleft$ (1) By 4.32(9), A is regular $\Longleftrightarrow$ A_M is regular for all $M \in \max(R)$. In addition, semiprime Artinian rings are regular. Therefore, (1) follows from the fact that orthogonally finite regular rings are Artinian.

(2) is directly verified.

(3) The following assertion (*) holds (see 1.2): If X is an ideal of a ring Y, then X is left vanishing $\Longleftrightarrow$ $NX \neq N$ for each nonzero right module N_Y. Let N be a nonzero right A-module. It follows from (*) that $NS \neq N$. In addition, $NS = NSA$. Therefore $NSA \neq N$. It follows from (*) that SA is left vanishing. Analogously, SA is right vanishing. $\triangleright$

5.38 Let a ring A be a finitely generated module over its unitary central subring R.

(1) For each $M \in \max(R)$, the ring of quotients A_M is a finitely generated module over its local central subring R_M.

(2) Every prime factor ring of A is an order in a simple Artinian ring.

(3) $A_M J(R_M) \subseteq J(A_M)$ for all $M \in \max(R)$.

(4) The ring A_M is module-finite over its centre, semilocal, and orthogonally finite for all $M \in \max(R)$.

(5) A is right distributive $\Longleftrightarrow$ for all $M \in \max(R)$, the ring of quotients A_M is a right Bezout ring, and $A_M/J(A_M)$ is a finite direct product of division rings.

$\triangleleft$ (1) is obvious. (2) follows from 5.34(2). (3) follows from 5.33(2) applied to R_M and A_M. (4) follows from (1), (3), and from the fact that $(A_M)/(A_M J(R_M))$ is a finite-dimensional algebra over the field $R_M/J(R_M)$. (5) follows from (4), 4.32(14), and 5.16(2). $\triangleright$

5.39 (1) A is a reduced right finite-dimensional right Bezout ring $\Longleftrightarrow$ A is a finite direct product of right uniform right semihereditary right Bezout domains.

(2) A is a semilocal right distributive right nonsingular ring $\Longleftrightarrow$ A is a right nonsingular right Bezout ring, and $A/J(A)$ is a finite direct product of division rings $\Longleftrightarrow$ A is a reduced ring, all cyclic right A-modules are finite-dimensional, A is a finite direct product of right uniform right semihereditary right Bezout domains A_i, and all factor rings $A_i/J(A_i)$ are finite direct products of division rings.

(3) A is a semilocal reduced right Bezout ring $\Longleftrightarrow$ A is a finite direct product of semilocal right uniform right semihereditary right Bezout domains A_i, and all right cyclic A_i-modules are completely finite-dimensional.

$\triangleleft$ (1) It is sufficient to prove only $\Longrightarrow$. Since A is right finite-dimensional, there exists an essential right ideal $B = \oplus_{i=1}^n B_i$ such that all B_i are uniform principal right ideals. Since A is a right Bezout ring, $B = bA$. By 1.35(3), A is right nonsingular. Therefore $\ell(b) = 0$. By 1.35(2), $r(a) = 0$. Hence $A_A \cong B_A$, A is a finite direct sum of uniform right ideals. It follows from 1.35(2) that A is a finite direct product of right uniform right nonsingular rings $A_1, \ldots, A_n$. All A_i are right Bezout domains. By 4.28, all A_i are semihereditary domains.

(2) follows from (1), 5.16(2), and 3.8. (3) follows from (1) and 2.15(3). $\triangleright$

5.40 *Let a ring A be a finitely generated module over a unitary central subring R. Then the following conditions are equivalent.*

(1) A is a semiprime right distributive ring.

(2) A is a semiprime left distributive ring.

(3) For each $M \in \max(R)$, the ring of quotients A_M is a finite direct product of semihereditary uniform Bezout domains $A_i(M)$, and all factor rings $A_i(M)/J(A_i(M))$ are finite direct products of division rings for all i.

(4) A is a distributive reduced ring, all submodules of flat A-modules are flat, each prime factor ring of a ring A is a semihereditary order in a division ring, and there exists a positive integer m such that $Aa^m = a^m A$ for all $a \in A$.

$\triangleleft$ (3)$\Longrightarrow$(1) follows from 4.32(9) and 5.38(5). (4)$\Longrightarrow$(1) is obvious. (1)$\Longleftrightarrow$(2) follows from 5.36.

(1)$\Longrightarrow$(3) By 5.36, A is a distributive reduced ring. Let $M \in \max(A)$, and let $B \equiv R_M$. By 4.32(9) and 5.38(5), B is a reduced Bezout ring, and $B/J(B)$ is a finite direct product of division rings. The assertion follows now from 5.39(3).

$((1)+(2)+(3))\Longrightarrow(4)$ By 5.36, A is a distributive reduced ring, and there exists a positive integer m such that $Aa^m = a^m A$ for all $a \in A$. By 4.20, A is a pf-ring. All rings A_M are semihereditary. By 4.18, all submodules of flat A_M-modules are flat. By 4.36, all submodules of flat A-modules are flat. Let B be a prime factor ring of A. By 5.38(2), B is an order in a simple Artinian ring Q. By 3.8, B is a uniform domain, and Q is a division ring. By repeating for B the argument which were used for A, we obtain that all submodules of flat B-modules are flat. By 4.26, B is a semihereditary order in a division ring. $\triangleright$

5.41 Let a ring A be a finitely generated module over its unitary central subring, and let the prime radical N of A be a finitely generated right or left ideal. Then the following conditions are equivalent.

(1) A is right distributive, and A, A/N are rings with the maximum condition on right annihilators.

(2) A is left distributive, and A, A/N are rings with the maximum condition on left annihilators.

(3) A is a finite direct product of distributive uniform domains and uniserial Artinian rings.

$\triangleleft$ It is sufficient to prove $(1)\Longleftrightarrow(3)$.

$(3)\Longrightarrow(1)$ is obvious.

$(1)\Longrightarrow(3)$ By 5.33(3), N is a finitely generated left ideal of A. We may assume that the ring A is indecomposable. By 3.20, A is either a domain or an Artinian right uniserial ring. In the first case, it follows from 5.40 that A is a distributive domain. By 3.12(2), A is uniform. Therefore, we may assume that A is an Artinian right uniserial ring. By 5.33(4), A is a uniserial ring. $\triangleright$

5.42 *Let a ring A be a finitely generated module over its unitary central subring. Then the following conditions are equivalent.*

(1) A is a right Noetherian right or left distributive ring.

(2) A is a left distributive right or left Noetherian ring.

(3) A is a finite direct product of uniserial Artinian rings and invariant hereditary Noetherian domains.

$\triangleleft$ 5.42 follows from 5.41, 3.35(2), and 3.35(1). $\triangleright$

5.4 Commutative rings

5.43 *If M_A is a uniserial module over a right invariant ring A, then* $\mathrm{End}(M)$ *is a local ring.*

◁ Let $f \in \mathrm{End}(M)$, and let $g \equiv 1_M - f \in \mathrm{End}(M)$. It is necessary to prove that at least one of the endomorphisms f, g is an automorphism of M. From $f + g = 1_M$, we obtain $M = f(M) + g(M)$. Since M is uniserial, at least one of the modules $f(M)$, $g(M)$ coincides with M. Assume that $f(M) = M$. The following two cases are possible: $g(M) = M = f(M)$ and $g(M) \neq M = f(M)$.

Consider the case $g(M) = M$. Then f and g are epimorphisms. Since M is uniserial and $\mathrm{Ker}(f) \cap \mathrm{Ker}(g) = 0$, at least one of the epimorphisms f, g is an automorphism of M.

Consider case $g(M) \neq M = f(M)$. There exists $m \in M \setminus g(M)$. Set $N \equiv mA$. Since M is uniserial, $g(M) \subset N \neq g(M)$. Since $f \equiv 1 - g$ and $g(N) \subseteq N$, we obtain $f(N) \subseteq N$. Therefore f and g induce endomorphisms f_N and g_N of N such that $f_N + g_N = 1_N$. Since N is a cyclic uniserial right module over the right invariant ring A, the ring $\mathrm{End}(N)$ is isomorphic to a right uniserial factor ring of A. Therefore $\mathrm{End}(N)$ is local. Since $N \neq g(M)$, the endomorphism g_N is not an automorphism. Hence f_N is an automorphism of N. Therefore $\mathrm{Ker}(f_N) = 0$. Hence $N \cap \mathrm{Ker}(f) = 0$. Since $N \neq 0$ and M is uniserial, $\mathrm{Ker}(f) = 0$. Therefore f is an automorphism. ▷

5.44 Let T be a multiplicative subset of a unitary central subring R of a ring A, A_T be the ring of quotients of A with respect to T, F and G be right A-modules, $h \in \mathrm{Hom}(F_A, G_A)$, F_T and G_T be the modules of quotients, and let $f : A \to A_T$, $g : F \to F_T$ be the canonical homomorphisms.

(1) By the rule $h_T(g(m)f(t)^{-1} = g(h(m))f(t)^{-1}$, the homomorphism of right A_T-modules $h_T : F_T \to G_T$ is well defined. If $T = R \setminus M$, where $M \in \max(R)$, then we write h_M instead of h_T.

(2) If $u : G \to H$ is a homomorphism of right A-modules, then $(uh)_T = u_T h_T$.

(3) If $h_M \equiv 0$ for all $M \in \max(R)$, then $h \equiv 0$.

◁ (1) For each subset B of A, we write $\overline{B}$ instead of $f(B)$. Let $g(m_1)\overline{t_1}^{-1} = g(m_2)\overline{t_2}^{-1}$. There exist $a_1, a_2 \in A$ and $t \in T$ such that $\overline{t_i}^{-1} = \overline{at}^{-1}$. Then $g(m_1 a_1)t^{-1} = g(m_1)\overline{t_1}^{-1} = g(m_2)\overline{t_2}^{-1} = g(m_2 a_2)t^{-1}$. Therefore $g(m_1 a_1) = g(m_2 a_2)$. Hence $h_T(g(m_1)\overline{t_1}^{-1}) = h_T(g(m_1 a_1)\overline{t}^{-1}) = g(h(m_1 a_1))\overline{t}^{-1} = g(h(m_2 a_2))\overline{t}^{-1} = h_T(g(m_2)\overline{t_2}^{-1})$. Therefore, the map h_T is well defined. Since T is contained in the centre of A, it can analogously be verified that $h_T \in \mathrm{Hom}(M_T, N_T)$.

(2) is directly verified.

(3) Let $M \in \max(R)$. Then $(h(F))_M = h_M(F) = 0$. By 4.32(5), $h(F) = 0$. ▷

5.45 Assume that A is a commutative ring, all regular elements of A are invertible, M_A is a uniserial A-module, $f, g \in \operatorname{End}(M)$, and there exists $m \in M$ such that $r(m) = 0$.

Then $(fg - gf)(m) = 0$.

$\triangleleft$ Let $h \equiv fg - gf \in \operatorname{End}(M)$. We shall prove that $h(m) = 0$. Since M is uniserial, only the following cases are possible:

(1) $f(mA) \supseteq mA$ and $g(mA) \supseteq mA$.

(2) $f(mA) \subseteq mA$ and $g(mA) \subseteq mA$.

(3) $f(mA) \supseteq mA$ and $g(mA) \subseteq mA$.

(4) $f(mA) \subseteq mA$ and $g(mA) \supseteq mA$.

(1) Let $f(mA) \supseteq mA$, $g(mA) \supseteq mA$. There exist $a, b \in A$ such that $m = f(m)a = g(m)b$. Hence $r(a) \subseteq r(m) = 0$, $r(b) \subseteq r(m) = 0$, a and b are regular elements of A. By assumption, a and b are units. Hence $h(m) = h(m)ab(ab)^{-1} = ((fg)(m)ab - (gf)(m)ab)(ab)^{-1} = (m - m)(ab)^{-1} = 0$.

(2) Let $f(mA) \subseteq mA$, $g(mA) \subseteq mA$, $f(m) = ma$, and $g(m) = mb$, where $a, b \in A$. Then $h(m) = f(mb) - g(ma) = mab - mab = 0$.

(3) Let $f(mA) \supseteq mA$, $g(mA) \subseteq mA$. Then $m = f(m)a$ and $g(m) = mb$, where $a, b \in A$. Since $r(a) \subseteq r(m) = 0$, a is a regular element. By assumption, a is a unit. Therefore $f(m) = ma^{-1}$. Hence $h(m) = f(mb) - g(ma^{-1}) = ma^{-1}b - mba^{-1} = 0$.

(4) Let $f(mA) \subseteq mA$ and $g(mA) \supseteq mA$. This case can be considered analogously to the case (3). $\triangleright$

5.46 *If M is a uniserial module over a commutative ring A, then $\operatorname{End}(M)$ is a commutative local ring.*

$\triangleleft$ By 5.43, the ring $\operatorname{End}(M)$ is local. Let $\overline{f}, \overline{g} \in \operatorname{End}(M)$, and let $\overline{h} \equiv \overline{f}\overline{g} - \overline{g}\overline{f} \in \operatorname{End}(M)$. We have to prove that $\overline{h} \equiv 0$. Assume the contrary. There exists a nonzero $m \in M$ such that $\overline{h}(m) \neq 0$. Set $B \equiv A/r_A(m)$, $N \equiv \{x \in M \mid r_A(m) \subseteq r_A(x)\}$. Then N is a uniserial B-module, $r_B(m) = 0$, and N_A is fully invariant in M. The endomorphisms $\overline{f}$, $\overline{g}$, and $\overline{h}$ of M induce endomorphisms f, g, and h of the fully invariant submodule N of M_A. Let B_T be the ring of quotients of B with respect to the set T of all central regular elements of B, N_T be the module of quotients of N with respect to the set T, and let f_T, g_T, and h_T be the endomorphisms of B_T-module N_T induced by the endomorphisms f, g, and h in line with 5.44(1). Since N is a uniserial B-module, it can directly be verified that N_T is a uniserial B_T-module. In addition, the annihilator of m in B_T is equal to zero, and all regular elements of B_T are units. By 5.45, $f_T g_T - g_T f_T = 0$. By 5.44(2), $h_T \equiv 0$. Hence $h(m) = 0$; we have a contradiction. $\triangleright$

5.47 *If N is a distributive module over a commutative ring A, then* $\mathrm{End}(N)$ *is a commutative ring.*

◁ Let $M \in \max(A)$, $f, g \in \mathrm{End}(N)$, and let $h = fg - gf \in \mathrm{End}(M)$. By 4.31(4), N_M is a uniserial A_M-module. By 5.46, $f_M g_M - g_M f_M \equiv 0$. By 5.44(2), $h_M \equiv 0$. By 5.44(3), $h \equiv 0$. Therefore $\mathrm{End}(M)$ is commutative. ▷

5.48 *Let M be a distributive module over a commutative ring $A(0)$, and let n be any integer, $n \geq 0$. Denote by $A(n+1)$ the endomorphism ring of the module $M_{A(n)}$ over the ring $A(n)$.*

Then all the rings $A(n)$ are commutative, and M is a distributive $A(n)$-module.

◁ By 5.47, $A(1)$ is a commutative ring. Since the multiplication by any element of a commutative ring $A(0)$ is an endomorphism of $M_{A(0)}$, the ring $A(0)$ can be naturally embedded into the ring $A(1)$. Since $M_{A(0)}$ is distributive, $M_{A(1)}$ is distributive. Further, we use the induction on n. ▷

5.49 *Let F be a finitely generated distributive module over a commutative ring A.*

(1) $\mathrm{Hom}_A(F, G)$ is a distributive A-module for any distributive A-module G.

(2) $\mathrm{End}(F)$ is a commutative distributive ring which is a distributive A-module.

◁ (1) Let M be a maximal ideal of A, $B \equiv A_M$, $\overline{F} \equiv F_M$, $\overline{G} \equiv G_M$, $H \equiv \mathrm{Hom}_A(F, G)$, $\overline{H} \equiv H_M$, $E \equiv \mathrm{Hom}_B(\overline{F}, \overline{G})$, and let $W \equiv \mathrm{Hom}_B(B, \overline{G})$. Using 5.44, we can verify that $\overline{H}_B \cong E_B$. By 4.43, it is sufficient to prove that E_B is distributive. By 4.44(1), $\overline{F}_B$ is cyclic. Hence there exists a monomorphism $E_B \to W_B$, and $W_B \cong \overline{G}_B$. By 4.43, $\overline{G}_B$ is distributive. Therefore E_B is distributive.

(2) By 5.47, $\mathrm{End}(F)$ is commutative. By (1), $\mathrm{End}(F)$ is a distributive A-module. There exists the natural homomorphism $A \to \mathrm{End}(F)$. Therefore $\mathrm{End}(F)$ is a distributive ring. ▷

5.50 *For the classical ring of quotients Q of a commutative ring A, the following conditions are equivalent.*

(1) Q is a distributive ring.

(2) For any $m, n \in A$, there exist $a, b, c, d \in A$ such that $a + b$ is a regular element of A, $ma = nc$, and $nb = md$.

◁ (2) is equivalent to the following condition (*): for any $m, n \in A$, there exist $a, b, c, d, t \in A$ such that t is a regular element, $at^{-1} + bt^{-1} = 1$, $mt^{-1}at^{-1} = nt^{-1}ct^{-1}$, and $nt^{-1}bt^{-1} = mt^{-1}dt^{-1}$. If $q_1, \ldots, q_n \in Q$, then there exist $a_1, \ldots, a_n, t \in A$ such that t is a regular element, and $q_i = a_i t^{-1}$ for all i. Therefore, (*) is equivalent to the following condition (**): for any $p, q \in Q$, there exist $u, v, w, x \in Q$ such that $u + v = 1$, $pu = qw$, and $qv = px$. $(1) \Longleftrightarrow (**)$ follows from 2.4. ▷

5.51 Let A be a commutative uniserial ring, $M \equiv J(A)$, $0 \neq f \in \text{End}(M_A)$, $r(M) = 0$, and let each element of M be a zero-divisor in A.

(1) If f is an epimorphism, then f is an automorphism.

(2) For each $m \in M$, there exists $a \in A$ such that $f(m) = am$.

(3) $f(M)$ is not a cyclic module.

(4) There do not exist $a, b \in M$ such that $f(M) \subseteq bA$ and $af(m) = bm$ for any $m \in M$.

◁ (1) If $a \in \text{Ker}(f)$, then $Ma = f(M)a \subseteq f(Ma) = f(aM) \subseteq f(a)M = 0$, whence $a \in r(M) = 0$. Therefore f is an automorphism.

(2) Let $m \neq 0$. Since A is uniserial, either $f(m) \in mA$ or $m \in f(m)A$. In the first case, the assertion is obvious. Let $m \in f(m)A$. Then $m = f(m)a$, where $a \in A$. Let $r(a) = 0$. Since each element of M is a zero-divisor, a is a unit. Hence $f(m) = ma^{-1}$, as we wished to prove. Let $r(a) \neq 0$. Since A is uniserial, there exists $b \in A$ such that $0 \neq mb \in mA \bigcap r(a)$. Hence $mb = f(m)ab = f(mba) = 0$; we have a contradiction.

(3) Assume that $f(M)$ is cyclic. Since $f(M) \cong M/\text{Ker}(f)$, there exists $m \in M$ such that $M = mA + \text{Ker}(f)$. Since M is uniserial, $\text{Ker}(f)$ and mA are comparable. In addition, $\text{Ker}(f) \neq M$. Therefore $M = mA$. Hence $r(M) = r(m) \neq 0$; we have a contradiction.

(4) Assume the contrary. Let m be an arbitrary element of M. We have $(f(a) - b)m = f(a)m - af(m) = f(am) - f(am) = 0$, $f(a) - b \in r(M) = 0$, and $f(a) = b$. By assumption, $f(M) \subseteq bA = f(a)A \subseteq f(M)$. Therefore $f(M) = f(a)A$; we have a contradiction to (3). ▷

5.52 Let A be a commutative uniserial ring, $M \equiv J(A)$, $r(M) = 0$, each element of M be a zero-divisor, and let $\text{End}(M)$ be a right or left distributive ring.

(1) $\text{End}(M)$ is a commutative uniserial ring.

(2) If $f \in \text{End}(M)$ and f is not an automorphism, then there exists $d \in A$ such that $f(m) = dm$ for any $m \in M$.

(3) For each $g \in \text{End}(M)$, there exists $c \in A$ such that $g(m) = cm$ for any $m \in M$.

◁ The radical M is a uniserial module over the commutative uniserial ring A. Set $E \equiv \mathrm{End}(M)$.

(1) By 5.46, E is a commutative local ring. By assumption, E is distributive. By 2.8, E is uniserial.

(2) Let $f \neq 0$. Assume that f is not an automorphism. By 5.51(1), $f(M) \neq M$. Let $b \in M \setminus f(M)$. Since A is uniserial, $f(M) \subseteq bA$. Let $\bar{b}$ be the endomorphism of M such that $\bar{b}(x) = xb$. By (1), E is a commutative uniserial ring. Therefore, either $f \in E\bar{b}$ or $\bar{b} \in Ef$. Assume that $f = g\bar{b}$, where $g \in E$. By 5.51(2) applied to g and b, there exists $a \in A$ such that $g(b) = ab$. Then the equalities $f(m) = (g\bar{b})(m) = g(bm) = abm$ hold for any $m \in M$. Therefore, we can set $d \equiv ab$. Assume now that $\bar{b} = hf$, where $h \in E$. By 5.51(2) applied to h and b, there exists $a \in A$ such that $h(b) = ab$. Let $m \in M$. Since $f(m) \in bA$, $h(f(m)) = af(m)$. Hence $bm = \bar{b}(m) = hf(m) = af(m)$; we have a contradiction to 5.51(4).

(3) Let $0 \neq y \in M$. By 5.51(2), there exists $a \in A$ such that $g(y) = ay$. Define an endomorphism f of M as follows: $f(x) = g(x) - ax$ for any $x \in M$. Then $0 \neq y \in \mathrm{Ker}(f)$. Therefore f is not an automorphism. By 5.51(2), there exists $d \in A$ such that $f(m) = dm$ for any $m \in M$. Hence $g(m) = f(m) + am = (d + a)m$, and we can set $c \equiv d + a$. ▷

Example 5.53 *There exists a commutative uniserial ring A with Jacobson radical M such that the following conditions hold.*

(1) The module M_A is uniserial.

(2) There exists an endomorphism g of M_A such that g is not induced by multiplication by any element $b \in A$.

(3) $\mathrm{End}(M)$ is a commutative local ring which is not distributive. (In particular, $\mathrm{End}(M)$ is not a uniserial ring).

(4) M is a nil-ideal of A, and M is the unique proper faithful ideal of A.

(5) Every element of A is either a unit or a nilpotent element.

(6) A coincides with its classical ring of quotients.

◁ Let F be a field, T be the multiplicative monoid formed by all real numbers greater than, or equal to 1, N be the ideal of the monoid ring $F[T]$ formed by all elements whose supports consist of real numbers greater than, or equal to 2, A be the factor ring $F[T]/N$, and let $h : F[T] \to A$ be the natural epimorphism. Every $b \in A$ can be written in the form $b = \sum_{i=1}^{n} f_i h(t_i)$, where $1 \leq t_1 < t_2 < \ldots < 2$ and $f_1, \ldots, f_n \in F$. This form will be called here the canonical form of b. If $t_1 > 1$, then b is a nilpotent element. If $t_1 = 1$, then b is a unit. In addition, $bA = h(t_1)A$. If $t, u \in T$ and $t_1 \leq u_1$, then $h(t_1)A \supseteq h(u_1)A$. Therefore A is a commutative uniserial ring, $J(A) \equiv M$ is a nil-ideal, and $r(M) = 0$. Since each noninvertible element of A is nilpotent, A coincides with its classical ring of quotients. For each positive integer n, the element $h(1 + n^{-1})$ is denoted by a_n. Then $a_n A \subset a_{n+1} A$ for any n and $M = \bigcup_{n=1}^{\infty}$. Define an endomorphism g of M_A as multiplication by a formal series $\sum_{i=1}^{\infty} h(2 - i^{-1})$. Since $M = \bigcup_{n=1}^{\infty}$ and for any positive integer n, there exists a positive integer L such that $h(2 - i^{-1})a_n = 0$ for $i \geq L$, g is well defined. Assume that there exists an element b of A such that $g(x) = xb$ for all $x \in M$. Let $b = \sum_{i=1}^{m} f_i h(t_i)$ be the canonical form of b. Then the canonical form of each ba_n contains no more than m summands. On the other hand, the number of summands in the canonical form of $g(a_n)$ increases infinitely when n tends to infinity; we obtain a contradiction. By 5.52(3), $\mathrm{End}(M)$ is not a distributive ring. Let B be a proper ideal of A, $B \neq M$, and let $m \in M \setminus B$. Hence m is a nilpotent element. Since A is uniserial, $B \subseteq mA$. Hence B is a nilpotent ideal. Therefore $r(B) \neq 0$. $\triangleright$

5.54 *Let M be a direct limit of distributive (resp. Bezout) right modules $\{M_i\}_{i \in I}$ over a ring A.*

Then M_A is a distributive (resp. Bezout) module.

$\triangleleft$ Let M_A be distributive. The definition of a direct limit contains: a directed partially ordered subscript set I, homomorphisms of A-modules $h_{ij} : M_i \to M_j$ for $i \leq j$, $f_i : M_i \to M$ for $i, j \in I$. Let m and n be two elements of the module M. Then $m = f_i(m_i)$ and $n = f_j(n_j)$ for some $m_i \in M_i$ and $n_j \in M_j$. Let t be a subscript from I such that $t \geq i$ and $t \geq j$. Then $m = f_t(m_t)$ and $n = f_t(n_t)$, where $m_t \equiv h_{it}(m_i)$, $n_t \equiv h_{jt}(n_j)$. By 2.4 applied to the elements m_t and n_t of the distributive module M_t, there exist elements a and b in A such that $1 = a + b$, $m_t a \in n_t A$, and $n_t b \in m_t A$. Then $ma \in nA$ and $nb \in mA$. By 2.4, M is distributive. In the case of Bezout modules, the proof is analogous. $\triangleright$

5.55 [5]. *If M and N are two distributive modules over a commutative ring A, then $M \otimes_A N$ is a distributive A-module.*

$\triangleleft$ Let B be a maximal ideal of A, $A(B)$ be the ring of quotients of A with respect to the set $A \setminus B$, and let $M(B)$, $N(B)$ be the modules of quotients of M and N with respect to the set $A \setminus B$, respectively. Assume that $(M \otimes_A N)_B$ is the module of quotients of $M \otimes_A N$ considered as $A(B)$-module. By 1.53(14), $A(B)$ is a local ring. By 4.31(3), it is sufficient to prove that $(M \otimes_A N)_B$ is a distributive module over $A(B)$. It can directly be verified that $(M \otimes_A N)_B$ and $M(B) \otimes_{A(B)} N(B)$ are isomorphic $A(B)$-modules. Therefore, without loss of generality, we may assume that $A = A(B)$ is a local ring.

Let N' be any finitely generated submodule of N. By 4.31(4), the module N' is cyclic. Hence there exists the natural epimorphism $A_A \to N'$ which induces an epimorphism $M \otimes_A A \to M \otimes_A N'$. Since $M \cong M \otimes_A A$, $M \otimes_A A$ is a distributive module, whence $M \otimes_A N'$ is a distributive module. The module N is a direct limit of the set $\{N_i\}_{i \in I}$ of its finitely generated submodules N_i. Therefore $M \otimes_A N$ is isomorphic to the direct limit of modules $M \otimes_A N_i$. All modules $M \otimes_A N_i$ are distributive. By 5.54, $M \otimes_A N$ is distributive. $\triangleright$

Exercises 5.56 (1) If M is a right module over a right quasi-invariant ring and each finitely generated submodule of M is a direct sum of cyclic modules, then

M is distributive $\Longleftrightarrow$ M is a Bezout module.

(Hint: Use 2.26 and 2.35.)

(2) If M is a module over an invariant principal ideal ring A, then

M is distributive $\Longleftrightarrow$ M is a Bezout module.

(Hint: Use (1) and the fact that each finitely generated A-module is a direct sum of cyclic modules.)

(3) If M is a module over an invariant principal ideal domain A and Q is the classical division ring of quotients for A, then

M is distributive $\Longleftrightarrow$ M is a Bezout module $\Longleftrightarrow$ M is isomorphic to a subfactor of the module Q_A.

(Hint: Use (2) and 3.54.)

(4) If M is an Abelian group and Q is the additive group of rational numbers, then

M is distributive $\Longleftrightarrow$ M is a locally cyclic group $\Longleftrightarrow$ M is isomorphic to a subfactor of the group Q_A.

(Hint: Use (3).)

Chapter 6

Ring constructions

6.1 Endomorphism rings

6.1 A module M_A over a ring A is said to be *endodistributive (resp. endouniserial, endosimple, endo-Bezout)* if the module $_{\mathrm{End}(M)}M$ is distributive (resp. uniserial, simple, a Bezout module).

(1) A is a left distributive ring (resp. a left uniserial ring, a division ring, a left Bezout ring) $\Longleftrightarrow$ the module A_A is endodistributive (resp. endouniserial, endosimple, endo-Bezout).

(2) Let E_A and M_A be two modules over a ring A, F be a fully invariant submodule of E, N be a fully invariant submodule of M, and let $H \equiv \mathrm{Hom}(M, E)$.

Then $\mathrm{Hom}(M, F)$ is a submodule of the left module $_{\mathrm{End}(E)}H$, and $\mathrm{Hom}((M/N), E)$ is a submodule of a right module $H_{\mathrm{End}(M)}$.

(3) For any module M_A, there exists a natural $(\mathrm{End}(M), A)$-bimodule isomorphism $_{\mathrm{End}(M)}M_A \cong {}_{\mathrm{End}(M)}\mathrm{Hom}(A_A, M_A)$ which maps elements $m \in M$ to homomorphisms $\widehat{m}$ such that $\widehat{m}(a) = ma$ for $a \in A$.

(4) If each 2-generated submodule of a module M_A is a Bezout module over some unitary subring B of a ring A, then M_A is a Bezout module.

(5) Let $\{M_i\}_{i\in I}$ be some set of Bezout modules over rings A_i, respectively, and let A be the direct product of all the rings A_i.

Then $\prod_{i\in I} M_i$ and $\oplus_{i\in I} M_i$ are Bezout A-modules.

(6) If each 2-generated unitary subring of A is contained in some unitary subring of A which is a right Bezout ring, then A is a right Bezout ring.

(7) All direct products and factor rings of right Bezout rings are right Bezout rings.

(8) Let M be a nonzero module, $\{M_i\}_{i\in I}$ be the set of all nonisomorphic simple subfactors of M, N_i be the injective hull of M_i ($i \in I$), and let $N \equiv \prod_{i\in I} N_i$.

Then M is isomorphic to a submodule of a direct product E of some set of copies of N.

In addition, if M is injective, then M is isomorphic to a direct summand of E.

(9) Every module is isomorphic to a submodule of a direct product of injective hulls of simple modules.

(10) Every injective module is isomorphic to a direct summand of a direct product injective hulls of simple modules.

◁ (1), (2), and (3) can be directly verified.

(4) For any $m, n \in M$, there exist $a, b, c, d \in B$ such that $m = (ma + nb)c$ and $n = (ma + nb)d$. Since $a, b, c, d \in A$, we obtain that M_A is a Bezout module.

(5) Let $(m_i)_{i\in I}$, and let $(n_i)_{i\in I}$ be arbitrary elements of A-module $\prod_{i\in I} M_i \equiv M$. There exist $a_i, b_i, c_i, d_i \in A_i$ such that $m_i = (m_i a_i + n_i b_i)c_i$ and $n_i = (m_i a_i + n_i b_i)d_i$. Let $a \equiv (a_i)_{i\in I} \in A$, $b_i \equiv (b_i)_{i\in I} \in A$, $c_i \equiv (c_i)_{i\in I} \in A$, and let $d_i \equiv (d_i)_{i\in I} \in A$. Then $m = (ma + nb)c$ and $n = (ma + nb)d$. Therefore M_A is a Bezout module. Hence its submodule $\oplus_{i\in I} M_i$ is a Bezout module.

(6) and (7) can be proved analogously to (4) and (5), respectively. (8) is directly verified. (9) follows from (8). (10) follows from (9). ▷

6.2 Let M_A be a right module over a ring A, $\{N_i\}_{i\in I}$ be a set of right A-modules, $N_A \equiv \prod_{i\in I} N_i$, and let $R \equiv \prod_{i\in I} \text{End}(N_i)$.

(1) There exist a natural ring monomorphism $R \to \text{End}(N)$ and a natural $(\text{End}(N), \text{End}(M))$-bimodule isomorphism $\text{Hom}(M, N) \cong \prod_{i\in I} \text{Hom}(M, N_i)$.

(2) Every left $\text{End}(N)$-submodule of $_{\text{End}(N)}\text{Hom}(M, N)$ is a submodule of $_R\text{Hom}(M, N)$.

(3) If the module $_R\text{Hom}(M, N)$ is distributive, then $_{\text{End}(N)}\text{Hom}(M, N)$ is distributive.

(4) If $_R\text{Hom}(M, N)$ is a Bezout module, then $_{\text{End}(N)}\text{Hom}(M, N)$ is a Bezout module.

(5) If $_R N$ is distributive, then $_{\text{End}(N)} N$ is distributive.

(6) If $_R N$ is a Bezout module, then $_{\text{End}(N)} N$ is a Bezout module.

◁ (1) and (2) are directly verified. (3) and (4) follow from (2).

(5) and (6) By 6.1(3), there exists a natural isomorphism $N_A \cong \text{Hom}(A_A, N_A)$. Therefore, (5) and (6) follow from (3) and (4) for $M \equiv A_A$. ▷

6.3 Let M and E be two right modules over a ring A, $\{N_i\}_{i\in I}$ be a set of right A-modules, $N_A \equiv \oplus_{i\in I} N_i$, and let $R \equiv \prod_{i\in I} \operatorname{End}(N_i)$.

(1) There exists a natural $(\operatorname{End}(\oplus_{i\in I} N_i), \operatorname{End}(M))$-bimodule isomorphism $\operatorname{Hom}(M, \prod_{i\in I} N_i) \cong \prod_{i\in I} \operatorname{Hom}(M, N_i)$.

(2) There exists a natural $(\operatorname{End}(E), \operatorname{End}(\oplus_{i\in I} N_i))$-bimodule isomorphism $\operatorname{Hom}(\oplus_{i\in I} N_i, E) \cong \prod_{i\in I} \operatorname{Hom}(N_i, E)$.

(3) There exist a natural ring monomorphism $R \to \operatorname{End}(N)$ and a natural $(\operatorname{End}(M), \operatorname{End}(N))$-bimodule isomorphism $\operatorname{Hom}(N, M) \cong \prod_{i\in I} \operatorname{Hom}(N_i, M)$.

(4) Every right $\operatorname{End}(N)$-submodule of $\operatorname{Hom}(N, M)_{\operatorname{End}(N)}$ is a submodule of $\operatorname{Hom}(N, M)_R$.

(5) If $\operatorname{Hom}(N, M)_R$ is distributive, then $\operatorname{Hom}(N, M)_{\operatorname{End}(N)}$ is distributive.

(6) If $\operatorname{Hom}(N, M)_R$ is a Bezout module, then $\operatorname{Hom}(N, M)_{\operatorname{End}(N)}$ is a Bezout module.

(7) Let M_A be distributive, E be uniform, and let $f, g \in \operatorname{Hom}(M, E)$.

Then either $\operatorname{Ker}(f) \subseteq \operatorname{Ker}(g)$ or $\operatorname{Ker}(g) \subseteq \operatorname{Ker}(f)$.

$\triangleleft$ (1), (2), (3), and (4) are directly verified. (5) and (6) follow from (4).

(7) Assume the contrary. There exist $m, n \in M$ such that $f(n) = g(m) = 0$, $f(m) \neq 0$, and $g(n) \neq 0$. Since E is uniform, the intersection of its nonzero submodules $f(mA)$ and $g(nA)$ contains a nonzero element t. Let $f(mx) = g(ny) = t \neq 0$, where $x, y \in A$. By 2.4, for $mx, ny \in M$, there exist $a, b \in A$ such that $1 = a + b$ and $mxaA + nybA \subseteq mxA \cap nyA$. Hence $t = ta + tb = f(mxa) + g(nyb) \in f(mA \cap nA) + g(mA \cap nA) = 0$; we have a contradiction. $\triangleright$

6.4 Let M_A be a right module over a ring A, $V(M)$ be the class of all right modules E_A such that $\operatorname{Hom}_A(M, E)$ is a left distributive module over a ring $\operatorname{End}(E_A)$, and let $W(M)$ be the class of all right modules E_A such that $\operatorname{Hom}_A(M, E)$ is a left Bezout module over a ring $\operatorname{End}(E_A)$.

(1) All direct summands, direct products and fully invariant submodules of modules in $V(M)$ belong to $V(M)$.

(2) All direct products and fully invariant submodules of modules in $W(M)$ belong to $W(M)$.

$\triangleleft$ It follows from 6.1(2) that all fully invariant submodules of modules in $V(M)$ (resp. $W(M)$) belong to $V(M)$ (resp. $W(M)$). Let $\{N_i\}_{i\in I}$ be a set of modules in $V(M)$, and let $R \equiv \prod_{i\in I} \operatorname{End}(N_i)$.

Then each module $_{\text{End}(N_i)}\text{Hom}(M, N_i)$ can be considered as a left R-module. By 6.3(1), $\text{Hom}(M, \prod_{j \in J} N_j) \cong \prod_{j \in J} \text{Hom}(M, N_j)$. Therefore, using 2.5(3), we have that $\text{Hom}(M, \prod_{i \in I} N_i)$ is a left distributive R-module. By 6.2(3), $\text{Hom}(M, \prod_{i \in I} N_i)$ is a left distributive $\text{End}(\prod_{i \in I} N_i)$-module. Therefore, all direct products of modules in $V(M)$ belong to $V(M)$. Analogously, we can prove that all direct products of modules in $W(M)$ belong to $W(M)$. (We need only use 6.1(5) and 6.2(4) instead of 2.5(3) and 6.2(3).)

It is sufficient now to prove that all direct summands of modules in $V(M)$ belong to $V(M)$. Let $N \in V(M)$, and let $e = e^2 \in \text{End}(N)$. We may assume that $\text{End}(N) = e \cdot \text{End}(N) \cdot e$. In addition, we may assume that the left $e \cdot \text{End}(N) \cdot e$-module $e \cdot \text{Hom}(M, N)$ coincides with $\text{Hom}(M, e(N))$. Also, $_{\text{End}(N)}\text{Hom}(M, N)$ is distributive. By 1.5(11), $_{e \cdot \text{End}(N) \cdot e}(e \cdot \text{Hom}(M, N))$ is distributive. $\triangleright$

6.5 Let E_A be a right module over a ring A, $V^*(E)$ be the class of all right modules M_A such that $\text{Hom}_A(M, E)$ is a distributive right module over a ring $\text{End}(M_A)$, and let $W^*(E)$ be the class of all right modules M_A such that $\text{Hom}_A(M, E)$ is a right Bezout $\text{End}(M_A)$-module.

(1) All direct summands and direct sums of modules in $V^*(E)$ belong to $V^*(E)$. If $M \in V^*(E)$ and N is fully invariant in M, then $M/N \in V^*(E)$.

(2) All direct sums of modules in $W^*(E)$ belong to $W^*(E)$.

If $M \in W^*(E)$ and N is fully invariant in M, then $M/N \in W^*(E)$.

$\triangleleft$ If M belongs to $V^*(E)$ (resp. $W^*(E)$) and N is fully invariant in M, then $M/N \in V^*(E)$ (resp. $M/N \in W^*(E)$) by 6.1(2). Let $\{N_i\}_{i \in I}$ be some set of modules in $V^*(E)$, and let $R \equiv \prod_{i \in I} \text{End}(N_i)$. Each module $\text{Hom}(N_i, E)_{\text{End}(N_i)}$ can be considered as a right R-module. By 6.3(2), $\text{Hom}(\oplus_{j \in J} N_j, E) \cong \prod_{j \in J} \text{Hom}(N_j, E)$. Therefore, it follows from 2.5(3) that $\text{Hom}(\oplus_{i \in I} N_i, E)$ is a distributive right R-module. By 6.3(5), $\text{Hom}(\oplus_{i \in I} N_i, E)$ is a distributive right $\text{End}(\oplus_{i \in I} N_i)$-module. Therefore, all direct sums of modules in $V^*(M)$ belong to $V^*(M)$. Analogously, it can be proved that direct sums of modules in $W^*(M)$ belong to $W^*(M)$. (We need only use 6.1(5) and 6.3(6) instead of 2.5(3) and 6.3(5).)

It is sufficient to prove that all direct summands of modules in $V^*(M)$ belong to $V^*(M)$. Let $N \in V^*(M)$, and let $e = e^2 \in \text{End}(N)$. We may assume that $\text{End}(N)$ coincides with $e \cdot \text{End}(N) \cdot e$. In addition, we may assume that the right $e \cdot \text{End}(N) \cdot e$-module $e \cdot \text{Hom}(N, E)$ coincides with $\text{Hom}(e(N), E)$. The module $\text{Hom}(N, E)_{\text{End}(N)}$ is distributive. By 1.5(11), $(\text{Hom}(M, N) \cdot e)_{e \cdot \text{End}(N) \cdot e}$ is distributive. $\triangleright$

6.6 (1) Let $\mathcal{E}$ be the class of all endodistributive modules over a ring A.

Then all direct summands, direct sums, direct products, and fully invariant submodules of modules in $\mathcal{E}$ belong to $\mathcal{E}$.

If $M \in \mathcal{E}$ and N is fully invariant in M, then $M/N \in \mathcal{E}$.

(2) Let $\mathcal{F}$ be the class of all endo-Bezout modules over a ring A.

Then all direct sums, direct products, and fully invariant submodules of modules in $\mathcal{F}$ belong to $\mathcal{F}$.

If $M \in \mathcal{F}$ and N is fully invariant in M, then $M/N \in \mathcal{F}$.

(3) A ring A is left distributive $\iff$ all free right A-modules are left distributive modules over their endomorphism rings $\iff$ all projective right A-modules are left distributive modules over their endomorphism rings.

(4) A is a left Bezout ring $\iff$ all free right A-modules are left Bezout modules over their endomorphism rings.

(5) Let Q be an injective hull of a quasi-injective module E.

If Q is endodistributive (endo-Bezout), then E and Q/E are endodistributive (endo-Bezout).

◁ (1) and (2) follow from 6.4, 6.5, and from the fact that any fully invariant submodule $N \in \mathrm{Lat}(M_A)$ is contained in $\mathrm{Lat}(_{\mathrm{End}(M)}M)$.

(3), (4), and (5) follow from (1) and (2). ▷

6.7 Let M_A be a module, E_A be a quasi-injective module, $R \equiv \mathrm{End}(E)$, and let $H \equiv \mathrm{Hom}(M, E)$.

(1) If $f, g \in H$, then $\mathrm{Ker}(f) \subseteq \mathrm{Ker}(g) \iff g \in Rf$.

(2) $_RH$ is a cyclic module $\iff$ there exists a homomorphism $f \in H$ such that $\mathrm{Ker}(f) \subseteq \mathrm{Ker}(g)$ for any $g \in H$.

(3) $_RH$ is a uniserial module $\iff \mathrm{Ker}(f)$ and $\mathrm{Ker}(g)$ are comparable for any homomorphisms $f, g \in H$.

(4) If $x, y \in E$, then $r(x) \subseteq r(y) \iff y \in Rx$.

(5) If $x \in E$, then $E = Rx \iff r(x) = r(E)$.

(6) $_{\mathrm{End}(E)}E$ is a uniserial module $\iff$ for any $x, y \in E$, right ideals $r(x)$ and $r(y)$ of A are comparable.

(7) E is indecomposable $\iff E$ is uniform $\iff \mathrm{End}(E)$ is a local ring.

◁ (1) If $g \in Rf$, then $\mathrm{Ker}(f) \subseteq \mathrm{Ker}(g)$. Let $\mathrm{Ker}(f) \subseteq \mathrm{Ker}(g)$. By the rule $tf(m) = g(m)$, the epimorphism $t : f(M) \to g(M)$ is well defined. Since E is quasi-injective, t can be extended to an endomorphism $h \in R$. Hence $g = hf \in Rf$.

(2) and (3) follow from (1). (4), (5), and (6) follow from 6.1(3) and (1),(2),(3). (7) is proved in [236, 19.9]. ▷

6.8 Let E be a quasi-injective indecomposable right module over a ring A. Then

E is endodistributive $\iff$ E is an endo-Bezout module $\iff$ E is endouniserial $\iff$ for any $f, g \in E$, the right ideals $r(f)$ and $r(g)$ of A are comparable.

◁ By 6.7(7), End(E) is a local ring. Hence 6.8 follows grom 2.8 and 6.7(6). ▷

6.9 *For a module M_A, the following conditions are equivalent.*

(1) *M is distributive.*

(2) *For any indecomposable quasi-injective module E_A, the left* End(E)*-module* Hom(M, E) *is uniserial.*

(3) *For any module E_A which is a direct product of indecomposable quasi-injective right A-modules, the left* End(E)*-module* Hom(M, E) *is a distributive Bezout module.*

(4) *For any injective module E_A, the left* End(E)*-module* Hom(M, E) *is distributive.*

(5) *For any quasi-injective module E_A, the left* End(E)*-module* Hom(M, E) *is distributive.*

(6) *For any module E_A which is the injective hull of a direct sum of all representatives of classes of isomorphic simple right A-modules, the left* End(E)*-module* Hom(M, E) *is distributive.*

(7) *For any direct summand of any direct product E_A of quasi-injective right A-modules, the left* End(E)*-module* Hom(M, E) *is distributive.*

(8) *For any module E_A which is the injective hull of an arbitrary simple subfactor T of M, the left* End(E)*-module* Hom(M, E) *is distributive.*

(9) *For any module E_A which is the injective hull of an arbitrary simple subfactor T of M,* Hom(M, E) *is the left Bezout module over the ring* End(E).

(10) *For any module E_A which is a direct product of injective hulls of nonisomorphic simple subfactors of M,* Hom(M, E) *is a left Bezout module over the ring* End(E).

◁ (1)$\Longrightarrow$(2) follows from 6.7(3), 6.3(7), and from the fact that End(E) is a local ring by 6.7(7). (2)$\Longrightarrow$(3), (2)$\Longrightarrow$(7), (6)$\Longrightarrow$(8), and (9)$\Longleftrightarrow$(10) follow from 6.4. (3)$\Longrightarrow$(4) and (7)$\Longrightarrow$(4) follow from 6.4 and 6.1(10). (4)$\Longrightarrow$(5) follows from 6.1(2) and from the fact that each quasi-injective module is fully invariant in its injective hull. (5)$\Longrightarrow$(6) is directly verified. (8)$\Longleftrightarrow$(9) follows from 2.8 and from the fact that End(E) is a local ring by 6.7(7).

(8)$\Longrightarrow$(1) Assume that M is not distributive. By 2.32, there exist $N \in \text{Lat}(M)$ and $D \in \text{Lat}(N)$ such that $N/D \cong T \oplus T$, where T is a simple module. Therefore, there exist $N_1, N_2 \in \text{Lat}(N)$ such that $N = N_1 + N_2$, $N_1 \bigcap N_2 = D$, and $N_i \cong T$. Hence there exist epimorphisms $f_1, f_2 \in \text{Hom}(N, T)$ such that $\text{Ker}(f_i) = N_i$ and N_1, N_2 are not comparable. Let E be the injective hull of T, $R \equiv \text{End}(E)$, and let $H \equiv \text{Hom}(M, E)$. By 6.7(7), R is a local ring. By 2.8, $_R H$ is a uniserial module. Since E is injective, the homomorphisms f_1, f_2 can be extended to homomorphisms $g_1, g_2 \in H$, and modules $\text{Ker}(g_1)$, $\text{Ker}(g_2)$ are not comparable. Hence Rg_1 and Rg_2 are not comparable. Since $_R H$ is a uniserial module, we obtain a contradiction. $\triangleright$

6.10 *Let A be a ring. Then A is right distributive $\Longleftrightarrow$ all direct summands of all direct sums or direct products of quasi-injective right A-modules are endodistributive $\Longleftrightarrow$ all direct sums and all direct products of indecomposable quasi-injective right A-modules are endo-Bezout modules $\Longleftrightarrow$ all indecomposable quasi-injective right A-modules are endouniserial $\Longleftrightarrow$ the injective hull of the direct sum of all representatives of the classes of isomorphic simple right A-modules is endodistributive $\Longleftrightarrow$ injective hulls of all simple right A-modules are endodistributive $\Longleftrightarrow$ injective hulls of all simple right A-modules are endo-Bezout modules $\Longleftrightarrow$ the direct sum or the direct product of injective hulls of every set of nonisomorphic simple right A-modules is an endo-Bezout module.*

$\triangleleft$ By 6.1(3), $E_A \cong \text{Hom}(A_A, E_A)$ for any module E_A. Further, we apply 6.6(1),(2) and 6.9. $\triangleright$

6.11 Let E_A be a quasi-injective module, and let $\text{End}(E)$ be a division ring. Then
E is endodistributive $\Longleftrightarrow$ E is endosimple $\Longleftrightarrow$ $r(f) = r(E)$ for any nonzero $f \in E$.

$\triangleleft$ The first equivalence follows from the following fact: If $_S E$ is a vector space over a division ring S, then $_S E$ is distributive $\Longleftrightarrow$ $_S E$ is one-dimensional.

Since E is indecomposable, the second equivalence follows from 6.7(5) and from the fact that every simple module is generated by each nonzero element. $\triangleright$

6.12 Let M be a maximal right ideal of a ring A, and let $E \equiv A/M$. The following conditions are equivalent.
    (1) The module E is endodistributive.
    (2) E is an endo-Bezout module.
    (3) E is endosimple.
    (4) M is an ideal of A.

◁ Since E is simple, E is quasi-injective and $\operatorname{End}(E)$ is a division ring.

$(1)\Longleftrightarrow (2) \Longleftrightarrow (3)$ follow from 6.8 and 6.11.

$(4)\Longrightarrow(3)$ By (4), E is a cyclic module over the division ring A/M. Therefore E is endosimple.

$(3)\Longrightarrow(4)$ Let $h : A_A \to E$ be the natural epimorphism, and let $a \in A \setminus M$. By 6.11, $(a : M) = r(h(a)) = r(h(1)) = M$. Therefore $am \in M$ for any $a \in A$ and $m \in M$. ▷

6.13 *A ring A is right quasi-invariant $\Longleftrightarrow$ all simple right A-modules are endo-Bezout modules $\Longleftrightarrow$ all simple right A-modules are endodistributive $\Longleftrightarrow$ all direct products of semisimple right A-modules are left distributive Bezout modules over their endomorphism rings.*

◁ The first two equivalences follow from 6.12. The third equivalence follows from 6.12 and 6.6(1),(2). ▷

6.14 Let A be a semiregular ring.
(1) If A is normal, then $A/J(A)$ is strongly regular.
(2) A is right distributive $\Longleftrightarrow$ A is a normal right Bezout ring.

◁ (1) can directly be verified.
(2) By 2.54, we may assume that A is normal. By (1), we may assume that $A/J(A)$ is strongly regular. The assertion follows now from 5.15. ▷

6.15 Let M be a quasi-injective module.
(1) The ring $\operatorname{End}(M)$ is semiregular, and $J(\operatorname{End}(M)) = \operatorname{sg}(M)$.
(2) If M is finite-dimensional, then $\operatorname{End}(M)$ is semiperfect.
(3) If $\operatorname{End}(M)$ is normal, then the factor ring $\operatorname{End}(M)/J(\operatorname{End}(M))$ is strongly regular.
(4) $\operatorname{End}(M)$ is right (left) distributive $\Longleftrightarrow$ $\operatorname{End}(M)$ is a normal right (left) Bezout ring.
(6) If M is distributive, then $\operatorname{End}(M)$ is a normal semiregular ring, and the factor ring $\operatorname{End}(M)/J(\operatorname{End}(M))$ is strongly regular.

◁ (1) is proved in [236, 22.1]. (2) and (3) follow from (1). (4) follows from (1) and (2). (5) follows from (1), 6.14(1), and from the fact that the endomorphism ring of a distributive module is normal by 2.54. ▷

6.16 *Let M be a distributive quasi-injective right module.*
Then $\operatorname{End}(M)$ is a left distributive semiregular normal left Bezout ring, and the factor ring $\operatorname{End}(M)/J(\operatorname{End}(M))$ is strongly regular.

In addition, if M is indecomposable, then $\operatorname{End}(M)$ is a left uniserial ring.

◁ By 6.9, $\mathrm{End}(M)$ is left distributive. 6.16 follows now from 6.15(4),(5), 6.7(7), and 2.8. ▷

6.17 *Let A be a left distributive left self-injective ring.*
(1) *A is a right distributive semiregular normal right Bezout ring, and $A/J(A)$ is strongly regular.*
(2) *If A has no nontrivial idempotents, then A is right and left uniserial.*

◁ (1) follows from the symmetric analog of 6.16.
(2) By 6.16, A is right uniserial. Therefore A is local. Hence A is a local left distributive ring. By 2.8, A is left uniserial. ▷

6.18 Let E_A be a module, M_A be a quasi-projective module, and let $R \equiv \mathrm{End}(M)$.
(1) If $f, g \in \mathrm{Hom}(M, E)$, then
$f(M) \supseteq g(M) \Longleftrightarrow g \in fR.$
(2) $(\mathrm{Hom}(M, E))_R$ is cyclic $\Longleftrightarrow$ there exists a homomorphism $f \in \mathrm{Hom}(M, E)$ such that $f(M) \supseteq g(M)$ for any $g \in \mathrm{Hom}(M, E)$.
(3) $(\mathrm{Hom}(M, E))_R$ is uniserial $\Longleftrightarrow$ for any homomorphisms $f, g \in \mathrm{Hom}(M, E)$, the modules $f(M)$ and $g(M)$ are comparable.

◁ (1) If $g \in fR$, then $f(M) \supseteq g(M)$. Let $f(M) \supseteq g(M)$. Then $g \in \mathrm{Hom}(M, f(M))$. Since M is quasi-projective, there exists a homomorphism $h \in R$ such that $g = fh \in fR$.
(2) and (3) follow from (1). ▷

6.19 A module M is *strongly indecomposable* if the following two equivalent conditions hold.
(1) All factor modules of M are indecomposable.
(2) All proper submodules of M are superfluous in M.
Let E_A be a distributive module, M_A be a strongly indecomposable module, and let $f, g \in \mathrm{Hom}(M_A, E_A)$.
Then either $f(M) \subseteq g(M)$ or $g(M) \subseteq f(M)$.

◁ Let $\widehat{E} \equiv E/(f(M) \bigcap g(M))$, $u : E \to \widehat{E}$ be a natural epimorphism, $\widehat{f} \equiv uf, \widehat{g} \equiv ug \in \mathrm{Hom}(M, \widehat{E})$, $\widehat{N} \equiv \widehat{f}(M)$, and let $\widehat{T} \equiv \widehat{g}(M)$. We have $\widehat{N} \bigcap \widehat{T} = 0$. Therefore, by 2.23(1) applied to the distributive module $\widehat{E}$, the homomorphism $h \equiv \widehat{f} + \widehat{g}$ is an epimorphism of M onto $\widehat{N} + \widehat{T}$. Hence $M = h^{-1}(h(M)) = h^{-1}(\widehat{N}) + h^{-1}(\widehat{T})$. Since M is strongly indecomposable, at least one of the modules $h^{-1}(\widehat{N}), h^{-1}(\widehat{T})$ coincides with M. For example, let $h^{-1}(\widehat{N}) = M$. Then $\widehat{T} = \widehat{N} \bigcap \widehat{T} = 0$, whence $g(M) \subseteq f(M)$. ▷

6.20 Let M_A be a quasi-projective strongly indecomposable module.

(1) For any distributive module E_A, the right $\mathrm{End}(M)$-module $\mathrm{Hom}(M, E)$ is uniserial.

(2) If M is distributive, then $\mathrm{End}(M)$ is right uniserial.

◁ (1) follows from 6.18(3) and 6.19. (2) follows from (1). ▷

6.21 Let M_A be a quasi-injective module, and let $r(M) = r(m)$ for some $m \in M$.

(1) $_{\mathrm{End}(M)}M = \mathrm{End}(M)m$ is a faithful cyclic left $\mathrm{End}(M)$-module with the generator m.

(2) If $\mathrm{End}(M)$ is left invariant, then $_{\mathrm{End}(M)}M$ is a free cyclic left $\mathrm{End}(M)$-module with a free generator m.

(3) If A is right distributive and $\mathrm{End}(M)$ is left invariant, then $\mathrm{End}(M)$ is left distributive.

(4) If M is distributive and A is commutative, then $\mathrm{End}(M)$ and $A/r(M)$ are commutative distributive rings.

◁ (1) If $f \in \mathrm{End}(M)$ and $f(M) = 0$, then $f \equiv 0$. Hence $_{\mathrm{End}(M)}M$ is faithful. Let $x \in M$. There exists an epimorphism $h : mA \to xA$ such that $h(m) = x$. Since M is injective, h can be extended to $f \in \mathrm{End}(M)$. Hence $x = f(m) \in \mathrm{End}(M)m$.

(2) follows from (1) and from the fact that each cyclic faithful left module with generator x over a left invariant ring is a free cyclic module with a free generator x.

(3) By 6.10, $_{\mathrm{End}(M)}M$ is distributive. By (2), $_{\mathrm{End}(M)}M \cong {}_{\mathrm{End}(M)}\mathrm{End}(M)$.

(4) Since mA is distributive and $mA \cong A/r(m) = A/r(M)$, $A/r(M)$ is a distributive commutative ring. By 5.47, $\mathrm{End}(M)$ is commutative. Since M is a quasi-injective distributive $A/r(M)$-module, (3) shows us that $\mathrm{End}(M)$ is distributive. ▷

6.2   Maximal rings of quotients

6.22 Let N be a submodule of a right A-module M_A, and let E be the injective hull of M. We say that M is a *rational extension* of N (or N is a *rational submodule* of M) if the following two equivalent conditions hold.

(1) Given any $x, y \in M$, where $x \neq 0$, there exists $a \in A$ such that $xa \neq 0$ and $ya \in N$.

(2) $f(M) = 0$ for any $f \in \mathrm{End}(E)$ such that $f(N) = 0$.

◁ (1)$\Longrightarrow$(2) Assume that $f(M) \neq 0$. Then $M \bigcap f(M) \neq 0$. Let $0 \neq x = f(y) \in M \bigcap f(M)$, where $y \in M$. By (1), there exists $a \in A$ such that $xa \neq 0$ and $ya \in N$. Hence $0 = f(ya) = xa \neq 0$; we have a contradiction.

(2)$\Longrightarrow$(1) Assume the contrary. Then there exist $x, y \in M$ such that $x \neq 0$ and $(y : N) \subseteq r(x)$. Since $r(y) \subseteq (y : N) \subseteq r(x)$, there exists an epimorphism $g : yA \to xA$ such that $g(ya) = xa$ for all $a \in A$. Then $N \bigcap yA \subseteq \mathrm{Ker}(g)$, since $(y : N) \subseteq r(x)$. Therefore, there exists an epimorphism $h : (yA + N) \to xA$ such that $N \subseteq \mathrm{Ker}(h)$, and $h(ya) = xa$ for all $a \in A$. Since E is injective, h can be extended to $f \in \mathrm{End}(E)$. By assumption, the equality $f(N) = 0$ implies the equality $f(M) = 0$. Hence $x = f(y) = 0$; we have a contradiction. ▷

6.23 If M is an essential extension of a nonsingular module N, then M is a rational extension of N.

◁ Let E be the injective hull of M, $f \in \mathrm{End}(E)$, and let $f(N) = 0$. Since M is an essential extension of N and E is an essential extension of M, E is an essential extension of the nonsingular module N. Therefore E is nonsingular. By 1.29, $f \equiv 0$. Hence $f(M) = 0$, and M is a rational extension of N. ▷

6.24 Let M_A be a rational extension of a module N, and let E be the injective hull of M.

(1) Every submodule X of M is a rational extension of $X \bigcap N$.

(2) If M is a rational extension of a module L, then M is a rational extension of $L \bigcap N$.

(3) If N is a rational extension of a module L, then M is a rational extension of L.

(4) M is an essential extension of N.

(5) E is the injective hull of N.

(6) $g(N) \neq 0$ for any nonzero endomorphism $g \in \mathrm{End}(M)$.

◁ (1), (2), and (3) are directly verified.

(4) Let $0 \neq x = y \in M$. By the definition of a rational extension, there exists $a \in A$ such that $0 \neq xa = ya \in N$. Therefore M is an essential extension of N.

(5) follows from (4).

(6) Assume that $g(N) = 0$. Since E is injective, g can be extended to $f \in \mathrm{End}(E)$. Since $f(N) = 0$ and M is a rational extension of N, we obtain $f(M) = 0$. Hence $g \equiv 0$; we have a contradiction. ▷

6.25 Let E be the injective hull of a module M_A. A submodule Q of E is the *rational hull* of M in E if Q is a rational extension of M, and Q contains any rational extension of M which is contained in E.

(1) There exists precisely one rational hull Q of M in E, and Q coincides with the sum of all rational extensions of M in E.

(2) Let E_1 and E_2 be two injective hulls of M, and let Q_1 and Q_2 be rational hulls of M in E_1 and E_2 respectively.

Then $Q_1 \cong Q_2$.

(3) If E is a rational extension of M, then $Q = E$, and the ring $\mathrm{End}(Q)$ is semiregular.

(4) If M is injective, then $M = Q$.

(5) If M is nonsingular, then $Q = E$, and the ring $\mathrm{End}(Q)$ is regular.

(6) If N is a rational submodule of M, then Q is the rational hull of N.

(7) If X is a submodule of E and X is a rational extension of Q, then $X = Q$.

(8) If for an endomorphism g of E, there exists a rational submodule N of Q such that $g(N) \subseteq Q$, then $g(Q) \subseteq Q$.

(9) For any rational submodule N of Q and for each homomorphism $h : N \to Q$, there exists the unique endomorphism $\overline{h} \in \mathrm{End}(Q)$ coinciding with h on N.

(10) Every endomorphism g of any rational submodule N of Q can be extended to an endomorphism $\overline{g}$ of Q uniquely. By the rule $\varphi(g) = \overline{g}$, the injective ring homomorphism $\varphi : \mathrm{End}(N) \to \mathrm{End}(Q)$ is defined.

(11) If N is a rational fully invariant submodule of Q, then $\mathrm{End}(N) \cong \mathrm{End}(Q)$.

(12) If there exists a rational submodule N of M such that N is fully invariant in Q, then $\mathrm{End}(N) \cong \mathrm{End}(Q)$.

$\lhd$ (1) Let $\{Q_i\}_{i \in I}$ be the set of all submodules of E which are rational extensions of M, $Q \equiv \sum_{i \in I} Q_i$, and let f be an endomorphism of E such that $f(M) = 0$. Then $f(Q_i) = 0$ for all i. Therefore $f(Q) = 0$, and Q is a rational extension of M.

(2) follows from (1) and from the fact that any two injective hulls of the same module are isomorphic. (3) follows from 6.15(1) and from the definition of a rational extension. (4) follows from (3).

(5) By 6.23, E is a rational extension of M. By (3), $Q = E$. By 4.37, $\mathrm{End}(Q)$ is regular.

(6) follows from 6.24(3). (7) follows from the fact that X is a rational extension of M.

(8) Let $f \in \mathrm{End}(E)$ and $f(M) = 0$. Hence $fg(M \cap N) = 0$. By 6.24(2), Q is a rational extension of $M \cap N$. Therefore $fg(Q) = 0$. Hence $g(Q) \subseteq Q$.

(9) Since E is injective, h can be extended to $g \in \mathrm{End}(E)$. By (8), $g(Q) \subseteq Q$. Therefore g induces an endomorphism $\bar{h}$ of Q coinciding with h on N. Assume that there exists an endomorphism $\bar{h}_1$ of Q coinciding with h on N. Extend $\bar{h}_1$ to an endomorphism g_1 of the injective module E. Then $(g - g_1)(N) = 0$. Since Q is a rational extension of N, $(g - g_1)(Q) = 0$. Therefore $\bar{h}_1 = \bar{h}$.

(10) follows from (9).

(11) By (10), there exists an injective ring homomorphism $\varphi :$ $\mathrm{End}(N) \to \mathrm{End}(Q)$. Since N is fully invariant in Q, φ is an isomorphism.

(12) follows from (11) and (3). $\triangleright$

6.26 (1) If N is a rational right ideal of a ring A, then $\ell(N) = 0$.

(2) If N is an ideal of a ring A, then

N is a rational right ideal $\Longleftrightarrow \ell(N) = 0$.

(3) If ideals N_1 and N_2 of a ring A are rational right ideals, then the ideal $N_1 N_2$ is a rational right ideal.

$\triangleleft$ (1) follows from 6.24(6).

(2)

$\Longrightarrow$ follows from (1).

$\Longleftarrow$ Let $a, b \in A$, where $a \neq 0$. Since $\ell(N) = 0$, there exists $n \in N$ such that $an \neq 0$. Since N is an ideal, $bn \in N$.

(3) By (2), $\ell(N_1) = \ell(N_2) = 0$. Then $\ell(N_1 N_2) = 0$. By (2), $N_1 N_2$ is a rational right ideal. $\triangleright$

6.27 Let A be a ring, E be the injective hull of A_A, and let $Q \equiv \mathrm{End}(_{\mathrm{End}(E)}E)$. The ring Q is called the *maximal right ring of quotients* of A, and Q is denoted by $Q_{\max}(A)$. The *maximal left ring of quotients* $_{\max}Q(A)$ of A is defined analogously.

Since elements of Q are endomorphisms of the left module $_{\mathrm{End}(E)}E$, we write this endomorphisms to the right of arguments. Then E is a $(\mathrm{End}(E), Q)$-bimodule $_{\mathrm{End}(E)}E_Q$.

Associate to every element $a \in A$ its canonical image $\bar{a} \in Q$ as follows: $(x)\bar{a} = xa$ for $x \in E$. Let us associate with every element $q \in Q$ its canonical image $(1)q$, and let us associate with every endomorphism $f \in \mathrm{End}(E)$ its canonical image $f(1) \in E$.

(1) Given any $x \in E$, there exists an endomorphism $f \in \mathrm{End}(E_A)$ such that $x = f(1)$.

(2) The canonical map $f \to f(1)$ is an $\mathrm{End}(E)$-module epimorphism $_{\mathrm{End}(E)}\mathrm{End}(E) \to {}_{\mathrm{End}(E)}E$.

(3) The canonical map $a \to \bar{a}$ is a ring monomorphism $A \to Q$.

(4) The canonical map $q \to (1)q$ is an A-module monomorphism $u : Q_A \to E_A$.

(5) The image of the canonical monomorphism $u : Q_A \to E_A$ coincides with the rational hull $\overline{A}$ of A_A in E.

(6) For any module M_Q, each A-module homomorphism $f : M_A \to E_A$ is a Q-module homomorphism $f : M_Q \to E_Q$.

(7) $\mathrm{End}(E_A) = \mathrm{End}(E_Q)$, $\mathrm{End}(Q_A) = \mathrm{End}(Q_Q)$.

In addition, we may assume that $Q = \mathrm{End}(Q_A)$.

$\triangleleft$ (1) follows from 6.21(1). (2) follows from (1). (3) follows from the fact that for any $a \in A$ with $\overline{a} = 0$, we have $a = (1)\overline{a} = 0$.

(4) Let $q \in Q$, $(1)q = 0$, and let $x \in E$. By (1), there exists $f \in \mathrm{End}(E_A)$ such that $x = f(1)$. Hence $(x)q = (f(1))q = f((1)q) = f(0) = 0$ and $q \equiv 0$.

(5) Let $f \in \mathrm{End}(E)$, $f(A) = 0$. If $q \in Q$, then $f((1)q) = (f(1))q = (0)q = 0$. Therefore $Q \subseteq \overline{A}$. Let us prove that $\overline{A} \subseteq Q$. Let $m \in \overline{A}$. By (1), for any $x \in E$, there exists $f_x \in \mathrm{End}(E_A)$ such that $x = f_x(1)$. If f^* and f^{**} are two endomorphisms of E such that $f^*(1) = f^{**}(1)$, then $(f^* - f^{**})(A) = 0$ and $(f^* - f^{**})(m) = 0$. Therefore, by the rule $(x)q = (f_x(1))q = f_x(m)$, the element $q \in Q$ is well defined. If $x = 1$, then $f_x \equiv 1_E$. Therefore $(1)q = f_1(m) = m$ and $m \in u(Q)$.

(6) Let $m \in M$. Define a homomorphism $f_m : Q_A \to E_A$ as follows: $f_m(q) = f(mq) - f(m)q$. Then $f_m(A) = 0$. Since E_A is injective, f_m can be extended to an endomorphism of E_A. Since Q_A is a rational extension of A_A, we obtain $f_m(Q) = 0$. Therefore $f(mq) - f(m)q = 0$ for all $q \in Q$. Hence $f \in \mathrm{Hom}(M_Q, E_Q)$.

(7) follows from (6). $\triangleright$

6.28 Let A be a ring, E be the injective hull of A_A, Q be the maximal right ring of quotients of A, and let $\overline{A}$ be the rational hull of A_A in E. We assume that A coincides with its canonical image in Q. Using 6.27(5),(6), we assume also that $Q = \mathrm{End}(Q_A)$ and $Q_A = \overline{A}$.

(1) E is a rational extension of $A_A \iff Q_A = E$.

(2) If A is right self-injective, then $A = Q$.

(3) For any rational submodule N of Q_A and for each homomorphism $h : N_A \to Q_A$, there exists the unique $\overline{h} \in Q_A$ such that $h(n) = \overline{h}n$ for all $n \in N$.

(4) For each endomorphism g and for any rational submodule N of Q_A, there exists the unique $\overline{h} \in Q$ such that $h(n) = \overline{h}n$ for all $n \in N$. By the rule $\varphi(g) = \overline{g}$, the injective ring homomorphism $\varphi : \mathrm{End}(N_A) \to Q$ is defined.

(5) If there exists a rational right ideal N of A such that N is a left ideal of Q, then $\mathrm{End}(N) \cong Q$.

(6) E_Q is the injective hull of Q_Q.

(7) Let N be a rational right ideal of A, $q \in Q$, and let $q^{-1}(N) \equiv \{a \in A \mid qa \in N\}$.

Then $q^{-1}(N)$ is a rational right ideal of A.

In particular, $q^{-1}(A)$ is a rational right ideal of A.

(8) If the intersection N of all rational right ideals of A is a rational right ideal of A, then N is a left ideal of Q, and $Q \cong \operatorname{End}(N_A)$.

(9) If the ring $\operatorname{End}(E)$ is left invariant, then $Q = E$.

(10) If the left annihilator of each proper right ideal of A is not equal to zero, then $A = Q$.

(11) If A is a local ring, and $\ell(J(A)) \neq 0$, then $A = Q$.

In particular, every local semiprimary ring coincides with its maximal right ring of quotients.

$\lhd$ (1), (2), (3), (4), and (5) follow from 6.25(3),(4),(9),(10),(12).

(6) Let $\overline{M}_Q$ be a submodule of some Q-module M_Q, and let $\overline{f} \in \operatorname{Hom}(\overline{M}_Q, E_Q)$. Then $\overline{f} \in \operatorname{Hom}(\overline{M}_A, E_A)$. Since E_A is injective, $\overline{f}$ can be extended to an A-homomorphism $f : M_A \to E_A$. By 6.27(6), $f \in \operatorname{Hom}(M_Q, E_Q)$. Therefore E_Q is injective.

(7) Let h be an endomorphism of E_A such that $h(q^{-1}(N)) = 0$. Let $n_1, n_2 \in N$, and let $a_1, a_2 \in A$ be elements such that $n_1 + qa_1 = n_2 + qa_2$. Then $q(a_1 - a_2) \in N$, $a_1 - a_2 \in q^{-1}(N)$, and $h(a_1) = h(a_2)$. Therefore, by the rule $f(n + qa) = h(a)$, the homomorphism $f : N + qA \to A$ is well defined. Then $f(N) = 0$. Since E is injective, f can be extended to an endomorphism $\overline{f}$ of E_A. Since $\overline{f}(N) = 0$ and N is a rational right ideal, $\overline{f}(A) = 0$. By 6.27(6), $\overline{f} \in \operatorname{End}(E_Q)$. Therefore $\overline{f}(q) = \overline{f}(1)q = 0$ for all $q \in Q$. Hence $h(1) = f(q) = \overline{f}(q) = 0$. Therefore $h(A) = h(1)A = 0$.

(8) Let $q \in Q$, and let $q^{-1}(N) \equiv \{a \in A \mid qa \in N\}$. By (7), $q^{-1}(N)$ is a rational right ideal of A. Then $N \bigcap q^{-1}(N)$ is a rational right ideal, and $N \bigcap q^{-1}(N) \subseteq N$. Since N is the intersection of all rational right ideals, $N \bigcap q^{-1}(N) = N$. Therefore $qN \subseteq N$, and N is a left ideal of Q. By (5), $Q \cong \operatorname{End}(N_A)$.

(9) Let $x \in E$, and let f be an endomorphism of E_A such that $f(A) = 0$. By 6.21(1), there exists $g \in \operatorname{End}(E)$ such that $x = g(1)$. Since $\operatorname{End}(E)$ is left invariant, there exists $h \in \operatorname{End}(E)$ such that $fg = hf$. Hence $f(x) = fg(1) = hf(1) = 0$. Therefore $f(E) = 0$, E is a rational extension of A_A, and $E = Q$.

(10) Let $q \in Q$, and let $N \equiv \{a \in A \mid qa \in A\}$. By (7), N is a rational right ideal. By 6.26(1), $\ell(N) = 0$. It follows from the assumption that $N = A$ and $1 \in N$. Therefore $q = q \cdot 1 \in A$, $A = Q$.

(11) Every proper right ideal N of A is contained in $J(A)$. Therefore $\ell(N) \neq 0$. By (10), $A = Q$. $\rhd$

6.29 Let Q be the maximal right ring of quotients of A, and let E be the injective hull of A_A. The following conditions are equivalent.

(1) Q is right self-injective.

(2) Q is an injective right A-module.

(3) $Q = E$.

(4) E is a rational extension of A_A.

(5) E is a rational extension of A_A, $Q_A = E$, $Q = \operatorname{End}(Q_A)$ is a semiregular ring, and $_QQ \cong {}_{\operatorname{End}(E)}E$.

◁ (1)$\Longleftrightarrow$(2) follows from 6.28(6). (2)$\Longleftrightarrow$(3) and (3)$\Longleftrightarrow$(4) follow from the definition of Q. (5)$\Longrightarrow$(4) is obvious. (4)$\Longrightarrow$(5) follows from 6.25(3). ▷

6.30 *Let Q be the maximal right ring of quotients of a ring A.*

(1) *The centre of A is contained in the centre of Q.*

(2) *If B is a nonzero right ideal of A, then BQ is a nonzero right ideal of Q, $A \bigcap BQ$ is a nonzero right ideal of A, and $(BQ)_A$, $A \bigcap BQ$ are essential extensions of B_A.*

(3) *If B is a closed right ideal of A, then $B = A \bigcap BQ$.*

(4) *If B and C are two right ideals of A such that $B \bigcap C = 0$, then $BQ \bigcap CQ = 0$.*

(5) *If $B_A = \oplus_{i \in I} B_i$ is a right ideal of A, then $(BQ)_Q = \oplus_{i \in I}(B_i Q)$.*

(6) *A is right uniform $\Longleftrightarrow$ Q is right uniform.*

(7) *A is right finite-dimensional$\Longleftrightarrow$ Q is right finite-dimensional.*

(8) *If R is an essential right ideal of Q, then $R \bigcap A$ is an essential right ideal of A.*

(9) *If B is an essential right ideal of A, then BQ is an essential right ideal of Q.*

(10) *$\operatorname{Sing}(Q_A) = \operatorname{Sing}(Q_Q)$ and $\operatorname{Sing}(A_A) = A \bigcap \operatorname{Sing}(Q_Q)$.*

◁ (1) Let $c \in C(A)$ and $q \in Q$. Assume that $qc - cq \neq 0$. Since Q_A is a rational extension of A_A, there exists $a \in A$ such that $(qc - cq)a \neq 0$ and $qa = b \in A$. Hence $0 \neq (qc - cq)a = qca - cb = qac - bc = 0$; we have a contradiction.

(2) It is sufficient to prove that $(BQ)_A$ is an essential extension of B_A. Let $0 \neq x = \sum_{i=1}^{n} b_i q_i \in BQ$, where $b_i \in B$ and $q_i \in Q$. Since Q_A is a rational extension of A_A, there exists $a_1 \in A$ such that $0 \neq xa_1 = \sum_{i=1}^{n} b_i q_i a_1 \in BQ$, $q_1 a_1 \in A$, and $b_1 q_1 a_1 \in B$. Analogously, there exists $a_2 \in A$ such that $0 \neq xa_1 a_2 = \sum_{i=1}^{n} b_i q_i a_1 a_2 \in BQ$ and $q_2 a_1 a_2 \in A$. In addition, $b_1 q_1 a_1 a_2 \in B$. Therefore, there exist $a_1, \ldots, a_n \in A$ such that $0 \neq xa_1 \cdots a_n \in B$.

(3) follows from (2).

(4) Let $\overline{B}$ and $\overline{C}$ be closures of right ideals B and C in A, respectively. Then $\overline{B} \bigcap \overline{C} = 0$. By (3), $\overline{B} = A \bigcap \overline{B}Q$ and $\overline{C} = A \bigcap \overline{C}Q$. Hence

$0 = (A \cap \overline{B}Q) \cap (A \cap \overline{C}Q) = A \cap (\overline{B}Q \cap \overline{C}Q)$. Since Q_A is an essential extension of A_A, we have $\overline{B}Q \cap \overline{C}Q = 0$. Therefore $BQ \cap CQ = 0$.

(5) follows from (4) and from the fact that $(\sum_{i \in J} B_i)Q = \sum_{i \in J}(B_i Q)$ for any subset $J \subseteq I$. (6) follows from (4) and from the fact that Q_A is an essential extension of A_A. (7) follows from (5) and from the fact that Q_A is an essential extension of A_A. (8) follows from the fact that Q_A is an essential extension of A_A. (9) follows from (4). (10) follows from (8) and (9). $\triangleright$

6.31 *Let Q be the maximal right ring of quotients of a right distributive ring A, E be the injective hull of A_A, N be the prime radical of Q, U be the group of units of Q, and let T be the set of all nilpotent elements of Q.*

(1) E is a left distributive $\mathrm{End}(E)$-module, and Q is isomorphic to the endomorphism ring of the distributive module $_{\mathrm{End}(E)}E$.

(2) T is a nil-subring of $J(Q)$, T is the sum of all nilpotent ideals of the nil-ring T, and $UT = TU$ is an ideal of $J(Q)$.

(3) The prime radical N is the largest right nil-ideal and the largest left nil-ideal of Q.

(4) If the nil-ring T is right or left vanishing, then T coincides with the prime radical N.

(5) If Q is right self-injective, then Q is a semiregular normal left distributive left Bezout ring.

(6) If Q is a self-injective ring, then Q is a distributive semiregular normal Bezout ring.

(7) If A is right nonsingular, then Q is a right self-injective strongly regular ring.

$\triangleleft$ (1) follows from 6.10 and the isomorphism $Q \cong \mathrm{End}(_{\mathrm{End}(E)}E)$. (2), (3), and (4) follow from (1) and 2.67.

(5) By 6.29, $Q_A = E$. By 6.27(7), we may assume that $Q = \mathrm{End}(E)$. It follows from (1) that Q is left distributive. By 2.50(3), Q is normal. By 6.15(1), the right self-injective ring Q is semiregular. By 5.16(1), Q is a left Bezout ring.

(6) follows from (5) and 6.17.

(7) By 6.25(5), Q is a regular ring, and Q_A is injective. By 6.29, Q is right self-injective. By (5), Q is left distributive. By 5.16(1), the left distributive regular ring Q is strongly regular. $\triangleright$

6.32 Let A be a right uniserial ring, $M \equiv J(A)$, $\ell(M) = 0$, and let each element of M be a right zero-divisor of A.

(1) $Q_{\mathrm{max}}(A) \cong \mathrm{End}(M) = \mathrm{Hom}(M_A, A_A)$.

(2) If $r(a) = \ell(a)$ for all $a \in A$, then the rings $Q_{\mathrm{max}}(A)$ and $\mathrm{End}(M_A)$ are local.

◁ (1) By 6.26(2), M is a rational right ideal. Let N be a proper right ideal of A, and let $N \neq M$. There exists $m \in M \setminus N$. Since A is right uniserial, $N \subseteq mA$. By assumption, $am = 0$, where $0 \neq a \in A$. Hence $aN = 0$. By 6.26(1), N is not a rational right ideal. Hence M is the least rational right ideal. By 6.28(8), $Q_{\max}(A) \cong \operatorname{End}(M)$. Let $f \in \operatorname{Hom}(M_A, A_A)$ and $f(M) \not\subseteq M$. Hence $f(M) = A$, $M_A \cong A_A$, and M_A is a free cyclic module with a free generator x. This is impossible, since by assumption x is a zero-divisor. Therefore $\operatorname{End}(M) = \operatorname{Hom}(M_A, A_A)$.

(2) By (1), it is sufficient to prove that $\operatorname{End}(M_A)$ is a local ring. Let $f, g \in \operatorname{End}(M_A)$, and let $f + g = 1$. Since $\operatorname{Ker}(f) \bigcap \operatorname{Ker}(g) = 0$ and A is right uniserial, at least one of the endomorphisms f, g is a monomorphism. For example, let f be a monomorphism. It is sufficient to prove that $f(M) = M$. Assume the contrary. There exists $m \in M \setminus f(M)$. Since A is right uniserial, $f(M) \subseteq mA$. Since $r(m) = \ell(m)$, $r(m)$ is an ideal. Hence $0 \neq r(m) \subseteq r(f(M)) = r(M) = \ell(M) = 0$; we have a contradiction. ▷

6.33 *Let Q be the maximal right ring of quotients of a commutative ring A, and let E be the injective hull of A_A.*

(1) Q is a commutative ring.

(2) If E_A is distributive, then Q is a distributive commutative ring, and the module Q_A is distributive.

◁ (1) Let $q_1, q_2 \in Q$, and let $N_i \equiv \{a \in A \mid q_i a \in A\}$ $(i = 1, 2)$. By 6.28(7), M_i are rational ideals of A. By 6.26(3), $M_1 M_2$ is a rational ideal. Let $m_i \in M_i$. Then $q_i m_i \in A$, $q_1 q_2 m_1 m_2 = q_1 (q_2 m_2) m_1 = (q_1 m_1)(q_2 m_2) = q_2 m_2 (q_1 m_1) = q_2 q_1 m_1 m_2$. Therefore $(q_1 q_2 - q_2 q_1) m_1 m_2 = 0$. Hence $(q_1 q_2 - q_2 q_1) M_1 M_2$. Since $M_1 M_2$ is a rational ideal, $(q_1 q_2 - q_2 q_1) Q = 0$. Therefore $q_1 q_2 = q_2 q_1$.

(2) By 5.47, $\operatorname{End}(E)$ is commutative. By 6.28(9), $E = Q_A$. Hence the module Q_A is distributive. Therefore, the ring Q is distributive. ▷

6.34 Let A be a commutative uniserial ring, $M \equiv J(A)$, $Q \equiv Q_{\max}(A)$, $E \equiv \operatorname{End}(M)$, $\ell(M) = 0$, and let each element of M be a zero-divisor. The following conditions are equivalent.

(1) $A \cong Q_{\max}(A)$.

(2) $Q_{\max}(A)$ is right or left distributive.

(3) $\operatorname{End}(M_A)$ is right or left distributive.

(4) $Q_{\max}(A)$ and $\operatorname{End}(M)$ are commutative uniserial rings.

(5) Given any $f \in \operatorname{End}(M)$, there exists $a \in A$ such that $f(m) = am$ for any $m \in M$.

(6) Given any homomorphism $f : M_A \to A_A$, there exists $a \in A$ such that $f(m) = am$ for any $m \in M$.

◁ (1)$\Longrightarrow$(2) and (4)$\Longrightarrow$(2) are directly verified. (2)$\Longleftrightarrow$(3) and (5)$\Longleftrightarrow$(6) follow from the fact that, by 6.32(1), $Q_{\max}(A) \cong \mathrm{End}(M) = \mathrm{Hom}(M_A, A_A)$.

(3)$\Longrightarrow$(4) By 5.52(1), $\mathrm{End}(M)$ is a commutative uniserial ring. By 6.32(1), $Q_{\max}(A) \cong \mathrm{End}(M)$. Therefore $Q_{\max}(A)$ is a commutative uniserial ring.

(3)$\Longrightarrow$(5) follows from 5.52(3).

(5)$\Longrightarrow$(1) There exists the natural homomorphism $h : A \to \mathrm{End}(M)$. Since $\ell(M) = 0$, h is a monomorphism. It follows from (5) that h is an isomorphism. By 6.32(1), $Q_{\max}(A) \cong \mathrm{End}(M)$. Therefore $Q_{\max}(A) \cong A$. ▷

Example 6.35 *There exists a commutative uniserial ring A such that A coincides with its classical ring of quotients, and $Q_{\max}(A)$ is a commutative local nondistributive ring.*

◁ By 5.53, there exists a commutative uniserial ring A such that A coincides with its classical ring of quotients, $J(A) \equiv M$ is a nil-ideal, $\ell(M) = 0$, and $\mathrm{End}(M_A)$ is a commutative nondistributive local ring. By 6.32(1), $Q_{\max}(A) \cong \mathrm{End}(M)$. ▷

Example 6.36 Let $F[x, y]$ be the polynomial ring of two variables x, y with coefficients from a field F, S be the ideal of $F[x, y]$ generated by x^2, y^2, xy, $A \equiv F[x, y]/S$, $h : F[x, y] \to A$ be a natural epimorphism, and let M be an ideal of A formed by images of polynomials with zero constant terms.

Then $M^2 = 0$, $A/M \cong F$, A is a commutative local Artinian ring, $M = J(A)$, $M_A \cong (A/M)_A \oplus (A/M)_A$ is a direct sum of two simple A-modules, and A is not a uniform ring.

By 6.7(7), local self-injective rings are uniform. Therefore A is not self-injective. By 6.28(11), $A = Q_{\max}(A)$.

Consequently, the maximal ring of quotients can be non-self-injective.

6.3 Quaternion algebras

6.37 Let a and b be two units of a commutative ring A. We denote by $(a, b/A)$ the (*generalized*) *quaternion algebra* over A, i.e., the free A-module with the canonical basis $\{1, i, j, k\}$, in which multiplication is A-bilinear, and is defined on the canonical basis such that 1 is the common identity element of the rings A and $(a, b/A)$, and $i^2 = a$, $j^2 = b$, $k^2 = -ab$, $ij = -ji = k$, $ik = -ki = aj$, and $kj = -jk =$

bi. For $a = b = -1$, the quaternion algebra $(-1, -1/A)$ is called the *Hamiltonian quaternion algebra* over A.

Each quaternion algebra is a finitely generated module over a central subring A.

If the equality $x^2 - ay^2 - bz^2 = 0$ is possible in A only for $x = y = z = 0$, then A is called (a, b)-*ring*.

We denote by q_0, q_1, q_2, q_3, q^*, and $|q|$ the canonical basis coefficients of q, the *conjugate* for q, and the *norm* of q, respectively, i.e., $q_i \in A$, $q = q_0 + q_1 i + q_2 j + q_3 k$, $q^* \equiv q_0 - q_1 i - q_2 j - q_3 k \in (a, b/A)$, and $|q| \equiv q_0^2 - aq_1^2 - bq_2^2 + abq_3^2 \in A$.

For any two quaternions $q, t \in (a, b/A)$, the following assertions are verified by means of direct calculations.

(1) $|q| = qq^* = q^*q$.

(2) $|qt| = |q||t|$.

(3) $|t| = |t^*|$.

(4) $|0| = 0$.

(5) $(qt)^* = t^*q^*$.

(6) $(q + t)^* = q^* + t^*$.

(7) $q^{**} = q$.

(8) $d \in A$, $d^* = d$, and $|d| = d^2$.

(9) $[i, q] = 2aq_3 j + 2q_2 k$, $[j, q] = -2bq_3 i - 2q_1 k$, and $[k, q] = 2bq_2 i - 2aq_1 j$.

(10) If $x, y, z \in A$, then $(x + yi + zj)(x - yi - zj) = |x + yi + zj| = x - ay^2 - bz^2$.

6.38 Let a and b be two units of a commutative ring A, $Q \equiv (a, b/A)$, and let $q \equiv q_0 + q_1 i + q_2 j + q_3 k \in Q$.

(1) $q \in U(Q) \iff q \in U(Q)$, and $q^{-1} = |q|^{-1}q^* \iff q^* \in U(Q) \iff |q| \in U(A)$.

(2) q is regular in $Q \iff q^*$ is regular in $Q \iff |q|$ is regular in A.

(3) If q belongs to the centre of Q, then $2q = 2q_0 \in A$ (i.e., $2q_1 = 2q_2 = 2q_3 = 0$).

(4) $4q_1 = [[k, q], i](ak)^{-1} \in QqQ$, $4q_2 = [[i, q], j](bi)^{-1} \in QqQ$, $4q_3 = [k, [j, q]](abj)^{-1} \in QqQ$, $4q_0 = 4(q - q_1 - q_2 - q_3) \in QqQ$.

(5) If M is an ideal of A, $h : A \to A/M$ is a natural epimorphism, then the factor ring Q/MQ is isomorphic to the quaternion algebra $(h(a), h(b)/h(A))$ over the ring $h(A)$.

(6) If $M \in \max(A)$, then the ring of quotients Q_M is isomorphic to the quaternion algebra $(a_M, b_M/A_M)$ over the ring of quotients A_M.

(7) If q_0 is a unit of the ring A and $q_1, q_2, q_3 \in J(A)$, then q is a unit of the algebra Q.

(8) If $q_0, q_1, q_2, q_3 \in J(A)$, then $1 - q$ is a unit of Q.

(9) $J(A)Q \subseteq J(Q)$.

(10) Q is right distibutive (resp. right invariant, reduced, regular, strongly regular) $\Longleftrightarrow$ for all $M \in \max(A)$, the quaternion algebra $(a_M, b_M/A_M)$ is right distibutive (resp. right invariant, reduced, regular, strongly regular).

(11) If for all $M \in \max(A)$, the algebra $(a_M, b_M/A_M)$ is right quasi-invariant, then the algebra Q is right quasi-invariant.

(12) The norm $|q|$ of every nonzero quaternion $q \in Q$ is nonzero $\Longleftrightarrow A$ is an (a, b)-ring.

(13) If A is a reduced (a, b)-ring, then Q is reduced.

(14) Q is a domain $\Longleftrightarrow A$ is a domain, and A is an (a, b)-ring.

(15) Q is a division ring $\Longleftrightarrow A$ is a field, and A is an (a, b)-ring.

$\lhd$ (1) and (2) follow from 6.37(1). (3) follows from 6.37(9). (4), (5), (6) are directly verified.

(7) Since $q_0^2 \in U(A)$ and $-aq_1^2 - bq_2^2 + abq_3^2 \in J(A)$, we obtain $|q| = q_0^2 - aq_1^2 - bq_2^2 + abq_3^2 \in U(A)$. By (1), $q \in U(Q)$.

(8) follows from (7). (9) follows from (8) and from the fact that $J(A)Q$ is an ideal of Q. (10) follows from 6.38(6), 4.32(9), 4.32(14), and 4.32(11). (11) follows from 6.38(6) and 4.32(10).

(12)

$\Longrightarrow$ follows from 6.37(10).

$\Longleftarrow$ Let $q \equiv q_0 + q_1 i + q_2 j + q_3 k$ be a nonzero quaternion with the nonzero norm $|q|$, $x \equiv q_0 q_1 + b q_2 q_3 \in A$, $y \equiv q_1^2 - b q_3^2 \in A$, and let $z \equiv q_0 q_3 + q_1 q_2 \in A$. It is verified directly that $x^2 - a y^2 - b z^2 = y|q| = 0$. Since A is an (a, b)-ring, we have $y = q_1^2 - b q_3^2 = 0$. Since A is an (a, b)-ring, the equality $y = 0$ implies the equalities $q_1 = q_3 = 0$. Since $0 = |q| = q_0^2 - a q_1^2 - b q_2^2 + a b q_3^2$ and $q_1 = q_3 = 0$, we have $q_0^2 - b q_2^2 = 0$. We have $q_0 = q_2 = 0$, since A is an (a, b)-ring. Then $q = 0$; we have a contradiction.

(13) If $q \in Q$ and $q^2 = 0$, then $|q|^2 = |q^2| = 0$ by 6.37(2). Since A is reduced, $|q| = 0$. By assumption, $q = 0$.

(14) $\Longrightarrow$ follows from (13) and from the fact that a subring of a domain is a domain.

$\Longleftarrow$ Let $0 \neq q, t \in Q$. Since A is an (a, b)-ring without zero-divisors, it follows from (12) and 6.37(5) that $|qt| = |q||t| \neq 0$. By (12), $qt \neq 0$.

(15) $\Longleftarrow$ follows from (14) and from the fact that a finite-dimensional algebra over a field without zero-divisors is a division ring.

$\Longrightarrow$ By (14), A is an (a, b)-ring. Every nonzero element $a \in A$ has the inverse element a^{-1} in the division ring Q. It is verified directly that $a^{-1} \in A$. Therefore A is a field. $\rhd$

6.39 Let a and b be two units of a commutative ring A, $2^{-1} \in A$, and let $Q \equiv (a, b/A)$ be the quaternion algebra.

(1) The centre of Q coincides with the ring A.

(2) The ideal (q) generated by q in Q contains the ideal C_q of A generated by all basis coefficients q_0, q_1, q_2, and q_3 of q.

Consequently, $C_q C_t = C_t C_q \subseteq (q)(t) \bigcap (t)(q)$ for any $t \in Q$.

(3) If R and S are two nonzero ideals of Q, then for any ideal T of Q such that $RS \subseteq T$, there exist two nonzero ideals B and D of A such that $BD \subseteq T$, $B \subseteq R$, and $D \subseteq T$ (in particular if $RS = 0$, then $BD = 0$).

(4) Q is prime $\Longleftrightarrow$ A is a domain.

(5) Every nonzero nilpotent ideal of Q contains a nonzero nilpotent ideal of A.

(6) Q is semiprime $\Longleftrightarrow$ A is reduced.

(7) Q is reduced $\Longleftrightarrow$ A is a reduced (a, b)-ring.

$\lhd$ (1) and (2) follow from 6.38(3) and 6.38(4). (3) follows from (2). (4) and (5) follow from (3). (6) follows from (5).

(7)

$\Longleftarrow$ follows from 6.38(13).

$\Longrightarrow$ By (6), A is reduced. Let $x, y, z \in A$, $d \equiv x^2 - ay^2 - bz^2 = 0$, and let B be the ideal of A generated by x, y, z. If $q \equiv x + yi + zj \in Q$, then $qq^* = d = 0$. By 1.35(2), $(q)(q^*) = 0$. By (2), $B^2 \subseteq (q)(q^*) = 0$. Since A is semiprime, $B = 0$ and $x = y = z = 0$. $\rhd$

6.40 Let a and b be two units of a commutative ring A, $2^{-1} \in A$, and let $Q \equiv (a, b/A)$ be the quaternion algebra.

(1) Q is a simple ring $\Longleftrightarrow$ A is a field, Q is a simple four-dimensional algebra over A $\Longleftrightarrow$ A is a field.

(2) Q is a division ring $\Longleftrightarrow$ A is a field, and Q is reduced $\Longleftrightarrow$ A is a field, and Q is right quasiinvariant.

(3) If A is local and $h : A \to A/J(A)$ is the natural epimorphism, then $\mathrm{char}(A/J(A)) \neq 2$, $J(A)Q = J(Q)$, and the factor ring $Q/J(Q)$ is isomorphic to the quaternion algebra $(h(a), h(b)/h(A))$.

(4) For any $M \in \max(A)$, the factor ring $Q_M/J(Q_M)$ is a simple Artinian ring, and $2^{-1} \in Q_M$.

(5) If N is a right Q-module, then

N is distributive $\Longleftrightarrow$ N_M is a uniserial Q_M-module for all $M \in \max(A)$.

$\lhd$ (1) Since the centre of a simple ring is a field, then in view of 6.39(1) we may assume that A is a field. By 6.39(1) and 6.39(4), Q is a central four-dimensional prime algebra over A. Therefore Q is a simple algebra.

(2) By 6.38(15), we may assume that A is a field. By (1), we may assume, that Q is a simple Artinian ring. In this case, the assertion is directly verified.

(3) Since $2^{-1} \in A$, we have $\mathrm{char}(A/J(A)) \neq 2$. Set $T \equiv (h(a), h(b)/h(A))$. By 6.38(5) and 6.38(9), $Q/J(A)Q \cong T$, $J(A)Q \subseteq J(Q)$. By (1) applied to the quaternion algebra T over the field $A/J(A)$, T is a simple Artinian ring. Therefore $J(Q/J(A)Q) = 0$. Then, taking into account the inclusion $J(A)Q \subseteq J(Q)$, we obtain the equality $J(A)Q = J(Q)$.

(4) follows from (3), 6.38(6), 4.32(2), and 4.32(12). (5) follows from (4) and 4.32(15). $\triangleright$

6.41 *Let a and b be two units of a commutative local ring A, and let $2^{-1} \in A$. The following conditions are equivalent.*

(1) The quaternion algebra $(a, b/A)$ is a local ring.

(2) $(a, b/A)$ is right quasi-invariant.

(3) For any $x, y, z \in A$, the inclusion $x^2 - ay^2 - bz^2 \in J(A)$ is possible only for $x \in J(A), y \in J(A), z \in J(A)$.

$\triangleleft$ (1)$\Longrightarrow$(2) is obvious.

(2)$\Longrightarrow$(3) Since Q is right quasi-invariant, $Q/J(Q)$ is right quasi-invariant. (3) follows now from 6.40(2), 6.40(3).

(3)$\Longrightarrow$(1) Let $Q \equiv (a, b/A)$, $h : A \to A/J(A)$ be the natural epimorphism onto the field $A/J(A)$, and let $T \equiv (h(a), h(b)/h(A))$. By 6.40(3), $\mathrm{char}(h(A)) \neq 2$. It follows from 6.38(15), that condition (3) is equivalent to the following condition (3*): $(h(a), h(b)/h(A))$ is a division ring. (3*) $\Longrightarrow$(1) follows from the fact that $T \cong Q/J(Q)$ by 6.40(3). $\triangleright$

6.42 *Let a and b be two units of a commutative ring A and let $2^{-1} \in A$. The following conditions are equivalent.*

(1) The algebra $(a, b/A)$ is right quasi-invariant.

(2) For any $M \in \max(A)$, the algebra $(h(a), h(b)/h(A))$ is local.

(3) For any $M \in \max(A)$ and for any $x, y, z \in A$, the inclusion $x^2 - ay^2 - bz^2 \in M$ is possible only for $x \in M, y \in M, z \in M$.

$\triangleleft$ (3)$\Longrightarrow$(2) By 4.32(2) and 4.32(12), for all $M \in \max(A)$, the ring A_M is local, $J(A_M) = M_M$, and $2^{-1} \in A$. Therefore, (2) follows from 6.41.

(2)$\Longrightarrow$(1) follows from 4.32(10) and 6.38(6).

(1)$\Longrightarrow$(3) Let $M \in \max(A)$, and let $h : A \to A/M$. Let $Q \equiv (a, b/A)$ and $T \equiv (h(a), h(b)/h(A))$. By 6.38(5), $Q/MQ \cong T$, whence T is right quasi-invariant if Q is right quasi-invariant. Further, we apply 4.32(13) and 6.41. $\triangleright$

6.43 [200]. *Let a and b be two units of a commutative local ring A, $Q \equiv (a, b/A)$, and let $2^{-1} \in A$. The following conditions are equivalent.*

   (1) *Q is right invariant.*

   (2) *Q is right distributive.*

   (3) *Q is right uniserial.*

   (4) *Q is right quasi-invariant, and A is uniserial.*

   (5) *Q is a uniserial invariant ring.*

   (6) *A is uniserial, and for any quaternion $q \in Q$, the equality $Qq = qQ = q_i Q = Qq_i$ holds at least for one of the canonical basis coefficients $q_i \in A$.*

   (7) *A is uniserial, and for any $x, y, z \in A$, the inclusion $x^2 - ay^2 - bz^2 \in J(A)$ is possible only for $x \in J(A), y \in J(A), z \in J(A)$.*

◁ (3)$\Longrightarrow$(4), (6)$\Longrightarrow$(5), (5)$\Longrightarrow$(3), (3)$\Longrightarrow$(2), and (5)$\Longrightarrow$(1) are directly verified. (4)$\Longleftrightarrow$(7) follows from 6.41. (2)$\Longrightarrow$(3) follows from 6.40(5).

(1)$\Longrightarrow$(4) Let $x, y \in A$. By assumption, $(x + yi)Q$ is an ideal of Q. Therefore, 6.39(2) shows us that there exist coefficients $g, h, m, n \in A$ such that $x = (x + yi)(m + gi + hj + nk)$. Equating the coefficients of the identity element first, and then the coefficients of i in this equality, we obtain the equality $x = xm + yag$, $0 = ym + xg$. Then $x(1 - m) \in yA$, $ym \in xA$, and in view of the fact that A is local, at least one of the m, $1 - m$ is a unit of A. Therefore, either $x \in yA$ or $y \in xA$. Hence A is uniserial.

(7)$\Longrightarrow$(6) The ring A is uniserial. Therefore, among the basis coefficients q_0, q_1, q_2, q_3 of q, there exists a coefficient d such that $dA \supseteq q_n A$ for $0 \leq n \leq 3$. Since $qQ = qiQ = qjQ = qkQ$, in view of the invertibility of i, j, k, we may assume that $d = q_0$. Then there exist $t_0 = 1$ and $t_1, t_2, t_3 \in A$ such that $q_n = dt_n$ for $0 \leq n \leq 3$. Set $t \equiv t_0 + t_1 i + t_2 j + t_3 k$. By 6.41, the ring Q is local and $J(Q) = J(A)Q$, whence $t \in U(Q)$. Hence $Qq = Qtd = Qd = dQ = dtQ = qQ$. ▷

6.44 [197], [198]. *Let a and b be two units of a commutative ring A, and let $2^{-1} \in A$. Then the following conditions are equivalent.*

   (1) *$(a, b/A)$ is right distributive.*

   (2) *$(a, b/A)$ is right invariant.*

   (3) *For any $M \in \max(A)$, the algebra $(a_M, b_M/A_M)$ is right uniserial.*

   (4) *For any $M \in \max(A)$, the algebra $(a_M, b_M/A_M)$ is right invariant.*

   (5) *For any $M \in \max(A)$, the algebra $(a_M, b_M/A_M)$ is a uniserial invariant ring.*

(6) *For any $M \in \max(A)$, the ring of quotients A_M is a uniserial ring, and for any $x, y, z \in A$, the inclusion $x^2 - ay^2 - bz^2 \in M$ is possible only for $x \in M, y \in M, z \in M$.*

(7) *A is a distributive ring, and for all $M \in \max(A)$ and for any elements x, y, z of the factor ring A/M, the equality $x^2 - ay^2 - bz^2 = 0$ is possible only for $x = y = z = 0$.*

◁ $(5) \Longleftrightarrow (6)$ follows from 4.32(2) and 6.43. $(6) \Longleftrightarrow (7)$ follows from 3.22. $(4) \Longleftrightarrow (2)$ and $(3) \Longleftrightarrow (1)$ follow from 6.38(5) and 6.40(5). $(5) \Longrightarrow (4)$ and $(5) \Longrightarrow (3)$ are directly verified. $(3) \Longrightarrow (5)$ and $(4) \Longrightarrow (5)$ follow from 6.43 and 4.32(12). ▷

6.45 Let A be a commutative ring of characterisic 2, H be the a direct product of two cyclic groups of order 2, and let Q be the ring which is a free A-module with the basis $1, i, j, k$, in which multiplication is defined by basis relations $i^2 = j^2 = k^2 = 1^2 = 1$, $ij = ji = 1k = k1 = k$, $jk = kj = 1i = i1 = i$, and $ik = ki = 1j = j1 = j$. Let, further, $t \equiv 1 + i + j + k$, $u \equiv 1 + i$, $v \equiv j + k$, $x \equiv i + j$, $y \equiv 1 + k$, and $w \equiv 1 + j$ be elements of Q.

(1) Q is a commutative ring of characterisic 2 which is isomorphic both to the group ring $A[H]$ and the quaternion algebras $(-1, -1/A)$, $(1, 1/A)$.

(2) $uA + vA = uQ = vQ$ is a nonzero ideal in Q whose square is equal to zero.

(3) $xA + yA = xQ = yQ \subseteq uQ + wQ$ and $xQ \bigcap uQ = xQ \bigcap wQ = tQ = tA \neq xQ$.

(4) $xQ \bigcap (uQ + wQ) \neq xQ \bigcap uQ + xQ \bigcap wQ$.

(5) The ring Q is neither distributive nor semiprime.

◁ (1), (2), and (3) are directly verified. (4) follows from (3). (5) follows from (4). ▷

6.46 Let A be a unitary subring of R, and let $_A A$ be isomorphic to a direct summand of the module $_A R$ (e.g., this is the case if $_A R$ is free).

(1) There exists an A-module epimorphism $t : {}_A R \to {}_A A$ with $t(x) = x$ for any $x \in A$.

(2) If R is right distributive, then A is right distributive.

(3) If $a, b \in A$ and $a \in bR$, then $a \in bA$ (in particular, if $a \in a^2 R$, then $a \in a^2 A$).

(4) If an element $a \in A$ is right invertible in R, then a is right invertible in A.

(5) If R is local, then A is local.

(6) If R is right uniserial, then A is right uniserial.

(7) If R is strongly regular, then A is strongly regular.

(8) If R is regular and A is a subring of the centre of R, then A is regular.

(9) If R is right invariant, then A is right invariant.

$\lhd$ (1) By assumption, $_AR = {_AF} \oplus {_AB}$, where $_AF$ is a free cyclic module with a free generator f. Let $h : {_AR} \to {_AF}$ be the projection with kernel B, $g : {_AF} \to {_AA}$ be an isomorphism such that $g(f) = 1$, and let $t \equiv gh \in \mathrm{Hom}(_AR, {_AA})$. Then $t(x) = x$ for any $x \in A$.

(2) Since R is right distributive, 2.4 shows us that there exist $\bar{a}, \bar{b}, \bar{c}, \bar{d} \in R$ such that the following equalities (*) hold: $1 = \bar{a} + \bar{b}$, $m\bar{a} = n\bar{c}$, $n\bar{b} = m\bar{d}$. Let $a \equiv t(\bar{a})$, $b \equiv t(\bar{b})$, $c \equiv t(\bar{c})$, and let $d \equiv t(\bar{d})$. Since $t(x) = x$ for any $x \in A$, it follows that applying t to (*), we obtain the following equalities $1 = a + b$, $ma = nc$, and $nb = md$. By 2.4, A is right distributive.

(3) Let $a = bx$, where $x \in R$, and let $d \equiv t(x) \in A$. Then $a = t(a) = t(bx) = bd \in bA$.

(4) Since $1 \in aR$, we obtain $1 \in aA$ by (3).

(5) follows from (4) and from the fact that an arbitrary ring is local $\Longleftrightarrow$ that for any element x of this ring, at least one of the elements x and $1 - x$ is right invertible. (6), (7), and (8) follow from (3).

(9) For any $a, b \in A$, it is sufficient to prove the existence of $c \in A$ such that $ba = ac$. Since R is right invariant, there exists $m \in R$ such that $ba = am$. Let $t : {_AR} \to {_AA}$ be the epimorphism of (1). Hence $ba = t(ba) = t(am) = at(m) = ac$, where $c \equiv t(m) \in A$. $\rhd$

6.47 Let A be a commutative ring, G be the quaternion group of order 8 with generators α, β and relations $\alpha^4 = 1$, $\alpha^2 = \beta^2$, and $\beta^{-1}\alpha\beta = \alpha$.

(1) If $\mathrm{char}(A) = 2$, then the group ring $A[G]$ is not right distributive.

(2) If $2^{-1} \in A$, then $A[G]$ is isomorphic to the direct product of the Hamiltonian quaternion algebra $(-1, -1/A)$ and four isomorphic copies of A.

(3) If either $(-1, -1/A)$ or $A[G]$ is right distributive, then $2^{-1} \in A$ and A is distributive.

(4) Either $2^{-1} \in A$, or there exists a maximal ideal M of A such that A/M is a field of characterisic 2, in which the equation $x^2 + y^2 + z^2 = 0$ has the solution $\{x = 1, y = 1, z = 0\}$, and the Hamiltonian quaternion algebra over the field A/M is not a semiprime ring and is isomorphic to a factor ring of the algebra $(-1, -1/A)$.

$\lhd$ (1) Assume the contrary. The quaternion group G has a factor group H which is isomorphic to a direct product of two cyclic groups of order 2. The group ring $A[H]$ is isomorphic to a factor ring of $A[G]$ and therefore is right distributive; we have a contradiction to 6.45(1), 6.45(4).

(2) Let $R \equiv A[G]$, $t \equiv 2^{-1}(1-\alpha^2)$, $u \equiv 8^{-1}(1+\alpha+\alpha^2+\alpha^3)(1+\beta)$, $x \equiv 8^{-1}(1+\alpha+\alpha^2+\alpha^3)(1-\beta)$, $y \equiv 8^{-1}(1-\alpha+\alpha^2-\alpha^3)(1+\beta)$, $z \equiv 8^{-1}(1-\alpha+\alpha^2-\alpha^3)(1-\beta)$, $i \equiv \alpha t$, $j \equiv \beta t$, and let $k \equiv \alpha\beta t$.

It is verified directly that t, u, x, y, and z are central orthogonal idempotents of R, and the rings uR, xR, yR, and zR are isomorphic to R, $i^2 = j^2 = k^2 = -t$, $ij = -ji = k$, $jk = -kj = i$, and $ki = -ik = j$. Therefore $tR \cong (-1, -1/A)$.

(3) Since $(-1, -1/A)$ and $A[G]$ are free A-modules, 6.46(2) shows us that A is distributive. Assume that $A \neq 2A$. Let $\overline{A} \equiv A/2A$, $\overline{Q} \equiv (-1, -1/\overline{A})$. Since the rings $\overline{Q}$ and $\overline{A}[G]$ are isomorphic to factor rings of the rings $(-1, -1/A)$ and $A[G]$, respectively, it follows that either $\overline{Q}$ or $\overline{A}[G]$ is right distributive. The first case contradicts 6.45(1) and 6.45(4). Since the direct product H of two cyclic groups of order 2 is a homomorphic image of the quaternion group G, the ring $A[H]$ is isomorphic to a factor ring of $A[G]$. Therefore, in the second case, $A[H]$ is right distributive; we have a contradiction to 6.45(1) and 6.45(4).

(4) Assume that $2A \neq A$. Then $2A \subseteq M \in \max(A)$. Therefore, the equation $x^2 + y^2 + z^2 = 0$ has a solution $x = 1$, $y = 1$, $z = 0$ in A/M. The remaining assertions follow from 6.45(5) and 6.38(5). $\triangleright$

6.48 *For a commutative ring A, the following conditions are equivalent.*

(1) The Hamiltonian quaternion algebra $(-1, -1/A)$ is right distributive.

(2) The group ring $A[G]$, where G is the quaternion group of order 8, is right distributive.

(3) The ring of quotients A_M with respect to any $M \in \max(A)$ is uniserial, and the equation $x^2 + y^2 + z^2 = 0$ has the unique solution $x = y = z = 0$ in the field A/M.

(4) A is distributive, and for all $M \in \max(A)$, the equation $x^2 + y^2 + z^2 = 0$ has the unique solution $x = y = z = 0$ in the field A/M.

$\triangleleft$ (1)$\Longleftrightarrow$(2) follows from 6.47(2), 6.47(3). (3)$\Longleftrightarrow$(4) follows from 4.31(4).

(1)$\Longleftrightarrow$(4) If $2^{-1} \in A$, then (1)$\Longleftrightarrow$(4) follows from 6.44. If one of conditions (1), (4) holds, then the inclusion $2^{-1} \in A$ follows from 6.47(3) and 6.47(4). $\triangleright$

6.49 *Let a and b be two units of a commutative ring A, and let $2^{-1} \in A$.*

(1) *For any $M \in \max(A)$, the quaternion algebra $(a_M, b_M/A_M)$ is a matrix-local ring, and in particular, $(a_M, b_M/A_M)$ is an indecomposable ring.*

(2) *$(a, b/A)$ is regular $\iff$ $(a_M, b_M/A_M)$ is a simple Artinian ring for all $M \in \max(A)$ $\iff$ A is regular.*

(3) *$(a, b/A)$ is strongly regular $\iff$ $(a_M, b_M/A_M)$ is a division ring for all $M \in \max(A)$ $\iff$ A is regular, and for each $M \in \max(A)$ and for any elements x, y, z of the factor ring A/M, the equality $x^2 - ay^2 - bz^2 = 0$ is possible only for $x = y = z = 0$.*

$\triangleleft$ (1) follows from 6.38(6) and 6.40(4).

(2) The first equivalence follows from 6.49(1) and 5.38(5). Let us prove the second equivalence. By 6.46(8), the regularity of $(a, b/A)$ implies the regularity of A. If A is regular, then A_M is a field and, by 4.32(12), 6.38(6), and 6.40(1), $(a_M, b_M/A_M)$ is a simple Artinian ring for any $M \in \max(A)$.

(3) Since strongly regular rings are distributive, the assertion follows from 6.49(2) and 6.44. $\triangleright$

6.50 *Let A be a commutative ring.*

(1) *The Hamiltonian quaternion algebra $(-1, -1/A)$ is regular $\iff$ $(a_M, b_M/A_M)$ is a simple Artinian ring for all $M \in \max(A)$ $\iff$ A is regular.*

(2) *$(-1, -1/A)$ is strongly regular $\iff$ $(a_M, b_M/A_M)$ is a division ring for all $M \in \max(A)$ $\iff$ A is regular, and for every $M \in \max(A)$ and for any elements x, y, z of A/M, the equality $x^2 + y^2 + z^2 = 0$ is possible only for $x = y = z = 0$.*

(3) *Assume that for all $M \in \max(A)$, the field A/M is isomorphic to a subfield of the field of real numbers.*

If A is a distributive (resp. regular) ring, then $(-1, -1/A)$ is distributive (resp. strongly regular).

$\triangleleft$ (1) and (2) follow from 6.49(2), 6.49(3), and 6.47(4). (3) follows from (1),(2), and 6.48. $\triangleright$

6.4 Polynomial rings

6.51 Let φ be an injective endomorphism of a ring A. We denote by $A_\ell[[x, \varphi]]$ the *left skew (power) series ring* consisting of formal series $\sum_{i=0}^{\infty} a_i x^i$ of the variable x with canonical coefficients $a_i \in A$, where addition is defined naturally and multiplication is defined by the rule $x^i a = \varphi^i(a) x^i$.

The *right skew (power) series ring* $A_r[[x, \varphi]]$ consists of series $\sum_{i=0}^{\infty} x^i a_i$, and their multiplication is defined by the rule $ax^i = x^i \varphi^i(a)$.

The *left skew polynomial ring* $A_\ell[x, \varphi] \subset A_\ell[[x, \varphi]]$ and *right skew polynomial ring* $A_r[x, \varphi] \subset A_r[[x, \varphi]]$ are the subrings of skew power series rings $A_\ell[[x, \varphi]]$ and $A_r[[x, \varphi]]$, respectively, consisting of the series with a finite number of nonzero coefficients.

Let φ be an automorphism of a ring A. Analogously, we define the *left skew Laurent series ring* $A_\ell((x, \varphi))$ and the *right skew Laurent series ring* $A_r((x, \varphi))$ consisting of the series $f \equiv \sum_{i=m}^{\infty} a_i x^i$ and $g \equiv \sum_{i=n}^{\infty} x^i a_i$, where $m = m(f), n = n(g)$ are (maybe, negative) integers, and $x^i a = \varphi^i(a) x^i$ in the left-side case, $ax^i = x^i \varphi^i(a)$ in the right-side case. It follows from the two last equalities that the set $T \equiv \{x^i\}_{i=0}^{\infty}$ is a right and left denominator set both in the ring $A_\ell[[x, \varphi]]$ and in $A_r[[x, \varphi]]$, and isomorphisms $(A_\ell[[x, \varphi]])_T \cong A_\ell((x, \varphi))$, $(A_r[[x, \varphi]])_T \cong A_r((x, \varphi))$ are directly verified.

The *left skew Laurent polynomial ring* $A_\ell[x, x^{-1}, \varphi] \subset A_\ell((x, \varphi))$ and the *right skew Laurent polynomial ring* $A_r[x, x^{-1}, \varphi] \subset A_r((x, \varphi))$ are the subrings of skew Laurent series rings $A_\ell((x, \varphi))$ and $A_r((x, \varphi))$, respectively, consisting of the series with a finite number of nonzero coefficients.

It follows from the equalities $x^i a = \varphi^i(a) x$ (for $A_\ell((x, \varphi))$) and $ax^i = x\varphi^i(a)$ (for $A_r((x, \varphi))$) that the set $T \equiv \{x^i\}_{i=0}^{\infty}$ is a two-sided Ore subset both in the rings $A_\ell[x, x^{-1}, \varphi]$ and $A_r[x, x^{-1}, \varphi]$, and isomorphisms $(A_\ell[x, \varphi])_T \cong A_\ell[x, x^{-1}, \varphi]$, $(A_r[x, \varphi])_T \cong A_r[x, x^{-1}, \varphi]$, $(A_\ell[[x, \varphi]])_T \cong A_\ell((x, \varphi))$, $(A_r[[x, \varphi]])_T \cong A_r((x, \varphi))$ are directly verified.

If the series f belongs to one of the rings $A_\ell[[x, \varphi]]$, $A_r[[x, \varphi]]$, $A_\ell((x, \varphi))$, $A_r((x, \varphi))$, then we denote by f_i and by $C(f)$ the coefficient $a_i \in A$ of x^i in the canonical form of f and the *content* of f, (i.e., the ideal of A generated by all coefficients f_i), respectively. If a polynomial g belongs to one of the rings $A_\ell[x, \varphi]$, $A_r[x, \varphi]$, $A_\ell[x, x^{-1}, \varphi]$, and $A_r[x, x^{-1}, \varphi]$, then we denote by $\deg(g)$ the *degree* of g.

6.52 Let φ be an injective endomorphism of a ring A. Then the following conditions are equivalent.

(1) The ring $A_\ell[[x, \varphi]]$ is reduced.

(2) $A_\ell[x, \varphi]$ is reduced.

(3) $A_r[[x, \varphi]]$ is reduced.

(4) $A_r[x, \varphi]$ is reduced.

(5) A is reduced, and $a\varphi^n(a) \neq 0$ for any nonzero $a \in A$ and for any $n \geq 0$.

(6) A is reduced, and $\varphi^n(a)a \neq 0$ for any nonzero $a \in A$ and for any $n \geq 0$.

◁ It is sufficient to prove the equivalence of conditions (1), (2), (5), and (6).

(1)$\Longrightarrow$(2) follows from the fact that $A_\ell[x, \varphi] \subseteq A_\ell[[x, \varphi]]$.

(2)$\Longrightarrow$(5) Since $A \subseteq A_\ell[x, \varphi]$, A is reduced. Let $a \in A$, and let $a\varphi^n(a) = 0$. Then $(ax^n)^2 = a\varphi^n(a)x^{2n} = 0$. Therefore $ax^n = 0$ and $a = 0$.

(5)$\Longleftrightarrow$(6) follows from 1.35(2).

(5)$\Longrightarrow$(1) Assume the contrary. There exist $n \geq 0$ and a series $f \in A_\ell[[x, \varphi]]$ with the nonzero constant term a such that $(fx^n)^2 = 0$. Compare coefficients of x^{2n} in both sides of this equality. Then $0 = a\varphi^n(a)$; we have a contradiction. ▷

6.53 Let φ be an automorphism of a ring A. Then the following conditions are equivalent.

(1) The ring $A_\ell((x, \varphi))$ is reduced.

(2) $A_\ell[x, x^{-1}, \varphi]$ is reduced.

(3) $A_r((x, \varphi))$ is reduced.

(4) $A_r[x, x^{-1}, \varphi]$ is reduced.

(5) A is reduced, and $a\varphi^n(a) \neq 0$ for any nonzero $a \in A$ and for any integer n.

(6) A is reduced, and $\varphi^n(a)a \neq 0$ for any nonzero $a \in A$ and any integer n.

◁ The proof of 6.53 is similar to the proof of 6.52. ▷

6.54 Let φ be an injective endomorphism of a ring A, and let $R \equiv A_\ell[x, \varphi]$.

(1) If A is a domain, then the rings R, $A_r[x, \varphi]$, $A_\ell[[x, \varphi]]$, and $A_r[[x, \varphi]]$ are domains.

(2) If for any $a \in A$, the right ideal $(a + x)R$ of R is an ideal of this ring, then φ is the identity automorphism, and A is commutative.

(3) If A is a division ring, and one of the rings R, $A_\ell[[x, \varphi]]$ is right uniform, then φ is an automorphism, and R is a right and left principal ideal domain.

(4) If A is a division ring and R is right distributive, then A is a field, and φ is the identity automorphism.

◁ (1) follows from the fact that the constant term of a product of two series is equal to the product of the constant terms of these series.

(2) Let $0 \neq a, b \in A$. Since $(a + x)R$ is an ideal of R, there exists a nonzero polynomial f of the degree $n \geq 0$ depending on a and b such that $b(a + x) = (a + x)f$. We take the polynomial $(a + x)f$ in the form $(a + x)f = \varphi(f_n)x^{n+1} + g$, where $\varphi(f_n) \neq 0$ and $\deg(g) \leq n$.

It follows from the equality $b(a + x) = \varphi(f_n)x^{n+1} + g$ that $n = 0$ and $f = f_0 \equiv c(a, b) \in A$. We obtain the equality (*): $ba + bx = ac + \varphi(c(a, b))x$. Substituting in (*) the identity element for a, we obtain that for any nonzero $b \in A$, there exists a nonzero $d \equiv c(1, b)$ such that $b + bx = d + \varphi(d)x$. Then $b = d = \varphi(d)$, whence it follows that $b = \varphi(b)$, $\varphi \equiv 1$, and the equality (*) is turned into the equality (**): $ba + bx = ac(a, b) + c(a, b)x$. Equating in (**) the coefficient of x first and then the constant terms, we obtain equalities $b = c(a, b)$ and $ba = ab$. Therefore A is commutative.

(3) Let a be a nonzero element of a division ring A. If one of the rings R and U is right uniform, then in both cases, for ax and x, there exist series $f, g \in R$ such that $axf = xg \neq 0$. Cancelling by the appropriate degree of x, we may assume that one of the series f and g has a nonzero constant term. Then it follows from the equality $axf = xg$ and (1) that $f_0 \neq 0$ and $g_0 \neq 0$. Equating in both sides of the equality $axf = xg$ the coefficients of x, we obtain the equality $a\varphi(f_0) = \varphi(g_0)$. Since A is a division ring, we have $a = \varphi(g_0)\varphi(f_0)^{-1} = \varphi(g_0 f_0^{-1})$. Therefore φ is an automorphism. Then U is a Euclidean ring, and in particular, U is a right and left principal ideal domain.

(4) By (1), R is a right distributive domain. By 3.12(2), R is right uniform. By (3), R is a right distributive right Noetherian domain. By 2.72, R is right invariant. By (2), $\varphi \equiv 1$ and A is a field. $\triangleright$

6.55 Let φ be an injective endomorphism of a ring A, $R \equiv A_\ell[x, \varphi]$, and $R(n) \equiv R/Rx^n$.

(1) $A \cong R/Rx$.

(2) Rx^n is an ideal of R for any positive integer n.

(3) $R(n)(x + Rx^n)$ is a nilpotent ideal of the ring $R(n)$, and $A \cong R(n)/R(n)(x + Rx^n)$ for any n.

(4) Set $T \equiv \{x^n\}_{n=0}^{\infty}$.

Then all elements of T are regular, and T is a left Ore set in R.

If φ is an automorphism of A, then the left ring of quotients $_T R$ is isomorphic to the left skew Laurent polynomial ring $A_\ell[x, x^{-1}, \varphi]$.

(5) If A is strongly regular, then $R(n)/J(R(n))$ is strongly regular.

(6) If A is a strongly regular ring, then

$R(n)$ is right distributive $\iff R(n)$ is a right Bezout ring.

$\triangleleft$ (1), (2), and (3) are directly verified.

(4) The equality $x^n a = \varphi^n(a)x^n$ implies that T is left permutable. It is verified directly that all elements of T are regular.

(5) follows from (3). (6) follows from (5) and 5.15. $\triangleright$

6.56 For a reduced ring A, the following conditions are equivalent.

(1) A is right or left Rickartian.

(2) A is Rickartian.

(3) For any finitely generated ideal B of A, there exists a central idempotent $e \in A$ such that $r(B) = \ell(B) = eA$.

(4) Every element of A is a product of a central idempotent and a regular element.

◁ By 1.35(1), the reduced ring A is normal. Therefore, the equivalence of conditions (1), (2), and (4) follows from 3.36. (3)$\Longrightarrow$(2) follows from 1.35(2).

(2)$\Longrightarrow$(3) Let $B = \sum_{i=1}^{n} Ab_i A$, e_i be central idempotents of A such that $r(b_i) = e_i A$, and let $e \equiv e_1 \cdot \ldots \cdot e_n$. It follows from 1.35(2) that $r(B) = \ell(B) = \bigcap(e_i A) = eA$. ▷

6.57 Let φ be an injective endomorphism of a reduced ring A such that $a\varphi^n(a) \neq 0$ for any nonzero $a \in A$ and each $n \geq 0$.

(1) Rings $A_\ell[x, \varphi]$, $A_r[x, \varphi]$, $A_\ell[[x, \varphi]]$, $A_r[[x, \varphi]]$, the left ring of quotients of $A_\ell[[x, \varphi]]$ with respect to the left denominator set $T \equiv \{x^n\}_{n=0}^{\infty}$, and the right ring of quotients of $A_r[[x, \varphi]]$ with respect to a right denominator set T are all reduced.

(2) If $a \in A$, then $r_A(a) = \ell_A(a) = \{b \in A \mid (b) \bigcap (\sum_{i=0}^{\infty} A\varphi^i(a)A) = 0\}$, and consequently, $r(a) \subseteq r(\varphi(a))$, $r(\varphi^n(a)) \subseteq \bigcap_{j=n}^{\infty} r(\varphi^j(a))$ for any $n \geq 0$.

(3) Let $a, b \in A$. Then
$ab = 0 \iff 0 = (\sum_{i=0}^{\infty} A\varphi^i(a)A)(\sum_{i=0}^{\infty} A\varphi^i(b)A) = (\sum_{i=0}^{\infty} A\varphi^i(b)A)(\sum_{i=0}^{\infty} A\varphi^i(a)A)$.

(4) Let either $f, g \in A_\ell[[x, \varphi]]$ or $f, g \in A_r[[x, \varphi]]$. Then
$fg = 0 \iff C(f) \bigcap C(g) = 0$.

(5) $A_\ell[x, \varphi]$, $A_r[x, \varphi]$, $A_\ell[[x, \varphi]]$, and $A_r[[x, \varphi]]$ are normal rings, and all their idempotents belong to A.

Consequently, if X is any one of these rings, then X is indecomposable $\iff X$ has no nontrivial idempotents $\iff A$ is indecomposable.

(6) $\varphi(e) = e$ for any idempotent $e \in A$.

In addition, if $\overline{A}$ is a factor ring of A with respect to the ideal B generated by some set of idempotents E, then φ induces an injective endomorphism $\overline{\varphi}$ of $\overline{A}$.

◁ (1) follows from 6.52 and 1.52. (2) follows from 1.35(2) and from the fact that $a\varphi^n(a) \neq 0$ for any $a \in A$ and for each nonnegative integer n. (3) follows from (2).

(4) Consider only the case $f, g \in A_\ell[[x, \varphi]]$. If $C(f) \bigcap C(g) = 0$, then $f_i g_j = 0$ for all i, j. By (3), $f_i \varphi^i(g_j) = 0$ for all i, j. Therefore $fg = 0$. Assume now that $fg = 0$. To prove the equality $C(f) \bigcap C(g)$, it is sufficient to prove that $f_i g_j = 0$ for $0 \leq i, j \leq \infty$.

First, let us prove by the induction on j that $f_0 g_j = 0$ for any j. For $j = 0$, the assertion follows from the fact that $f_0 g_0 = (fg)_0 = 0$. Let $f_0 g_j = 0$ for $0 \leq j \leq n$. In view of (3), $f_0 f_i \varphi^i(g_j) = 0$ for all i for $0 \leq j \leq n$. Hence $0 = f_0(fg)_{n+1} = f_0(\sum_{i+j=n+1} f_i \varphi^i(g_j)) = f_0^2 g_{n+1}$. Therefore $(f_0 g_{n+1} f_0)^2 = 0$, $f_0 g_{n+1} f_0 = 0$, $(f_0 g_{n+1})^2 = 0$, $f_0 g_{n+1} = 0$. It has been proved that equalities $f_i g_j = 0$ hold for $i = 0$ and for all j.

We use the induction on i. Assume that $f_i g_j = 0$ for $0 \leq i \leq n$ and $0 \leq j \leq \infty$. By (3), $f_i \varphi^i(g_j) = 0$ for $0 \leq i \leq n$. We take f in the form $f = \sum_{i=0}^n f_i x^i + t x^{n+1}$, where $t = \sum_{i=n+1}^\infty f_i x^{i-n-1}$. Then $fg = 0$ and $(\sum_{i=0}^n f_i x^i)g = 0$. Therefore $tg = 0$, and $t_0 = f_{n+1}$. By the induction hypothesis applied to t we obtain that $0 = t_0 g_j = f_{n+1} g_j$, where $0 \leq j \leq \infty$.

(5) Since the other cases can be treated analogously, it is sufficient to consider only the case of $A_\ell[[x, \varphi]] \equiv R$. Let $f = f^2 \in R$ and $g \equiv 1 - f$. We have to prove that $f_i = 0$ for $i \geq 1$. Since $fg = 0$, (4) shows us that $C(f) \bigcap C(g) = 0$ and $f = 1 - g$, $f_i = -g_i$ for $i \geq 1$. Therefore $f_i \in C(f) \bigcap C(g) = 0$. Since R is reduced, R is normal.

(6) Let $e = e^2 \in A$. By (5), A is normal. By (2), $r(e) = (1-e)A \subseteq r(\varphi(e))$. Therefore $\varphi(e) = e\varphi(e) \in eA$. Hence φ maps from any direct summand of A into itself. Denote by f the central idempotent $e(1 - \varphi(e))$. Then $\varphi(f) \in f(A)$ and $\varphi(f) = f\varphi(f) = e(1 - \varphi(e))\varphi(e)\varphi(1 - \varphi(e)) = 0$. Therefore $f = 0$ and $e = e\varphi(e) = \varphi(e)$. Since φ maps from each principal ideal generated by an idempotent into itself, $\varphi(B) \subseteq B$, whence φ induces an endomorphism $\overline{\varphi}$ of $\overline{A}$. Let $x + B \in \mathrm{Ker}(\overline{\varphi})$, where $x \in A$. Then $\varphi(x) \in B$ and $\varphi(x) = \sum_{i=1}^n e_i a_i$, where $e_i \in E$ and $a_i \in A$. Let $f \equiv \prod_{i=1}^n (1 - e_i)$. Then $\varphi(fx) = \varphi(f)\varphi(x) = f\sum_{i=1}^n e_i a_i = \sum_{i=1}^n f e_i a_i = 0$. Therefore $fx = 0$, $x = (1 - \prod_{i=1}^n (1 - e_i))x \in \sum_{i=1}^n e_i A \subseteq B$, and $\overline{\varphi}$ is a monomorphism. $\triangleright$

6.58 Let φ be an injective endomorphism of a reduced ring A such that $a\varphi^n(a) \neq 0$ for any nonzero $a \in A$ and each $n \geq 0$. Then the following conditions are equivalent.

(1) $A_\ell[x, \varphi]$ is right Rickartian.

(2) $A_\ell[x, \varphi]$ is left Rickartian.

(3) $A_r[x, \varphi]$ is right Rickartian.

(4) $A_r[x, \varphi]$ is left Rickartian.

(5) The annihilator of any finitely generated ideal in the reduced ring A is generated by a central idempotent.

(6) A is right or left Rickartian.

$\triangleleft$ It is sufficient to prove the equivalence of conditions (1), (2), (5), and (6). (5)$\Longleftrightarrow$(6) follows from 6.56.

Set $R \equiv A_\ell[x, \varphi]$. By 6.57(1), R is reduced. By 6.56, R is right (left) Rickartian $\iff$ any series $f \in R$ has the form $f = ed$, where e is a central idempotent, d is a regular element of R, and $e \in A$ by 6.57(5). The equivalence of conditions (1), (2), and (5) follows now from the fact that each finitely generated ideal of A has the form $C(g)$ for some polynomial $g \in R$. $\triangleright$

6.59 Let φ be an injective endomorphism of a reduced ring A such that $a\varphi^n(a) \neq 0$ for any nonzero $a \in A$ and each $n \geq 0$. Then the following conditions are equivalent.

(1) $A_\ell[[x, \varphi]]$ is right Rickartian.

(2) $A_\ell[[x, \varphi]]$ is left Rickartian.

(3) $S \equiv A_r[[x, \varphi]]$ is right Rickartian.

(4) $S \equiv A_r[[x, \varphi]]$ is left Rickartian.

(5) The annihilator of any countably generated ideal in a reduced ring A is generated by a central idempotent.

$\triangleleft$ It is sufficient to prove the equivalence of conditions (1), (2), (5), and (6). (5)$\iff$(6) follows from 6.56.

Set $R \equiv A_\ell[[x, \varphi]]$. By 6.57(1), R is reduced. By 6.56, R is right (left) Rickartian $\iff$ any series $f \in R$ has the form $f = ed$, where e is a central idempotent, d is a regular element of R and $e \in A$ by 6.57(5). The equivalence of conditions (1), (2), and (5) follows now from the fact that each countably generated ideal of A has the form $C(g)$ for some series $g \in R$. $\triangleright$

6.60 Let φ be an injective endomorphism of a reduced Rickartian ring A such that $\varphi(e) = e$ for any idempotent $e \in A$.

(1) $a\varphi^n(a) \neq 0$ and $\varphi^n(a)a \neq 0$ for any nonzero $a \in A$ and for each nonnegative integer n.

(2) $A_\ell[x, \varphi]$ and $A_r[x, \varphi]$ are Rickartian reduced rings.

(3) $A_\ell[[x, \varphi]]$ and $A_r[[x, \varphi]]$ are reduced rings.

(4) $A_\ell[[x, \varphi]]$ is Rickartian $\iff$ $A_r[[x, \varphi]]$ is Rickartian $\iff$ the annihilator of any countably generated ideal in A is generated by a central idempotent.

$\triangleleft$ (1) By 1.35(2), it is sufficient to prove that $a\varphi^n(a) \neq 0$. Assume the contrary. Since A is a normal Rickartian ring, $a = ed$, where e is a nonzero central idempotent of A, and the element d is regular in A. Hence $0 = a\varphi^n(a) = ed\varphi^n(d)$. Since the regularity of d implies the regularity of $\varphi^n(d)$, we have $a = 0$; this is a contradiction.

(2) follows from (1) and 6.58. (3) and (4) follow from 6.57(1) and 6.59. $\triangleright$

6.61 Let φ be an injective endomorphism of a ring A such that for any $a \in A$, there exists $b \in A$ such that $\varphi(a) = \varphi(a)ab$.

(1) $J(A) = 0$.

(2) If $r_A(a)$ is an ideal of A for any $a \in A$, then A is strongly regular, and for any idempotent $e \in A$, the equality $\varphi(e) = e$ holds, and e is central in $A_\ell[[x, \varphi]]$.

◁ (1) Let $a \in J(A)$. By assumption, there exists $b \in A$ such that $\varphi(a)(1 - ab) = 0$. Since $a \in J(A)$, we obtain that $1 - ab$ is a unit. Hence $\varphi(a) = 0$ and $a = 0$.

(2) By 4.58, A is strongly regular. By assumption, there exist $u, v \in A$ such that $\varphi(e) = \varphi(e)eu$, $\varphi(1 - e) = \varphi(1 - e)(1 - e)v$, and e, $1 - e$, $\varphi(e)$, $\varphi(1-e)$, $e\varphi(e)$, $(1-e)\varphi(1-e)$ are central idempotents of the ring A. Also, $1 = \varphi(e) + \varphi(1 - e)$. Multiplying this equality by the central idempotents $e, \varphi(e) \in A$, we obtain the equality $e = \varphi(e)$. Hence $ex = xe$, and e is a central idempotent of A. Therefore e is a central idempotent of R. ▷

6.62 *Let φ be an automorphism of a strongly regular ring A, and let $\varphi(e) = e$ for any idempotent $e \in A$.*

(1) $A_\ell[x, \varphi]$ is a reduced Bezout subring of the reduced ring $A_\ell[[x, \varphi]]$.

(2) If n is a positive integer, then $A_\ell[x, \varphi]/A_\ell[x, \varphi]x^n$ is a distributive Bezout ring.

◁ Set $R \equiv A_\ell[x, \varphi]$.

(1) By 6.57(1), R and $A_\ell[[x, \varphi]]$ are reduced rings. Let us prove that R is a right Bezout ring. Let $D \equiv fR + gR$ be an arbitrary 2-generated right ideal of R, and let $n \equiv \min(\deg(f), \deg(g))$. We use the induction on n. We may assume that $n = \deg(f) \leq \deg(g)$. Since A is strongly regular, $f_n = ea$, where e is a central idempotent of A, and a is a unit of A. Since R is reduced, e is a central idempotent of R. If $n = 0$, then $f = f_n = ea$. If we take $h \equiv e + g(1 - e)$, then we obtain that $f = hea$, $g = h(1 - e + ge)$, and $D = hR$.

Assume now that the right ideal $uR + vR$ is a principal right ideal for all polynomials $u, v \in R$ such that n greater than the least of $\deg(u)$, $\deg(v)$. Since $R = R(1 - e) \times Re$, it is sufficient to prove that $D(1 - e)$ and De are principal right ideals of $R(1 - e)$ and Re, respectively. Since $D(1 - e) = f(1 - e)R + g(1 - e)R$ and $(f(1 - e))_n = 0$, we have $n > \deg(f(1 - e))$. By induction, $D(1 - e)$ is a principal right ideal both in R and in $R(1 - e)$. Define $z \in Re$ by the equality $z \equiv fe\varphi^{-1}(g_n a^{-1}) - ge$. Since $z_n = (g_n a^{-1}fe)_n - g_n e = g_n a^{-1}f_n e - g_n e = 0$, we have $\deg(z) < n$, and $De = feR + zR$. By induction, De is a principal right ideal, and R is a right Bezout ring. Analogously, R is a left Bezout ring.

(2) By (1), $A_\ell[x,\varphi]/A_\ell[x,\varphi]x^n$ is a Bezout ring. By 6.55(6), $A_\ell[x,\varphi]/A_\ell[x,\varphi]x^n$ is distributive. $\triangleright$

6.63 Let φ be an injective endomorphism of a ring A, $R \equiv A_\ell[[x,\varphi]]$, $\pi : R \to R/Rx^2$ be the natural epimorphism, and let the ring $\pi(R)$ be right distributive. For any subset $Y \subseteq R$, we write $\overline{Y}$ instead of $\pi(Y)$.

(1) For any $a \in A$, there exists $b \in A$ such that $\varphi(a) = \varphi(a)ab$.

(2) φ is an automorphism, A is strongly regular, $\pi(R)$ is a distributive Bezout ring, and every idempotent e of A is central in R with $\varphi(e) = e$.

$\triangleleft$ (1) Let $a \in A$. By 2.4, for elements $\overline{a}$ and $\overline{x}$ of the right distributive ring $\overline{R}$, there exist series $f, g, h, s \in R$ such that $\overline{1} = \overline{f} + \overline{g}$, $\overline{af} = \overline{xh}$, and $\overline{xg} = \overline{as}$. Therefore, there exist series $u, v, z \in R$ such that $1 = f + g + ux^2$, $af = xh + vx^2$, and $xg = as + zx^2$. Then $\varphi(a)x = xa = xaf + xag + xaux^2 = \varphi(a)as + w$ and $w \equiv x^2h + xvx^2 + \varphi(a)zx^2 + xaux \in Rx^2$. Equating coefficients of x in the first and the last members of this sequence of equalities, we obtain $\varphi(a) = \varphi(a)as_1$. Therefore, we can set $b \equiv s_1$.

(2) By (1) and 6.61(1), $J(A) = 0$. The ring A is isomorphic to the factor ring $\overline{R}/\overline{Rx}$ of the right distributive ring $\overline{R}$. Hence A is right distributive. Since $J(A) = 0$, 2.54 shows us that A is reduced. By 1.35(2), all right annihilator ideals of A are ideals. By 6.61(2), A is strongly regular, and every idempotent e of A is central in R with $\varphi(e) = e$.

Let us prove that φ is an automorphism. It is necessary for any $a \in A$, to prove the inclusion $a \in \varphi(A)$. By 2.4, there exist series $f, g, h \in R$ such that $axf = xg + ux^2$ and $x(1 - f) = axh + vx^2$. Let b, c, and d be the constant terms of the series f, g, and h, respectively. Equating coefficients of x in equalities $axf = xg = ux^2$ and $x(1-f) = axh + vx^2$, we obtain the equalities (*) $a\varphi(b) = \varphi(c)$ and (**) $1 - \varphi(b) = a\varphi(d)$. Since A is strongly regular, $b = mt$, where m is a unit, and e is a central idempotent. Since $\varphi(e) = e$, (*) implies the equality $ae\varphi(m) = \varphi(c)$. Hence $ae = \varphi(cm^{-1})$. Let us prove the inclusion $a \in \varphi(A)$. By the equality $at = \varphi(cm^{-1})$, it is sufficient to prove the inclusion $a(1 - e) \in \varphi(A)$. Let $A_1 \equiv A(1-e)$, and let φ_1 be the restriction of φ to A_1. Then φ_1 is an injective endomorphism of A_1. For any $y \in A$, the element $y(1 - e)$ of A_1 will be denoted by y_1. Since $\varphi(b_1) = \varphi(0) = 0$, (**) implies the equality $1_1 = a_1\varphi_1(d_1)$. Therefore $\varphi_1(d_1)$ is a unit in A_1. Hence d_1 is a regular element of the strongly regular ring A_1. Then d_1 is a unit of A_1. Therefore, the equality $1_1 = a_1\varphi_1(d_1)$ implies that $a_1 = \varphi_1(d_1^{-1}) \in A_1$. Hence $a \in \varphi(A)$. It follows from 6.62(2) that R/Rx^2 is a distributive Bezout ring. $\triangleright$

6.64 Assume that φ is an injective endomorphism of a ring A, $R \equiv A_\ell[x, \varphi]$, R/Rx^2 is a right Bezout ring, and either A is a right quasi-invariant ring or right annihilators of all elements in A are ideals.

Then for any $a \in A$, there exists $b \in A$ such that $\varphi(a) = \varphi(a)ab$.

In addition, φ is an automorphism, A is strongly regular, R/Rx^2 is a distributive Bezout ring, and every idempotent e of A is central in R with $\varphi(e) = e$.

$\lhd$ Let $\pi : R \to R/Rx^2$ be the natural epimorphism. For any subset $Y \subseteq R$, we write $\overline{Y}$ instead of $\pi(Y)$. Since the ideal $\overline{Rx}$ of the ring $\overline{R}$ is nilpotent, $\overline{Rx} \subseteq J(\overline{R})$. In addition, $\overline{R}/\overline{Rx} \cong A$.

Assume that A is right quasi-invariant. Then $\overline{R}$ is a right quasi-invariant right Bezout ring. By 2.35, $\overline{R}$ is right distributive. The assertion follows now from 6.63.

Assume that all right annihilator ideals of A are ideals. Taking into account the above, it is sufficient to prove that A is right quasi-invariant. Let $a \in A$. By assumption, $\overline{aR + xR}$ is a principal right ideal of $\overline{R}$. Therefore, there exist $f, g, m, u \in R$ such that $\overline{(ag + xm)f} = \overline{a}$ and $\overline{(ag + xm)u} = \overline{x}$. Therefore, there exist $v, z \in R$ such that the following equalities hold; (*) $(ag + xm)f = a + vx^2$; (**) $(ag + xm)u = x + zx^2$. Comparing the constant terms in (*), we obtain $ag_0f_0 = a$. Comparing the constant terms and coefficients of x in (**), we obtain $ag_0u_0 = 0$ and $a[g_0u_1 + g_1\varphi(u_0)] + \varphi(m_0u_0) = 1$. Since $ag_0u_0 = 0$ and, by assumption, $r_A(ag_0)$ is an ideal of A, we have $am_0u_0 = ag_0t_0m_0u_0 = 0$. Hence $\varphi(a)\varphi(m_0u_0) = 0$. Let $b \equiv g_0u_1 + g_1\varphi(u_0)$. Then $\varphi(a) = \varphi(a)(ab + \varphi(m_0u_0)) = \varphi(a)ab$. By 6.61(2), A is strongly regular, and in particular, A is quasi-invariant. $\rhd$

6.65 *Let φ be an injective endomorphism of a ring A. Then the following conditions are equivalent.*

(1) $A_\ell[x, \varphi]$ is a right Bezout ring, and either A is right quasi-invariant, or all annihilator right ideals of A are ideals.

(2) A is strongly regular, φ is an automorphism, and $A_\ell[x, x^{-1}, \varphi]$, $A_\ell[x, \varphi]$ are semihereditary reduced Bezout rings.

(3) A is strongly regular, φ is an automorphism, and every idempotent e of A is central in $A_\ell[[x, \varphi]]$ with $\varphi(e) = e$.

$\lhd$ Set $R \equiv A_\ell[x, \varphi]$.

$(2) \Longrightarrow (1)$ is directly verified. $(1) \Longrightarrow (3)$ follows from 6.64.

$(3) \Longrightarrow (2)$ By 6.60(2) and 6.62(1), R is a reduced Rickartian Bezout ring. Hence R is semihereditary. The ring $S \equiv A_\ell[x, x^{-1}, \varphi]$ is isomorphic to the ring of quotients of R with respect to the two-sided denominator set $\{x^n\}_{n=0}^{\infty}$. By 1.52, S is reduced. By 4.30(14), S is semihereditary. By 4.45(2), S is a Bezout ring. $\rhd$

6.66 *For a ring A, the following conditions are equivalent.*
(1) $A[x, x^{-1}]$ *is a right distributive ring.*
(2) $A[x, x^{-1}]$ *is a right quasi-invariant right Bezout ring.*
(3) $A[x, x^{-1}]$ *is a commutative distributive Bezout ring.*
(4) A *is a commutative regular ring.*

◁ Set $R \equiv A[x, x^{-1}]$. By 2.35, any commutative Bezout ring is distributive. Hence (4)$\Longrightarrow$(3) follows from 6.65. (3)$\Longrightarrow$(2) is directly verified. (2)$\Longrightarrow$(1) follows from 2.35.

(1)$\Longrightarrow$(4) Elements x and $1 - x$ are central in R. Let $a \in A$. By 2.4, for elements a and $1 - x$ of the right distributive ring R, there exist Laurent polynomials $f, g, h, t \in R$ such that $1 = f + g$, $af = (1 - x)h$, and $(1 - x)g = at$. Then $a(1 - x) = a(1 - x)f + a(1 - x)g = (1 - x)^2 h + a^2 t$. Since Laurent polynomials h and t contain only a finite number of negative powers of x in the canonical form, there exists a positive integer n such that $hx^n \equiv z \in A[x]$ and $tx^n \equiv m \in A[x]$. Hence $a(1-x)x^n = (1-x)^2 z + a^2 m$, $(1-x)a = a(1-x)x^n + a(1-x)(1-x^n) = (1 - x)^2 z + a^2 m + a(1 - x)^2(1 + x + \ldots + x^{n-1}) = (1 - x)^2 w + a^2 m$, where $w \in A[x]$. Since $\{(1 - x)^j\}_{j=0}^{\infty}$ is a basis of $A[x]$ as a free right A-module, the equality $(1-x)a = (1-x)^2 w + a^2 m$ implies the equality $a = a^2 b$, where b is a coefficient of $1 - x$ in the decomposition of the polynomial m with respect to the basis $\{(1 - x)^j\}_{j=0}^{\infty}$. Therefore A is strongly regular. Hence A is a subdirect product of division rings D_j.

To prove that A is commutative, it is sufficient to prove that all the division rings D_j are fields. Fix $D_j \equiv D$. The ring $S \equiv D[x, x^{-1}]$ is isomorphic to a factor ring of the right distributive ring R, and therefore S is right distributive. The ring S is isomorphic to the ring of quotients of $D[x]$ with respect to the set $1, x, x^2 \ldots$. In addition, $D[x]$ is Noetherian. By 4.30(16), S is Noetherian. By 2.72, S is right invariant. For any nonzero $d \in D$ and for each polynomial $u \in D[x]$, there exists a Laurent polynomial $v \in S$ such that $u(d + x) = (d + x)v$. By 6.54(2), it is sufficient to prove the inclusion $v \in D[x]$. Assume the contrary. Then there exists a positive integer n such that $x^n v \equiv w \in D[x]$, and $w_0 \neq 0$. Therefore, the polynomial $(d + x)w$ has a nonzero constant term; we have a contradiction to the equality $(d + x)w = u(d + x)x^n$. ▷

6.67 *If φ is an injective endomorphism of a ring A, then*
$A_\ell[x, \varphi]$ *is a right distributive ring $\Longleftrightarrow$ A is a commutative regular ring and $\varphi \equiv 1$.*

◁ By 2.35, commutative Bezout rings are distributive. Therefore, $\Longleftarrow$ follows from 6.65.

$\Longrightarrow$ By 6.63(2), φ is an automorphism, A is strongly regular, and for any idempotent $e \in A$, the equality $\varphi(e) = e$ holds. Let M be an arbitrary maximal ideal of the strongly regular ring A, and let $\overline{A} \equiv A/M$ be a division ring. Each principal right ideal of A is generated by a central idempotent. Since φ is an automorphism and $\varphi(e) = e$ for any idempotent $e \in A$, we have $\varphi(M) = M$. Therefore φ induces an automorphism $\overline{\varphi}$ of the division ring $\overline{A}$. The ring $\overline{A}_\ell[x, \overline{\varphi}]$ is isomorphic to a factor ring of $A_\ell[x, \varphi]$, and therefore $\overline{A}_\ell[x, \overline{\varphi}]$ is right distributive. By 6.54(4), $\overline{A}$ is a field, and $\overline{\varphi}$ is the identity automorphism. Hence $(\varphi - 1)(A) \subseteq M$. In addition, the strongly regular ring A is a subdirect product of division rings. Hence $(\varphi - 1)(A) = 0$, and A is a subdirect product of fields. Therefore A is commutative. $\triangleright$

6.5 Series rings

6.68 Let φ be an injective endomorphism of a ring A, $R \equiv A_\ell[[x, \varphi]]$, $S \equiv A_r[[x, \varphi]]$, $U \equiv A_\ell[x, \varphi]$, $V \equiv A_r[x, \varphi]$, and let $T \equiv \{x^n\}_{n=0}^\infty$.

(1) For any series $f, g \in R$ (resp. $f, g \in S$), the series $\sum_{i=0}^\infty (fxg)^i$ is well defined, and this series is the inverse element for $1 - fxg$ in R (resp. in S).

(2) $R/Rx \cong U/Ux \cong A$ and $S/xS \cong V/xV \cong A$.

(3) For any positive integer n, the sets Rx^n, x^nS, Ux^n, and x^nV are ideals of the rings R, S, U, and V, respectively, and $R/Rx^n \cong U/Ux^n$, $S/x^nS \cong V/x^nV$.

(4) Let $R(n) \equiv R/Rx^n$, $S(n) \equiv S/x^nS$, $U(n) \equiv U/Ux^n$, and let $V(n) \equiv V/x^nV$.

Then $R(n)(x + Rx^n)$, $(x + x^nS)S(n)$, $U(n)(x + Ux^n)$, and $(x + x^nV)V(n)$ are nilpotent ideals of the rings $R(n)$, $S(n)$, $V(n)$, and $U(n)$, respectively, and $A \cong R(n)/R(n)(x + Rx^n) \cong S(n)/(x + x^nS)S(n) \cong U(n)/U(n)(x + Ux^n) \cong V(n)/(x + x^nV)V(n)$.

(5) $Rx = RxR \subseteq J(R)$ and $xS = SxS \subseteq J(S)$.

(6) If A is strongly regular, then factor rings $R/J(R)$, $S/J(S)$, $R(n)/J(R(n))$, and $S(n)/J(S(n))$ are strongly regular.

(7) If A is strongly regular, then

the ring R (resp. S, $R(n)$, $S(n)$) is right distributive $\Longleftrightarrow$ the ring R (resp. S, $R(n)$, $S(n)$) is a right Bezout ring.

(8) All elements of T are regular, T is a left Ore subset of R, and if φ is an automorphism of A, then the left ring of quotients $_TR$ is isomorphic to the ring $A_\ell((x, \varphi))$.

In addition, T is a right denominator subset of rings S and V, and if φ is an automorphism of A, then $S_T \cong A_r[[x, \varphi]]$ and $V_T \cong A_r[x, \varphi]$.

◁ (1), (2), (3), and (4) are directly verified. (5) follows from (1). (6) follows from (5) and (4). (7) follows from (6) and 5.15.

(8) is directly verified if we take into account that the equality $x^n a = \varphi^n(a)x^n$ (the equality $ax^n = x^n\varphi^n(a)$) implies that T is left (right) permutable. ▷

6.69 Let φ be an automorphism of a strongly regular ring A, $R \equiv A_\ell[[x, \varphi]]$, and let $\varphi(e) = e$ for any idempotent $e \in A$.

(1) For any positive integer n and for arbitrary series $u, v \in R$, there exist series $f, g, h, m, t, w, z \in R$ such that $1 = f + g + mx^n$, $uf = vh + wx^n$, and $vg = ut + zx^n$.

(2) If A is countably injective, then R is a distributive reduced Bezout ring, and all submodules of flat R-modules are flat.

(3) If A is countably injective and the annihilator of any countably generated ideal of A is generated by an idempotent, then R is a distributive reduced semihereditary Bezout ring.

◁ (1) By 6.68(3), $R/Rx^n \cong A_\ell[x, \varphi]/A_\ell[x, \varphi]x^n$. By 6.62(2), R/Rx^n is distributive. We apply 2.4 to this ring and pass to R.

(2) First, let us prove that R is left distributive. By 2.4, it is sufficient to prove that for any series $u, v \in R$, there exist series $f, g, h, k \in R$ such that $1 = f + g$, $fu = hv$, and $gv = ku$. It follows from the rule for multiplication of series that the system $\{1 = f + g, fu = hv, gv = ku\}$ of three equations in R is equivalent to a countable system of linear equations L in the module A_A of a countable number of unknowns f_i, g_i, h_i, k_i $(0 \le i < \infty)$ with coefficients from A depending on fixed elements $\{u_i, v_i \in A\}$, and each equation contains only a finite number of unknowns. Let N be a finite subsystem of the system L. There exists a positive integer n such that N contains no unknowns f_i, g_i, h_i, k_i for $i \ge n$. By 6.62(2), N is a solvable system. Hence L is a countable finitely solvable system, and 4.88 shows us that A is a right $\aleph_0$-algebraically compact ring. Therefore L is a solvable system, and R is left distributive. Analogously, R is right distributive. By 6.62(1) and 6.68(7), R is a reduced Bezout ring. By 4.21(1), all submodules of flat R-modules are flat.

(3) By (2) and 6.60(4), R is a Rickartian Bezout ring. Therefore R is semihereditary. ▷

6.70 If φ is an automorphism of a ring A and all 2-generated right ideals of $A_\ell[[x, \varphi]]$ are flat, then the ring A is regular.

◁ Let $a \in A$. Since $\varphi(a)x = xa$, 4.24(2) shows us that there exist series $f, h, t \in A_\ell[[x, \varphi]] \equiv R$ such that $\varphi(a)f = xh$ and $(1 - f)x = ta$. Let c be the constant term of f, and let b be the element of A such that $\varphi(b)$ is the coefficient of x of the series t. Equating the constant terms in the equality $\varphi(a)f = xh$ and coefficients of x in the equality $(1-f)x = ta$, we obtain equalities $\varphi(a) = c$, $1-c = \varphi(b)\varphi(a)$. Therefore $\varphi(a) = \varphi(a)(1-c) = \varphi(aba)$. Hence $a = aba$, and A is regular. $\triangleright$

6.71 *Let φ be an injective endomorphism of a ring A. Then the following conditions are equivalent.*

(1) $A_\ell[[x, \varphi]]$ is a right distributive ring.

(2) $A_\ell[[x, \varphi]]$ is a right Bezout ring, and either A is right quasi-invariant, or right annihilators of all elements in A are ideals.

(3) $A_\ell[[x, \varphi]]$ is either a right distributive ring or a right Bezout ring, and A is strongly regular.

(4) $A_\ell[[x, \varphi]]$ is a distributive reduced Bezout ring, and all submodules of flat A-modules are flat.

(5) A is a strongly regular countably injective ring, φ is an automorphism, and $\varphi(e) = e$ for any idempotent $e \in A$.

$\triangleleft$ (1)$\Longrightarrow$(3) and (2)$\Longrightarrow$(3) follow from 6.63(2) and 6.64. (5)$\Longrightarrow$(4) follows from 6.69(2). (4)$\Longrightarrow$(3), (3)$\Longrightarrow$(1), and (3)$\Longrightarrow$(2) are directly verified.

(3)$\Longrightarrow$(5) By 6.68(7), the ring $R \equiv A_\ell[[x, \varphi]]$ is right distributive. By 6.63(2), φ is an automorphism of the strongly regular ring A, and any idempotent e of A is central in R with $\varphi(e) = e$. Let $\{e_i\}_{i=0}^\infty$ be a countable set of central orthogonal idempotents of A, $\{a_i\}_{i=0}^\infty$ be a countable set of elements of A, $u \equiv \sum_{i=0}^\infty e_i x^i$, and let $v \equiv \sum_{i=0}^\infty a_i e_i x^{i+1}$. By 2.4, for the series u and v, there exist series $f, h, z \in R$ such that $uf = vh$ and $v(1 - f) = uz$. Multiplying these equalities by the central in R idempotents e_i, we obtain equalities $e_i x^i f = a_i e_i x^{i+1} h$ and $e_i x^i z = a_i e_i x^{i+1}(1 - f)$. Therefore $e_i x^{i+1} f = x e_i x^i f = x a_i e_i x^{i+1} h = e_i x^{i+2} g$, where $g \in R$. Hence $e_i x^i z = a_i e_i x^{i+1} - a_i e_i x^{i+2} g$. Equating coefficients of x^{i+1} in this equality, we obtain equalities $\varphi(z_1)e_i = a_i e_i$ for all i. By 4.88, A is countably injective. $\triangleright$

6.72 *Let φ be an automorphism of a normal ring A such that $\varphi(e) = e$ for any idempotent $e \in A$. Then the following conditions are equivalent.*

(1) All submodules of flat $A_\ell[[x, \varphi]]$-modules are flat.

(2) All 2-generated right ideals of $A_\ell[[x, \varphi]]$ are flat.

(3) All 2-generated left ideals of $A_\ell[[x, \varphi]]$ are flat.

(4) A is a strongly regular countably injective ring.

$\triangleleft$ (4)$\Longrightarrow$(1) follows from 6.71. (1)$\Longrightarrow$(2) and (1)$\Longrightarrow$(3) are directly verified.

$(2)\Longrightarrow(4)$ and $(3)\Longrightarrow(4)$ By 6.70, A is regular. Therefore, the normal regular ring A is strongly regular. Let $\{e_i\}_{i=0}^{\infty}$ be a countable set of central orthogonal idempotents of A, and let $\{a_i\}_{i=0}^{\infty}$ be a countable set of elements of A. Let $u \equiv \sum_{i=0}^{\infty} e_i x^i$, $v \equiv \sum_{i=0}^{\infty} a_i e_i x^{i+1}$, and let $w \equiv \sum_{i=0}^{\infty} e_i \varphi^i(a_i) x^{i+1}$. We have $uv = wu$. By 4.24(2), there exist series $f, g, h \in R$ such that $uf = wh$ and $(1-f)v = gu$. Multiplying these equalities by the central orthogonal idempotents e_i and taking into account that $\varphi(e_i) = e_i$ for all e_i, we obtain equalities $x^i e_i f = x^i a_i e_i x h$ and $(1-f)a_i e_i x^{i+1} = g e_i x^i$. Hence $g e_i x^i = (e_i - a_i e_i x h) a_i x^{i+1}$. Equating coefficients of x^{i+1} in this equality, we obtain that $e_i a_i = g_1 e_i$ for all i. By 4.88, A is countably injective. $\triangleright$

Example 6.73 *There exists a commutative regular ring A such that A is not countably injective, and the ring $A[[x]]$ is not right or left distributive.*

$\triangleleft$ Let T be a field, and let A be the ring formed by all eventually constant sequences $f = (f_n)_{n=0}^{\infty}$ of elements of T (for each $(f_n) \in A$, there exists a number $N = N(f)$ such that $f_n = f_N$ for all $n \geq N$). The ring A is commutative and regular. Let $\{e(i)\}_{i=0}^{\infty}$ be a countable set of orthogonal idempotents of A such that $e(i)_i = 1$ and $e(i)_j = 0$ for $j \neq i$, and let $\{a(i)\}_{i=0}^{\infty}$ be a countable set of elements in A such that $a(i) = 1_A$ for odd i and $a(i) = 0_A$ for even i. Assume that A is countably injective. By 4.88, there exists $b \in A$ such that $be(i) = a(i)e(i)$ for all i. Choose an even number $n \geq N(b)$. Then $b_n = (be(n))_n = 0_T$, $b_{n+1} = (be(n))_{n+1} = 1_T$. Since $b_n \neq b_{n+1}$, we have a contradiction. Hence A is not countably injective. If $A[[x]]$ is distributive, then 6.71 shows us that A is countably injective; we have a contradiction. $\triangleright$

6.74 *Let φ be an automorphism of a normal ring A such that $\varphi(e) = e$ for any idempotent $e \in A$. Then the following conditions are equivalent.*

(1) *$A_\ell[[x, \varphi]]$ is a right semihereditary ring.*
(2) *$A_\ell[[x, \varphi]]$ is a left semihereditary ring.*
(3) *All 2-generated right ideals of $A_\ell[[x, \varphi]]$ are projective.*
(4) *All 2-generated left ideals of $A_\ell[[x, \varphi]]$ are projective.*
(5) *A is a strongly regular countably injective ring, and the annihilator of any countably generated ideal in A is generated by a central idempotent.*

$\triangleleft$ It is sufficient to prove $(1)\Longleftrightarrow(3)\Longleftrightarrow(5)$.

$(5)\Longrightarrow(1)$ follows from 6.69(3). $(1)\Longrightarrow(3)$ is directly verified. $(3)\Longrightarrow(5)$ follows from 6.72 and 6.57(6). $\triangleright$

Example 6.75 (1) There exists a commutative regular ring D such that D is a factor ring of a commutative regular self-injective ring A, D has a countably generated ideal B such that $r_D(B)$ is not generated (as an ideal) by an idempotent, D is not a Baer ring, and D is not self-injective.

(2) There exists a commutative regular countably injective ring D such that $D[[x]]$ is a commutative distributive reduced Bezout ring, all submodules of flat $D[[x]]$-modules are flat, $D[[x]]$ is not a Rickartian ring, and a classical ring of quotients of $D[[x]]$ is not regular.

(3) There exists a commutative regular countably injective ring D such that D is not self-injective, and D is a factor ring of a commutative regular self-injective ring A.

◁ (1) Let A be a direct product of an infinite set of fields A_i, $T \equiv \oplus_{i=1}^{\infty} A_i$, and let $D \equiv A/T$. By 4.90(2), A and D are commutative regular rings, and D is not self-injective. By 4.84(5), A is self-injective. By 4.89, D is the required ring.

(2) By (1), there exists a commutative regular ring D which is a factor ring of a commutative regular self-injective ring A and has a countably generated ideal $\overline{B}$ such that the ideal $r_D(\overline{B})$ is not generated by an idempotent. By 4.88, all factor rings of A are countably injective. Therefore D is countably injective. By 6.71, $D[[x]]$ is a commutative distributive reduced Bezout ring, and all submodules of flat $D[[x]]$-modules are flat. By 6.59, $D[[x]]$ is not Rickartian. By 4.56, a classical ring of quotients of the distributive non-Rickartian ring $D[[x]]$ cannot be regular.

(3) Let A be a direct product of an infinite number of fields A_i, and let D be the factor ring of A modulo the direct sum of fields A_i. By (1), A is a commutative regular self-injective ring, D is a commutative regular ring, and D is not self-injective. By 4.88, D is countably injective. ▷

6.76 If A is the division ring of real quaternions, then the ring $A[[x]]$ is distributive, and the ring $A[x]$ is not right or left distributive.

If A is the commutative regular ring constructed in 6.73, then $A[x]$ is distributive, and $A[[x]]$ is not right or left distributive.

◁ 6.76 follows from 6.71, 6.73, and 6.67. ▷

6.77 Assume that a ring A has an injective endomorphism φ such that $\varphi(a)$ is a unit of A for any nonzero $a \in A$. Set $R \equiv A_r[[x, \varphi]]$.

(1) A and R are domains.

(2) $x^n R = a x^n R$ and $(a + xf)R = aR$ for any nonzero $a \in A$, for each positive integer n, and for any series $f \in R$.

(3) Let M and N be nonzero principal right ideals of R.

Then there exist two nonzero principal right ideals D and E of A and nonnegative integers m, n such that $M = x^m DR$, $N = x^n ER$, and the equality $M = N$ is equivalent to the equalities $m = n$ and $D = E$.

In addition, $M \subseteq N \iff$ either $m < n$ or $m = n, D \subseteq E$.

(4) If A is right uniserial (resp. right distributive, a ring with the maximum condition on principal right ideals), then R is right uniserial (resp. right distributive, a ring with the maximum condition on principal right ideals).

(5) If A is a right uniserial right Noetherian domain, then R is a right uniserial right Noetherian principal right ideal domain.

(6) If a is a nonzero noninvertible element of the domain A, then $Ra \bigcap Rx = 0$.

(7) If the ring R is either left distributive or left finite-dimensional or a left Bezout ring, then A is a division ring.

$\triangleleft$ (1) It follows by assumption that $\varphi(A)$ is a domain. Hence A is a domain. Since A is a domain and $(fg)_0 = f_0 g_0$ for any $f, g \in R$, we get R is a domain.

(2) From $ax = x\varphi(a)$ and $\varphi(a) \in U(A)$ we obtain $xR \subseteq axR$. Hence $x^n R \subseteq ax^n R$. Set $g \equiv 1 + x\varphi(a)^{-1}f$. Then $ag = a + xf$ and $g \in U(R)$. Therefore $(a + xf)R = aR$.

(3) follows from (2) and from the fact that elements x^n are regular and are not units of R. (4) can be verified with the use of (3). (5) follows from (4).

(6) Assume that $fa = gx$, where f and g are two nonzero series from R. Let $0 \neq g_i \in A$. Equating coefficients of x^{i+1}, we obtain $f_{i+1}a = \varphi(g_i) \in U(A)$. Therefore $a \in U(A)$; we have a contradiction.

(7) By 3.13, R is a left uniform domain. The assertion follows now from (6). $\triangleright$

Example 6.78 There exists a commutative uniserial principal ideal domain A which is not a field, and A has an injective ring endomorphism φ such that $\varphi(a) \in U(A)$ for all $a \in A \setminus 0$.

In addition, the ring $A_r[[x, \varphi]] \equiv R$ possesses the following properties.

(1) R is a right uniserial principal right ideal domain, and therefore R is a right distributive right Noetherian right hereditary right Bezout domain.

(2) R is not left distributive, R is not left finite-dimensional, and R is not a left Bezout ring.

(3) There exist two nonzero distinct completely prime ideals M and N of R such that $N = MN \subset M \subseteq J(R)$, $N \neq NM \subset N$, and

R/NM is a right uniserial right Noetherian principal right ideal ring which contains a nonzero nonmaximal nilpotent completely prime ideal N/NM.

(4) Multiplication of completely prime ideals of the right uniserial principal right ideal domain R is not commutative.

In addition, there exists a completely prime ideal N such that N is not a finitely generated left ideal, and R has an indecomposable right uniserial right Noetherian factor ring $\widehat{R}$ which is neither right Artinian nor semiprime.

◁ Let k be a field, F be the field of rational functions over k of countable number variables $\{t_i\}_{i=0}^{\infty}$, and let $t \equiv t_0$. There exists an injective endomorphism α of F such that $\alpha(t_i) = t_{i+1}$ for all i, and t is transcedental over $\alpha(F)$. Let A be a ring which is formed by all fractions $f/(1 + yg)$, where f, g are arbitrary polynomials from the polynomial ring $F[y]$ of one variable y. The rule $\varphi(y) = t$ allows us to extend α to an injective endomorphism φ of A. It can directly be verified that A is a commutative uniserial principal ideal domain, and $\varphi(a) \in U(A)$ for all $a \in A \setminus 0$. Set $R \equiv A_r[[x, \varphi]]$. By 6.77(5) and 6.77(7), R is a right uniserial principal right ideal domain, R is not left distributive, R is not left finite-dimensional, and R is not a left Bezout ring. Set $N \equiv xR$ and $M \equiv J(A)R$. Then M and N are nonzero completely prime ideals of R, $M \neq N \subset M \subseteq J(R)$, and $N \neq NM$ by Nakayama's Lemma. Since M is a principal right ideal which contains properly a completely prime ideal N, we obtain that $MN = N$. The remaining assertions can directly be verified if we substitute R/NM for $\widehat{R}$. ▷

6.79 Assume that φ is an injective endomorphism of a reduced ring A and for any $a \in A$, there exists $b \in A$ such that $a = a\varphi(a)b$.

(1) $r(a) = r(AaA) = r(\varphi^n(a)) = \ell(a) = \ell(AaA) = \ell(\varphi(a))$ for any $a \in A$ and for each nonnegative integer n.

(2) $a\varphi^n(a) \neq 0$ for any nonzero $a \in A$ and each $n \geq 0$.

(3) For any $a \in A$, there exists $b \in A$ such that $\varphi(a)b \equiv e$ is a central idempotent of A, $r(a) = \ell(a) = eA$, and $\varphi(a) = \varphi(a)e$.

(4) $\varphi(e) = e$ for any idempotent $e \in A$, $\varphi(a) \in U(A)$ for any regular element $a \in A$, and A is a Rickartian ring.

(5) If the ring A is indecomposable, then A is a domain, and for any $a \in A$, the element $\varphi(a)$ is a unit in A.

◁ By 1.35(2), $r(X) = r(\mathrm{Id}(X)) = \ell(\mathrm{Id}(X)) = \ell(x)$ for each subset X of A.

(1) It is sufficient to prove that $r(a) = r(\varphi(a))$. By assumption $a \in \mathrm{Id}(\varphi(a))$, whence $r(\varphi(a)) \subseteq r(a)$. Let $m \in r(a)$. Then $\varphi(m) \in$

$r(\varphi(a)) \subseteq r(a)$, whence $a\varphi(m) = 0$. By assumption, $m = m\varphi(m)d$ for some $d \in A$. Since $ma\varphi(m)d = 0$, 1.33 shows us that $0 = m\varphi(m)da = ma$. Then $am = 0$ and $m \in r(a)$.

(2) follows from (1).

(3) It follows from (1) and from the assumption that there exists $b \in A$ such that $1 - \varphi(a)b \in r(\varphi(a)) = \ell(\varphi(a))$. Hence $\varphi(a) = \varphi(a)b\varphi(a)$. Set $\varphi(a)b \equiv e$. Then $\varphi(a) = \varphi(a)e$ and $r(a) = \ell(a) = eA$.

(4) and (5) follow from (3). $\triangleright$

6.80 Let φ be an injective endomorphism of a ring A, and let $A_r[[x, \varphi]]/x^2 A_r[[x, \varphi]]$ be a right distributive ring.

(1) For any $a \in A$, there exists $b \in A$ such that $a = a\varphi(a)b$.

(2) $\varphi(e) = e$ for any idempotent $e \in A$, $\varphi(a) \in U(A)$ for any regular element $a \in A$, and A is a reduced Rickartian right distributive ring.

$\triangleleft$ Let $S \equiv A_r[[x, \varphi]]$, $h : S \to S/x^2 S$ be a natural epimorphism, and let $a \in A$.

(1) By 2.4, for elements $h(a)$ and $h(x)$ of the right distributive ring $h(S)$, there exist series $f, c, d \in S$ such that $h(a)h(f) = h(x)h(c)$ and $h(x)h(1 - f) = h(a)h(d)$. Therefore, there exist three series $u, v, w \in S$ such that the equalities $af = xc + x^2 v$ and $x(1 - f) = ad + x^2 w$ hold. Equating the constant terms in the first equality and coefficients of x in the second equality, we obtain equalities $af_0 = 0$ and $1 - f_0 = \varphi(a)d_1$, where f_0 is the constant term of the series f, and $d_1 \equiv b$ is the coefficient of x of the series d. Hence $a = a(1 - f_0) = a\varphi(a)b$.

(2) Since $A \cong h(S)/h(xS)$, A is right distributive. Let $a^2 = 0$. Then $(\varphi(a))^2 = 0$. By 2.54, $\varphi(a) \in J(A)$. In addition, (1) shows us that there exists $b \in A$ such that $a(1 - \varphi(a)b) = 0$. Since $1 - \varphi(a)b \in U(A)$, we obtain $a = 0$, whence A is reduced. By 6.79(4), $\varphi(e) = e$ for any idempotent $e \in A$, $\varphi(a) \in U(A)$ for any regular element $a \in A$, and A is Rickartian. $\triangleright$

6.81 *Let φ be an injective endomorphism of a ring A.*

(1) If A is an indecomposable ring, then

$A_r[[x, \varphi]]$ *is a right distributive ring* $\Longleftrightarrow$ A *is a right distributive domain, and* $\varphi(a) \in U(A)$ *for any nonzero $a \in A$.*

(2) $A_r[[x, \varphi]]$ is a right uniserial ring $\Longleftrightarrow$ A *is right uniserial, and* $\varphi(a)$ *is a unit of A for any nonzero $a \in A$.*

$\triangleleft$ (1)

$\Longleftarrow$ follows from 6.77(4).

$\Longrightarrow$ By 6.80(2), A is a reduced indecomposable right distributive ring. By 6.80(1) and 6.79(5), A is a domain, and $\varphi(a) \in U(A)$ for all $a \in A$.

(2) follows from (1), 6.77(4), and from the isomorphism $A \cong R/xR$. $\triangleright$

6.82 Let φ be an automorphism of a ring A, and let the ring $A_\ell((x, \varphi))$ be right uniserial.

(1) If N is an ideal of the ring A and $\varphi(N) = N$, then φ induces the automorphism $\overline{\varphi}$ of the ring A/N, and $(A/N)_\ell((x, \overline{\varphi}))$ is a right uniserial ring.

(2) A is a right uniserial ring.

(3) If A is a domain, then A is a division ring.

(4) If A has a completely prime ideal N such that $\varphi(N) = N$, then A/N is a division ring.

(5) If A is a ring with the maximum condition on principal right ideals, then A is a right uniserial right Artinian ring.

$\triangleleft$ Set $R \equiv A_\ell((x, \varphi))$.

(1) follows from the fact that $(A/N)_\ell((x, \overline{\varphi})) \cong R/N_\ell((x, \varphi))((x))$ and from the fact that all factor rings of any right uniserial ring are right uniserial.

(2) Let a and b be two elements of the ring A. Since R is a right uniserial ring, either $a \in bR$ or $b \in aR$. Assume that $a \in bR$. There is a series $f \in R$ such that $a = bf$. Let f_0 be the constant term of the series f. Then $a = bf_0$ and $aA \subseteq bA$. Therefore A is a right uniserial ring.

(3) Let a be a nonzero element of the domain A. It is sufficient to prove that a is a right invertible element. Consider elements a and $a + x$ of the ring R. Since R is a right uniserial ring, either $x + a \in aR$ or $a \in (x + a)R$.

Assume that $x + a \in aR$. Then $x = (x + a) - a \in aR$, whence there exists a series $f \in R$ such that $x = af$. Equaiting coefficients of x in the equality $x = af$, we obtain the equality $1 = af_1$, whence a is a right invertible element.

Assume that $a \in (x+a)R$. Then $x = (x+a) - a \in (x+a)R$, whence there exists a series $g \in R$ such that $x = (x + a)g$. Let $g = \sum_{i=m}^{\infty} g_i x^i$, and let $g_m \neq 0$. Since A is a domain, $ag_m \neq 0$. Also, $ag_m x^m$ is the lowest term of the series $(x + a)g = x$. Hence $m = 1$ and $ag_1 = 1$, whence the element a is right invertible.

(4) Let $\overline{A} \equiv A/N$, $\overline{\varphi}$ be the automorphism of the domain $\overline{A}$ induced by the automorphism φ, and let $\overline{R} \equiv \overline{A}_\ell((x, \overline{\varphi}))$. By (1), $\overline{R}$ is a right uniserial ring. By (3), $\overline{A}$ is a division ring.

(5) Since A is a right uniserial ring with the maximum condition on principal right ideals, A is right Noetherian. By 1.3(3), the prime radical N of A is nilpotent. Obviously, $\varphi(N) = N$. Since A is a right

Noetherian right uniserial ring and N is a semiprime ideal, 3.11 shows us that N is a completely prime ideal. By (4), A/N is a division ring. By 3.18, A is a right uniserial right Artinian ring. $\triangleright$

6.83 [218], [219]. *Let φ be an automorphism of a ring A. Then the following conditions are equivalent.*
   (1) *$A_\ell((x,\varphi))$ is a right uniserial ring.*
   (2) *$A_\ell((x,\varphi))$ is a right uniserial right Artinian ring.*
   (3) *A is a right uniserial right Artinian ring.*

   $\triangleleft$ Set $R \equiv A_\ell((x,\varphi))$.
   (2)$\Longrightarrow$(1) is obvious.
   (1)$\Longrightarrow$(3) By 6.82(2), A is right uniserial. By 6.82(5), it is sufficient to prove that A is a ring with the maximum condition on principal right ideals. Assume that A has the infinite properly ascending sequence $a_1 A \subset a_2 A \subset \ldots$ of principal right ideals $\{a_n A\}$. Let us decompose the sequence a_n into two subsequences $\{f_n\}_{i=0}^{\infty}$ and $\{g_n\}_{i=0}^{\infty}$ as follows:

$$\{f_n\} = a_1, a_4, a_5, a_6, a_7, a_{16}, \ldots, a_{2^{2i}}, a_{2^{2i}+1}, \ldots a_{2^{2i+1}-1}, \ldots,$$
$$\{g_n\} = a_2, a_3, a_8, \ldots, a_{2^{2i+1}}, a_{2^{2i+1}+1}, \ldots a_{2^{2i+2}-1}, \ldots$$

Define series f and g in R by the equalities $f = f_0 + f_1 x + f_2 x^2 + \ldots$ and $g = g_0 + g_1 x + g_2 x^2 + \ldots$. Since R is a right uniserial ring, either $f \in gR$ or $g \in fR$. Assume that $f = gh$, where $h = h_{-k}x^{-k} + h_{-k+1}x^{-k+1} + \ldots \in R$. For every positive integer n, we have

$$f_n = g_{n+k}\varphi^{n+k}(h_{-k}) + g_{n+k-1}\varphi^{n+k-1}(h_{-k+1}) + \ldots + g_0 h_n \in g_{n+k}A.$$

Set $u(i) \equiv 1 + 4 + 16 + \ldots + 2^{2i}$, $v(i) \equiv 2 + 8 + \ldots + 2^{2i+1} - 1$. Then $f_{u(i)} = a_{2^{2i}+2}$, $g_{v(i)} = a_{2^{2i}+2-1}$, and $v(i) - u(i) = 1 + 4 + 16 + \ldots + 2^{2i} - 1$. Let us take i such that $v(i) - u(i) > k$. Then $f_{u(i)} \in g_{u(i)+k}A \subset g_{v(i)}A$, whence $a_{2^{2i}+2} \in a_{2^{2i}+2-1}A$. This is a contradiction, since $a_{2^{2i}+2-1}A$ is properly contained in $a_{2^{2i}+2}A$. The case $g \in fR$ can be analogously considered.
   (3)$\Longrightarrow$(2) Let N be the prime radical of A. For every ideal B of A, denote by $B((x,\varphi))$ the subset of R consisting of series such that all their coefficients belong to the ideal B of A. Set $\overline{N} \equiv N((x,\varphi))$. Obviously, $\varphi(N) = N$, whence $N((x,\varphi))$ is an ideal of A. Let $\overline{\varphi}$ be the automorphism of A/N induced by the automorphism φ of A. Since A is a right uniserial right Artinian ring, N is a nilpotent ideal, A/N is a division ring, and $N = yA$ for some $y \in N$. Since $N = yA$, we have $N((x,\varphi)) = yR$. Then $N((x,\varphi))$ is a nilpotent ideal of R. The ring $R/N((x,\varphi))$ is isomorphic to the Laurent series ring $\overline{R} \equiv (A/N)_\ell((x,\overline{\varphi}))$ over the division ring A/N. Hence $R/N((x,\varphi))$ is a division ring. By 3.24(2), R is a right uniserial right Artinian ring. $\triangleright$

6.6 Pierce stalks and exchange rings

6.84 A ring A is an *exchange ring* if the following two equivalent conditions hold.

(1) For any two elements $a, b \in A$ with $a + b = 1$, there exists idempotents $e \in aA$ and $f \in bA$ with $e + f = 1$.

(2) For any two elements $a, b \in A$ with $a + b = 1$, there exists idempotents $e' \in Aa$ and $f' \in Aa$ with $e' + f' = 1$.

◁ It is sufficient only to prove (1)$\Longrightarrow$(2). Suppose that $e = ar$ and $f = bs$. By replacing r with re and s with sf, we may assume

$$rar = r, \; rbs = 0, \; sbs = s, \text{ and } sar = 0.$$

Let $r' \equiv 1 - sb + rb$, $s' \equiv 1 - ra + sa$, $e' \equiv r'a \in Aa$, and let $f' \equiv s'b \in Ab$. Then $r's = (1 - sb + rb)s = (s - sbs) + rbs = 0$ and $s'r = (1 - ra + sa)r = (r - rar) + sar = 0$. Since $ar + bs = e + f = 1$, we obtain

$$ar' = a(1 - sb) + arb = a(1 - sb) + (1 - bs)b = a(1 - sb) + b(1 - sb) = \\ (a + b)(1 - sb) = 1 - sb.$$

Therefore $r'ar' = r'$. Similarly, $bs' = 1 - ra$ and $s'bs' = s'$. Hence e' and f' are idempotents. Since $ab = ba$, we have $e' + f' = r'a + s'b = (a - sba + rba) + (b - rab + sab) = a + b = 1$. ▷

6.85 (1) All regular rings and all factor rings of exchange rings are exchange rings.

(2) A is an exchange ring without nontrivial idempotents $\Longleftrightarrow$ A is a local ring.

(3) Every normal exchange ring A is a quasi-invariant ring.

(4) Let A be a ring such that $A/J(A)$ is a normal exchange ring. Then A is a quasi-invariant ring, and $A/J(A)$ is a reduced ring. Consequently, $J(A)$ contains all nilpotent elements of A.

◁ (1) and (2) are directly verified.

(3) Assume that a maximal right ideal M of A is not an ideal. Then $A = M + AM$. There exist $m_1, m_2 \in M$ and $a \in A$ such that $m_1 + am_2 = 1$. Since A is a normal exchange ring, there exists a central idempotent $e \in m_1 A \subseteq M$ such that $1 - e \in am_2 A$. Therefore, there exists $m_3 \in m_2 A \subseteq M$ such that $1 - e = am_3$. Then $m_3 a$ is a central idempotent. Therefore

$$1 - e = a(m_3 a)m_3 = (m_3 a)am_3 \in M.$$

Then $1 = e + (1 - e) \in M$; this is a contradiction.

(4) By (3), $A/J(A)$ is a quasi-invariant ring. Then $A/J(A)$ is a subdirect product of division rings, whence $A/J(A)$ is a reduced ring. Since $A/J(A)$ is a right quasi-invariant ring, A is right quasi-invariant. $\triangleright$

6.86 (1) *Let A be normal exchange ring, and let M be a right (left) A-module. Then*

M is a distributive module $\Longleftrightarrow$ M is a Bezout module.

(2) *Let A be a ring such that $A/J(A)$ is a normal exchange ring, and let M be an A-module. Then*

M is a distributive module $\Longleftrightarrow$ M is a Bezout module.

$\triangleleft$ (1)

$\Longrightarrow$ Let $m, n \in M_A$. By 2.4, there exists $a \in A$ such that $maA + n(1 - a)A \subseteq mA \cap nA$. There exists a central idempotent e of A such that $e \in aA$ and $1 - e \in (1 - a)A$. Then $meA + n(1 - e)A \subseteq mA \cap nA$. Therefore $me \in neA$, $n(1 - e) \in m(1 - e)A$. Hence

$$N = Ne \oplus N(1 - e) = meA + neA + m(1 - e)A + n(1 - e)A = (ne + m(1 - e))A.$$

$\Longleftarrow$ By 6.85(3), A is a quasi-invariant ring. By 2.35, the Bezout module M_A is distributive.

(2)

$\Longrightarrow$ Let $m, n \in M$, $N \equiv mA + nA$, $J \equiv J(A)$, and let $h : N \to N/NJ$ be the natural epimorphism. It is sufficient to prove that N is a cyclic module. Since NJ is superfluous in N_A, it is sufficient to prove that the distributive right A/J-module $h(N)$ is cyclic. Therefore, we may assume that $J = 0$ and A is a normal exchange ring. In this case, the assertion follows from (1).

$\Longleftarrow$ By 6.85(3), $A/J(A)$ is a quasi-invariant ring. Hence A is a quasi-invariant ring. By 2.35, the Bezout module M_A is distributive. $\triangleright$

6.87 *Let A be a normal exchange ring, and let M be a maximal right or left ideal of A.*

Then there exists the two-sided ring of quotients $_M A_M$ which is a local ring, and the canonical ring homomorphism $f : A \to {}_M A_M$ is surjective.

$\triangleleft$ Let M be a maximal right ideal of A, and let $T \equiv A \setminus M$. By 6.85(3), M is an ideal. Then A/M is a division ring, whence T is a multiplicative set. Let us prove the following assertions.

(1) For every $t \in T$, there exist $u \in T$ and $m_1, m_2 \in M$ such that $tu = 1 - m_1 \in T$ and $ut = 1 - m_2 \in T$.

(2) T is a right permutable set.

(3) T is a right reversive set.

(4) There is the right ring of quotients A_M which is a local ring, and the canonical ring homomorphism $f : A \to A_M$ is surjective.

(1) follows from the fact that A/M is a division ring.

(2) Let $a \in A$, and let $t \in T$. By (1), there exist $u \in T$ and $m \in M$ with $tu = 1 - m \in T$. Since A is a normal exchange ring, there exists a central idempotent $e \in mA \subseteq M$ such that $1 - e \in (1 - m)A$. Therefore

$$a(1 - e) = (1 - e)a \in (1 - m)A = tuA \subseteq tA.$$

In addition, $1 - e \in T$, since $e \in M$. Therefore T is right permutable.

(3) Let $a \in A$, $t \in T$, and let $ta = 0$. By (1), there exist $u \in T$ and $m \in M$ with $ut = 1 - m \in T$. Then $(1 - m)a = 0$. Since A is a normal exchange ring, there exists a central idempotent $e \in Am \subseteq M$ such that $1 - e \in A(1 - m)$. Therefore $a(1 - e) = (1 - e)a \in A(1 - m)a = 0$. In addition, $1 - e \in T$, since $e \in M$. Therefore T is right reversive.

(4) It follows from (1), (2), and (3) that T is a multiplicative right permutable right reversive set. By 1.48 and 1.53(14), there exists the right ring of quotients A_M which is a local ring, and $J(A_M) = f(M)A_M$, where $f : A \to A_M$ is the canonical ring homomorphism. Then $f(A)$ is an exchange ring, and $f(A)$ has no nontrivial idempotents, since $f(A)$ is a subring of the local ring A_M. Hence $f(A)$ is a local ring. In addition, all elements of $f(M)$ are not invertible in $f(A)$, since $J(A_M) = f(M)A_M$. Therefore $f(M) = J(f(A))$. Then all elements of $f(T)$ are invertible in $f(A)$. Therefore $A_M = f(A)$ and $f : A \to A_M$ is a surjective homomorphism.

Now 6.87 follows from (4) and 1.58. ▷

6.88 *For an exchange ring A, the following conditions are equivalent.*

(1) *A is a right distributive ring.*

(2) *A is a normal right Bezout ring.*

(3) *For every maximal right or left ideal M of A, there exists the two-sided ring of quotients ${}_M A_M$ which is a right uniserial ring, and the canonical ring homomorphism $f : A \to {}_M A_M$ is surjective.*

◁ (1)$\Longleftrightarrow$(2) follows from 6.86 and from the fact that all right distributive rings are normal rings by 2.50(3). (3)$\Longrightarrow$(1) follows from 3.22.

(1)$\Longrightarrow$(3) By 6.87, there exists the two-sided ring of quotients ${}_M A_M$ which is a local ring, and the canonical ring homomorphism $f : A \to {}_M A_M$ is surjective. By 3.22, ${}_M A_M$ is a right uniserial ring. ▷

6.89 Let A be a ring, $B(A)$ be the set of all central idempotents of A, and let $S(A)$ be a (non-empty) set of all proper ideals of A generated by central idempotents.

An ideal $P \in S(A)$ is said to be a *Pierce ideal* of A if P is a maximal (with respect to inclusion) element of the set $S(A)$. The set of all Pierce ideals of A is denoted by $\mathcal{P}(A)$.

If P is a Pierce ideal of A, then the factor ring A/P is said to be a *Pierce stalk* of A.

(1) Assume that an ideal Q of A is generated by central idempotents and is not a Pierce ideal.

Then there exists a central idempotent e of A such that $Q + eA$ and $Q + (1-e)A$ are proper ideals of A which properly contain the ideal Q.

In particular, the ring A/Q is isomorphic to a direct product of nonzero rings $A/(Q + eA)$, $A/(Q + (1-e)A)$ and is not an indecomposable ring.

(2) Let a proper ideal P of A is generated by central idempotents e_i $(i \in I)$.

(i)   If A/P is an indecomposable ring, then P is a Pierce ideal of A.

(ii)   If d is a right (left) regular element of A, then $d + P$ is a right regular (left) element of A/P.

(iii) If A is a reduced ring, then A/P is a reduced ring.

(iv) If A is a *pf*-ring, then A/P is a *pf*-ring.

(3) Every ideal from $S(A)$ is contained in at least one Pierce ideal of A.

(4) For every ideal $S \in S(A)$, all idempotents of $A/S(A)$ can be lifted to idempotents of A.

(5) For any indecomposable A-module M_A, there exists a Pierce ideal P of A such that $P \subseteq r(M)$.

(6) If M_A is a module, then $\bigcap_{P \in \mathcal{P}(A)} MP = 0$.

In particular if $M \neq 0$, then there exists a Pierce ideal P of A such that $M \neq MP$.

(7) If N is a submodule of a module M_A, then for every Pierce ideal P of A, there exists the natural A-module isomorphism $MP/NP \to (M/N)P$.

(8) If X and Y are two submodules of a module M_A, then $(X + Y)P = XP + YP$ and $(X \cap Y)P = XP \cap YP$ for every Pierce ideal P of A.

(9) If X and Y are two submodules of a module M_A such that $XP = YP$ for every Pierce ideal P of A, then $X = Y$.

In particular if N is a proper submodule of a module M, then there exists a Pierce ideal P of A such that $MP/NP \neq 0$.

(10) Every ring is a subdirect product of their Pierce stalks.

◁ (1) Since the ideal Q of A is generated by central idempotents and is not a Pierce ideal, there exists a central idempotent $e \in A \setminus P$ such that $P + eA$ is a proper ideal of A. Then $P + (1-e)A$ is a proper ideal of A, since otherwise $e \in e(P + (1-e)A) \subseteq P$. The ideal $P + (1-e)A$ properly contains P, since otherwise $1 = (1-e) + e \in P + eA$.

(2)

(i) follows from (1).

(ii) Let d be a right regular element, and let a be an element of A such that $da \in P$. Then there exists a finite subset J of the set I such that $da \in \sum_{i \in J} e_i A$. Since J is a finite set, there exists a central idempotent e of A such that $\sum_{i \in J} e_i A = eA$. Then $da = dae$. Therefore $da(1-e) = 0$. Since d is a right regular element, $a(1-e) = 0$. Then

$$a = ae = ea \in \sum_{i \in J} e_i A \subseteq P.$$

Therefore $d + P$ is a right regular element of the ring A/P.

(iii) Let a be an element of A such that $a^2 \in P$. Then there exists a finite subset J of the set I such that $a^2 \in \sum_{i \in J} e_i A$. Since J is a finite set, there exists a central idempotent e of A such that $\sum_{i \in J} e_i A = eA$. Since $A/eA \cong (1-e)A$, the ring A/eA is reduced and $a^2 \in eA$. Therefore $a \in eA \subseteq P$, whence A/P is a reduced ring.

(iv) Let $h : A \to A/P$ be the natural epimorphism, and let a, b be two elements of A such that $h(a)h(b) = 0$. Then $ab \in P$ and there exists a finite subset J of the set I such that $ab \in \sum_{i \in J} e_i A$. Since J is a finite set, there exists a central idempotent e of A such that $\sum_{i \in J} e_i A = eA$. Then $a(1-e)b = 0$. Therefore, there exist elements x and y of the pf-ring A such that $x + y = 1$, $a(1-e)x = 0$, and $y(1-e)b = 0$. Then

$$h(x) + h(y) = h(1), \ h(a)h(x) = 0, \text{ and } h(y)h(b) = 0.$$

Therefore A/P is a pf-ring.

(3) By Zorn's Lemma, it is sufficient to prove that if S is the union of an ascending chain of ideals $S_i \in S(A)$ ($i \in I$), then $S \in S(A)$. Since all ideals S_i are generated by central idempotents, S is generated by central idempotents. It remains to prove that $S \neq A$. Assume the contrary. Then $1 \in S = \bigcup_{i \in I} S_i$. Therefore 1 belongs to some ideal S_i; this is a contradiction.

(4) Let $h : A \to A/S$ be the natural epimorphism, and let $h(a)$ be an idempotent of A/S, where $a \in A$. Then $a - a^2 \in S$. Therefore, there exist central idempotents $e_1, \ldots, e_n \in S$ such that $a - a^2 \in \sum_{i=1}^{n} e_i A$.

There is a central idempotent $e \in B(A)$ such that $eA = \sum_{i=1}^{n} e_i A$. Then $a(1-e) - (a(1-e))^2 \in eA \bigcap (1-e)A = 0$. Therefore $a(1-e)$ is an idempotent of A and $h(a(1-e)) = h(a)$.

(5) Let P be an ideal of A generated by the set of all central idempotents of A which are contained in the ideal $r(M)$ of A. It is sufficient to prove that $P + eA = A$ for an arbitrary central idempotent $e \in A \backslash P$. It follows from the definition of the ideal P that $e \in A \backslash r(M)$. Therefore Me is a nonzero submodule of the indecomposable module $M = Me \oplus M(1-e)$. Then $M(1-e) = 0$ and $1 - e \in r(M)$. Hence $1 - e \in P$. Then $A = (1-e)A + eA \subseteq P + eA$.

(6) The module M is a subdirect product of subdirectly indecomposable modules M/N_i, where $\bigcap_{i \in I} N_i = 0$. Therefore $r(M) = \bigcap_{i \in I} r(M/N_i)$. By (5), there exist Pierce ideals $P_i \subseteq r(M/N_i)$ $(i \in I)$. Hence $MP_i \subseteq N_i$ for all i. Then

$$\bigcap_{P \in \mathcal{P}(A)} MP \subseteq \bigcap_{i \in I}(MP_i) \subseteq \bigcap_{i \in I} N_i = 0.$$

(7) Let $P = \sum_{i \in I} e_i A$, where e_i are nonzero central idempotents of A. Let $h : M \to M/N$ be the natural epimorphism. Define the epimorphism $f : MP \to (M/N)P$ by the rule $f(\sum me_i a_i) = \sum h(me_i a_i)$. It can be directly verified that the epimorphism f is well defined and $N = \mathrm{Ker}(f)$.

(8) can be directly verified.

(9) Let P be an arbitrary Pierce ideal of A. By (8),

$$(X + Y)P = XP + YP = XP = XP \bigcap YP = (X \bigcap Y)P.$$

By (7), $((X + Y)/(X \bigcap Y))P \cong (X + Y)P/(X \bigcap Y)P = 0$. By (6), $(X + Y)/(X \bigcap Y) = 0$. Therefore $X + Y = X \bigcap Y$, whence $X = Y$.

(10) follows from (6). $\triangleright$

6.90 Let A be a ring, and let $f_1, \ldots, f_k$ be polynomials with integral coefficients of non-commuting indeterminates $x_1, \ldots, x_m, y_1, \ldots, y_n$, $a_1, \ldots, a_m$ be elements of A. In 6.90, a factor ring A/Q of A is called a *special* ring if there exist elements $\bar{b}_1, \ldots, \bar{b}_n$ of the ring A/Q such that $f_i(\bar{a}_1, \ldots, \bar{a}_m, \bar{b}_1, \ldots, \bar{b}_n) = 0$ for $i = 1, \ldots, k$, where $\bar{x}$ is a natural image of an element x in the ring A/Q. Then the following conditions are equivalent.

(1) A be a special ring.

(2) Every factor ring of A is a special ring.

(3) Every Pierce stalk A/P of A is a special ring.

(4) Every indecomposable factor ring A/P of A is a special ring.

$\triangleleft$ (1)$\Longrightarrow$(2), (2)$\Longrightarrow$(3), and (2)$\Longrightarrow$(4) can be directly verified.

$(3)\Longrightarrow(1)$ Let $\mathcal{E}$ be the set of all proper ideals E of A such that E is generated by central idempotents, and the ring A/E is not special. Assume that A is not a special ring. Then $0 \in \mathcal{E}$ and the set $\mathcal{E}$ is not empty. It can be directly verified that the union of every ascending chain of ideals from $\mathcal{E}$ belongs $\mathcal{E}$. By Zorn's Lemma, the set $\mathcal{E}$ contains a maximal element P. By condition (3), it is sufficient to prove that P is a Pierce ideal. Assume the contrary. By 6.89(1), there exists a central idempotent e of A such that $P+eA$ and $P+(1-e)A$ are proper ideals of A which properly contain the ideal P, and $A \cong (A/(Q+eA) \times (A/(Q+(1-e)A))$. Since ideals $P+eA$ and $P+(1-e)A$ do not belong to $\mathcal{E}$ and are generated by central idempotents, $A/(Q+eA)$ and $A/(Q+(1-e)A)$ are special rings. Since $A \cong (A/(Q+eA) \times (A/(Q+(1-e)A))$, it can be directly verified that A is special.

$(4)\Longrightarrow(1)$ Let $\mathcal{E}$ be the set of all proper ideals E of A such that the ring A/E is not special. Assume that A is not a special ring. Then $0 \in \mathcal{E}$ and the set $\mathcal{E}$ is not empty. It can be directly verified that the union of every ascending chain of ideals from $\mathcal{E}$ belongs $\mathcal{E}$. By Zorn's Lemma, the set $\mathcal{E}$ contains a maximal element P. By condition (4), it is sufficient to prove that A/P is an indecomposable ring. Assume the contrary. Then there exist proper ideals Q and R of A such that $A \cong A/Q \times A/R$ and ideals Q, R properly contain ideal P. Since ideals $P+eA$ and $P+(1-e)A$ do not belong to $\mathcal{E}$, rings A/Q and A/R are special. Since $A \cong A/Q \times A/R$, it can be directly verified that A is a special ring. $\triangleright$

6.91 For a ring A, the following conditions are equivalent.

(1) A is a normal ring.

(2) A/S is a normal ring for every ideal S generated by central idempotents of A.

(3) All Pierce stalks of A have no nontrivial idempotents.

(4) All Pierce stalks of A are normal rings.

$\triangleleft$ $(1)\Longleftrightarrow(2)$ follows from 6.89(4) and from the isomorphism $A \cong A/0$.

$(1)\Longrightarrow(3)$ Let A/P be a Pierce stalk of the normal ring A, and let $\bar{e}$ be a nonzero idempotent of the ring A/P. By 6.89(4), $\bar{e}$ can be lifted to the nonzero central idempotent e of the normal ring A. Then the ideal $P + eA$ is generated by central idempotents and properly contains the Pierce ideal P of A. Therefore $P + eA = A$, whence $\bar{e}$ is the identity element of the ring A/P.

$(3)\Longrightarrow(4)$ can be directly verified.

$(4)\Longrightarrow(1)$ Let e be an idempotent of A, a be an arbitrary element of A, $f_1(X_1) \equiv X_1 - X_1^2$ be the polynomial of an indeterminate X_1, $f_2(X_1, X_2) \equiv X_1 X_2 - X_2 X_1$ be the polynomial of two non-commuting indeterminates X_1, X_2, A/P be any Pierce stalk of A, and let $h : A \to A/P$ be the natural epimorphism. By assumption, $h(e)$ is a central idempotent of the normal ring A/P. Therefore $f_1(h(e)) = 0$ and $f_2(h(e), h(a)) = 0$. By 6.90, $0 = f_2(e, a) = ea - ae$. Therefore e is a central idempotent, A is a normal ring. $\triangleright$

6.92 For a ring A, the following conditions are equivalent.
(1) A is an exchange ring.
(2) All Pierce stalks of A are exchange rings.
(3) All indecomposable factor rings of A are exchange rings.

$\triangleleft$ (1)$\Longrightarrow$(2) and (1)$\Longrightarrow$(3) follow from the fact that every factor ring of an exchange ring is an exchange ring.

(2)$\Longrightarrow$(1) and (3)$\Longrightarrow$(1) Let a be an arbitrary element of A, $f_1(X_1, Y_1, Y_2) \equiv Y_1 - X_1 Y_2$ and $f_2(X_1, Y_1, Y_3) \equiv 1 - Y_1 - (1 - X_1)Y_3$ be polynomials of non-commuting indeterminates X_1, Y_1, Y_2, Y_3, and let $f_3(Y_1) \equiv Y_1 - Y_1{}^2$ be the polynomial of an indeterminate Y_1. In the case of condition (2), let A/P be an arbitrary Pierce stalk of A. In the case of condition (3), let A/P be an arbitrary indecomposable factor ring of A. Let $h : A \to A/P$ be the natural epimorphism. Since $h(A)$ is an exchange ring, there exist elements $\bar{e}, \bar{b}, \bar{c} \in h(A)$ such that $0 = \overline{(e)} - \bar{e}^2 = f_1(\bar{e})$, $0 = \bar{e} - h(a)\bar{b}$, and $0 = 1 - \bar{e} = (1 - h(a))\bar{c}$. By 6.90, there exist elements $e, b, c \in A$ such that $0 = e - e^2 = f_1(e)$, $0 = e - ab$, and $0 = 1 - e - (1 - a)c$. Then e is an idempotent, $e \in aA$, and $1 - e \in (1 - a)A$. Therefore A is an exchange ring. $\triangleright$

6.93 A is a normal exchange ring $\Longleftrightarrow$ all Pierce stalks of A are local rings.

$\triangleleft$ $\Longleftarrow$ follows from 6.92 and from the fact that every local ring is an exchange ring.

$\Longrightarrow$ Let R be a Pierce stalk of A. By 6.91 and 6.92, R is an exchange ring without nontrivial idempotents. Hence R is a local ring. $\triangleright$

6.94 For a ring A, the following conditions are equivalent.
(1) A is a right distributive ring.
(2) All Pierce stalks of A are right distributive rings.
(3) All indecomposable factor rings of A are right distributive.

$\triangleleft$ (1)$\Longrightarrow$(2) and (1)$\Longrightarrow$(3) follow from the fact that every factor ring of a right distributive ring is a right distributive ring.

(2)$\Longrightarrow$(1) and (3)$\Longrightarrow$(1) Assume that m and n are arbitrary elements of A and $Y_1 + Y_2 - 1$, $X_1 Y_1 - X_2 Y_3$, and $X_2 Y_2 - X_1 Y_4$ are polynomials of non-commuting indeterminates X_1, X_2, Y_1, Y_2, Y_3, and Y_4. In the case of condition (2), let A/P be an arbitrary Pierce stalk of A. In the case of condition (3), let A/P be an arbitrary indecomposable factor ring of A. Let $h : A \to A/P$ be the natural epimorphism. Since $h(A)$ is a right distributive ring, it follows from 2.4 that there exist elements $\bar{a}, \bar{b}, \bar{c}, \bar{d} \in h(A)$ such that

$$0 = \bar{a} + \bar{b} - h(1),\ 0 = h(m)\bar{a} - h(n)\bar{c},\ \text{and}\ 0 = h(n)\bar{b} - h(n)\bar{d}.$$

By 6.90, there exist $a, b, c, d \in A$ with $0 = a + b - 1$, $0 = ma - nc$, and $0 = nb - md$. By 2.4, A is right distributive. $\triangleright$

6.95 *A is a right distributive exchange ring $\Longleftrightarrow$ all Pierce stalks of A are right uniserial rings.*

$\triangleleft$ $\Longleftarrow$ follows from 6.94 and from the fact that all right uniserial rings are right distributive.

$\Longrightarrow$ By 2.50(3), A is a normal ring. Let A/P be a Pierce stalk of A. Since A is a normal right distributive exchange ring, A/P is a local right distributive ring by 6.93 and 6.94. By 2.8, A/P is a right uniserial ring. $\triangleright$

Exercises 6.96 Let B be a unitary subring of a ring A, $\{A_i\}_{i=1}^{\infty}$ be the countable set of copies of the ring A, D be the direct product of all the rings A_i, and let R be the subring of D generated by the ideal $\oplus_{i=1}^{\infty} A_i$ and by the subring $B' \equiv \{(b, b, b, \ldots) \mid b \in B\}$.

(1) $R = \{(a_1, \ldots, a_n, b, b, b, \ldots) \mid a_i \in A, b \in B\}$, where the integer n depends on the element $(a_1, \ldots, a_n, b, b, b, \ldots)$ of R.

(2) The rings B and B' are isomorphic to the factor ring $R/(\oplus_{i=1}^{\infty} A_i)$ of the ring R.

(3) Each nonzero right ideal M of the ring R contains the nonzero right ideal $\oplus_{i=1}^{\infty} M A_i$ of the ring D.

(4) For any element $x \in R$ there exists an element $b' \in B' \bigcap xR$ such that $x - b' \in xR \bigcap (\oplus_{i=1}^{\infty} A_i)$.

(5) For any finite set X of elements of the ring R, there exists a unitary subring S of the ring R such that $X \subseteq S$ and $S = \times A_1 \times \ldots \times A_n \times B^*$, where $B^* \cong B$ and B^* is the set of all the elements $s \in S$ such that $se_i = 0$ for all $i \leq n$ and $se_i = se_{n+1} \in B$ for all $i > n$.

(6) R is a semiprimitive ring $\Longleftrightarrow$ A is a semiprimitive ring.

(7) R is an exchange ring $\Longleftrightarrow$ A and B are exchange rings.

(8) R is a regular ring $\Longleftrightarrow$ A and B are regular rings.

(9) If A is the field of rational numbers and B is the ring of integers, then R is a commutative reduced semiprimitive ring, all nonzero

(right or left) ideals of R contain nonzero idempotents, and R is not an exchange ring.

(10) If A is the field of rational numbers and B is the ring of all rational numbers with odd denominators, then R is a commutative reduced semiprimitive exchange ring, and R is not regular.

Exercise 6.97 If A is the ring of all rational numbers with odd denominators, then the domain $A((x))$ is isomorphic to the ring of quotients of the local domain $A[[x]]$ with respect to the set $\{1, x, x^2, \ldots\}$ and is not a local ring.

6.98 (Open questions)

(1) If a right distributive ring A has right and left Krull dimensions, is A left distributive?

(2) If A is a distributive ring, is A localizable?

(3) If A is a distributive reduced ring, are all right ideals of A flat?

(4) If M is a distributive module over a regular ring, is M is a Bezout module?

(5) If A is a right distributive ring which is module-finite over its centre, is A left distributive?

(6) If A is a right distributive ring which is integral over its centre, is A left distributive?

(7) If A is a Noetherian semidistributive ring, is A a direct product of an Artinian ring and a semiprime ring?

(8) If A is a semiprime Noetherian semidistributive ring, is it hereditary?

(9) If all right modules over a ring A are endo-distributive, are all left A-modules endo-distributive?

(10) If all right modules over a ring A are endo-distributive, is A right or left Noetherian?

(11) If all right modules over a ring A are endo-distributive, is A a finite direct product of uniserial Artinian rings?

(12) When is every finitely generated (resp. finitely presented, flat) right module over a ring A endo-distributive?

(13) When is every finitely generated (resp. finitely presented, flat) right module over a ring A semidistributive?

(14) When is a generalized quaternion algebra over commutative rings right serial (resp. right semidistributive)?

(15) When is a skew Laurent series ring right distributive (resp. right serial, right semidistributive)?

(16) When is a group (monoid) ring right semidistributive (resp. right serial, right distributively generated)?

Bibliography

[1] H. Achkar, "Sur les anneaux arithmétiques à gauche," *C. R., Acad. Sci.*, **278**, No. 5, A307–A309 (1974).

[2] H. Achkar, " Sur les propriétés des anneaux arithmétiques à gauche," *C. R., Acad. Sci.*, **284**, No. 17, A993–A995 (1977).

[3] H. Achkar, "Anneaux arithmétiques à gauche," *C. R., Acad. Sci.*, **286**, No. 20, A871–A873 (1978).

[4] H. Achkar, "Anneaux arithmétiques à gauche: propriétés de transfert," *C. R., Acad. Sci.* **288**, No. 11, A583–A586 (1979).

[5] T. Albu and C. Năstăsescu, "Modules arithmétiques," *Acta. Math. Acad. Sci. Hung.*, **25**, No. 3-4, 299–311 (1974).

[6] D. D. Anderson, "Multiplication ideals, multiplication rings and the ring $R(X)$," *Canad. J. Math.*, **28**, No. 4, 760–768 (1976).

[7] D. D. Anderson, "Some remarks on multiplication ideals," *Math. Jpn.* **25**, No. 4, 463–469 (1980).

[8] D. D. Anderson, D. F. Anderson, and R. Markanda, "The rings $R(X)$ and $R\langle X\rangle$," *J. Algebra*, **95**, No. 1, 96–115 (1985).

[9] F. Anderson and K. Fuller, *Rings and Categories of Modules*, Springer, Berlin (1973).

[10] A. Barnard, "Distribitive extensions of modules," *J. Algebra*, **70**, No. 2, 303–315 (1981).

[11] A. Barnard, "Multiplication modules," *J. Algebra*, **71**, No. 1, 174–178 (1981).

[12] E. A. Behrens, "Algebren mit Vorgegebenem Endlichem, Distributivem Idealverband," *Math. Ann.*, **133**, 79–90 (1957).

[13] E. A. Behrens, "Eine Charakteriesierung der T-Moduln mit Distributivem Untermodulevervand bei Halbprimärem T," *Arch. Math.*, **8**, 265–273 (1957).

[14] E. A. Behrens, "Distributive Darstellbare Ringe I," *Math. Z.*, **73**, No. 5, 409–432 (1960).

[15] E. A. Behrens, "Distributive Darstellbare Ringe II," *Math. Z.*, **76**, No. 4, 367–384 (1961).

[16] E. A. Behrens, "Die Halbgruppe der Ideale in Algebren mit Distributivem Idealverband," *Arch. Math.*, **13**, 251–266 (1962).

[17] E. A. Behrens, *Ring Theory*, Acad. Press, New York (1972).

[18] E. A. Behrens, "Noncommutative arithmetic rings," In: *Rings, Modules and Radicals*, Amsterdam, London, 67–71 (1973).

[19] K. I. Beidar, V. N. Latyshev, V. T. Markov, A. V. Mikhalev, L. A. Skornyakov, and A. A. Tuganbaev, "Associative rings," In: Progress of Science and Technology: Series on Algebra, Topology, and Geometry. Vol. 22, *Itogi Nauki i Tekhn. VINITI*, All-Union Institute for Scientific and Technical Information, Akad. Nauk SSSR, Moscow (1985), pp. 3–116.

[20] T. Belzner, "Towards self-duality of semidistributive artinian rings," *J. Algebra*, **135**, No. 1, 74–95 (1990).

[21] K. Bessenrodt, H. H. Brungs, and G. Törner, "Right chain rings. Part 1," Schriftenreihe des Fachbereich Mathematik. Universität Duisburg, 1990.

[22] K. Bessenrodt, H. H. Brungs, and G. Törner, "Right chain rings. Part 2a," Schriftenreihe des Fachbereich Mathematik. Universität Duisburg, 1992.

[23] K. Bessenrodt, H. H. Brungs, and G. Törner, "Right chain rings. Part 2b," Schriftenreihe des Fachbereich Mathematik. Universität Duisburg, 1992.

[24] R. L. Blair, "Ideal lattice and the structure of rings," *Trans. Amer. Math. Soc.*, **75**, No. 1, 136–153 (1953).

[25] H. H. Brungs, "Subkommutative Dedekindringe," *Publ. Math. Debrecen*, **18**, 89–94 (1972).

[26] H. H. Brungs, "Rings with distributive lattice of right ideals," *J. Algebra*, **40**, No. 2, 392–400 (1976).

[27] H. H. Brungs, "Bezout domains and rings with a distributive lattice of right ideals," *Can. J. Math.* **38**, No. 2, 286–303 (1986).

[28] H. H. Brungs, "Ideal theory in rings with a distributive lattice of right ideals," *Wiss. Beitr. M.-Luther-Univ. Halle. Wittenberg*, No. 48, 73–80 (1987).

[29] H. H. Brungs and J. Gräter, "Valuation rings in finite-dimensional division algebras," *J. Algebra*, **120**, 90–99 (1989).

[30] H. H. Brungs and J. Gräter, "Value groups and distributivity," *Can. J. Math.* **43**, No. 6, 1150–1160 (1991).

[31] H. H. Brungs and J. Gräter, "Noncommutative Prüfer and Valuation rings," In: *Proc. Int. Conf. Algebra Dedicat. Mem. A.I.Mal'cev* Novosibirsk, (1989), Pt. 1. Providence (R.I.), 253–269 (1992).

[32] H. S. Butts and R. C. Phillips, "Almost multiplication Rings," *Canad. J. Math.*, **17**, 267–277 (1965).

[33] C. E. Caballero, "Self-duality and ℓ-hereditary semidistributive rings," *Comm. Algebra* **14**, No. 10, 1821–1843 (1986).

[34] V. Camillo, "Distributive modules," *J. Algebra*, **36**, No. 1, 16–25 (1975).

[35] R. Camps and W. Dicks, "On semilocal rings," *Isr. J. Math.*, **81**, 203–221 (1993).

[36] R. Camps, A. Facchini, and G. E. Puninski, "Chain rings and serial rings that are endomorphism rings of artinian modules," In "Rings and Radicals," Pitman Series, Longman, Sci & Tech., 1996.

[37] Qinghua Chen, "On graded distributive modules," *Northeast. Math. J.* **10**, No. 2, 273–278 (1994).

[38] Chang Woo Choi, "Multiplication modules and endomorphisms," *Math. J. Toyama Univ.* **18**, 1–8 (1995).

[39] A. W. Chatters, "Serial rings with Krull dimension," *Glasgow Math. J.*, **32**, 71–78 (1990).

[40] L. G. Chouinard, B. R. Hardy, and T. S. Shores, "Arithmetical semihereditary semigroup rings, *Comm. Algebra*, **8**, No. 17, 1593–1652 (1980).

[41] P. Cohn, *Free Rings and Their Relations*, Academic Press, London (1971).

[42] T. M. K. Davison, "Distributive homomorphisms of rings and modules," *J. Reine Angew Math.*, **271**, 28–34 (1974).

[43] Yu. A. Drozd, "On generalized uniserial rings," *Mat. Zametki*, **18**, No. 5, 707–710 (1975).

[44] N. I. Dubrovin, "Uniserial domains," *Vestn. MGU, Mat., Mekh.*, No. 2, 51–54 (1980).

[45] N. I. Dubrovin, "On uniserial rings," *Uspekhi Mat. Nauk*, **37**, No. 4, 139–140 (1982).

[46] N. I. Dubrovin, "The rational closure of group rings of left-ordered groups," In: *Gerhard Mercator Universität Duisburg Gesamthochschule*, (1994).

[47] D. Eisenbud and P. Griffith, "Serial rings, "J. Algebra," **17**, No. 3, 268–287 (1971).

[48] D. Eisenbud, P. Griffith, "The structure of serial rings," *Pacif. J. Math.*, **36**, 109–121 (1971).

[49] P. Eklof and I. Herzog, "Model theory of modules over a serial ring," *Ann. Pure Appl. Logic*, **72**, 145–176 (1995).

[50] Z. A. El-Bast and P. F. Smith, "Multiplication modules," *Comm. Algebra*, **16**, No. 4, 755–779 (1988).

[51] Z. A. El-Bast and P. F. Smith, "Multiplication modules and theorems of Mori and Mott," *Comm. Algebra*, **16**, N 4, 781–796 (1988).

[52] V. Erdoğdu, "Distributive modules," *Canad. Math. Bull.*, **30**, No. 2, 248–254 (1987).

[53] V. Erdoğdu, "Modules with locally linearly ordered distributive hulls," *J. Pure Appl. Algebra*, **47**, 119–130 (1987).

[54] V. Erdoğdu, "Multiplication modules which are distributive," *J. Pure Appl. Algebra*, **54**, 209–213 (1988).

[55] V. Erdoğdu, "Rings with principal distributive ideal," *Türk. Mat. Derg.*, **4**, No. 2, 120–124 (1990).

[56] A. Facchini and G. E. Puninski, "Σ-pure-injective modules over serial rings," In: "Abelian groups and modules". A. Facchini, C. Menini eds., Kluwer Acad. Publishers, Dodrecht. 1995, 145–162.

[57] A. Facchini and G. E. Puninski, "Classical localizations in serial rings," *Comm. Algebra*, **24**, No. 11, 3537–3559 (1996).

[58] A. Facchini and L. Salce, "Uniserial modules: sums and isomorphisms of subquotients, *Comm. Algebra*, **18**, No. 2, 499–517 (1990).

[59] C. Faith, *Algebra: Rings, Modules, and Categories I.*, Springer, Berlin (1973).

[60] C. Faith, *Algebra II.*, Springer, Berlin (1976).

[61] R. B. Feinberg, "Faithful distributive modules over incidence algebras," *Pacif. J. Math.*, **65**, No. 1, 35–45 (1976).

[62] M. Ferrero and G. Törner, "On the ideal structure of right distributive rings," *Comm. Algebra*, **21**, No. 8, 2697–2713 (1993)

[63] L. Fuchs, L. Salce, "Modules over valuation domains," *Lect. Notes Pure Appl. Math.*, **97**, (1985).

[64] K. R. Fuller, "Generalized uniserial rings and their Kupisch series," *Math. Z.*, **106**, No. 4, 248–260 (1976).

[65] K. R. Fuller, "Weakly symmetric rings of distributive module type," *Comm. Algebra*, **5**, No. 9, 997–1008 (1977).

[66] K. R. Fuller, "Rings of left invariant module type," *Comm. Algebra*, **6**, 153–167 (1978).

[67] K. R. Fuller, "On generalization of serial rings II," *Comm. Algebra*, **8**, No. 7, 635–661 (1980).

[68] K. R. Fuller, J. Haack, and H. Hullinger, "Stable equivalence of uniserial rings," *Proc. Amer. Math. Soc.*, **68**, No. 2, 153–158 (1978).

[69] R. W. Gilmer and J. L. Mott, "Multiplication rings as rings in whish ideals with prime radical are primary," *Trans. Amer. Math. Soc.* **114**, 40–52 (1965).

[70] K. R. Goodearl, *Von Neumann Regular Rings*, Pitman, London (1979).

[71] K. R. Goodearl, "Artinian and Noetherian modules over regular rings," *Comm. Algebra*, **8**, No. 5, 475–504 (1980).

[72] R. Gordon and J. C. Robson, "Krull dimension," *Mem. Amer. Math. Soc.* **133** (1973).

[73] J. Gräter, "Zur Theorie Nicht Kommutativer Prüfer Ringer," *Arch. Math.*, **41**, 30–36 (1983).

[74] J. Gräter, "Über die Distributivität des Idealverbandes Eines Kommutativem Ringes," *Monatsh. Math.*, **99**, No. 4, 267–278 (1985).

[75] J. Gräter, "On noncommutative Prüfer rings," *Arch. Math.*, **46**, No. 5, 402–407 (1986).

[76] J. Gräter, "Lokalinvariante Prüferringe," *Results Math.* **9**, 10–32 (1986).

[77] J. Gräter, "Ringe mit Distributivem Rechtidealverband," *Results Math.* **12**, 95–98 (1987).

[78] J. Gräter, "Strong right D-domains," *Monatsh. Math.* **107**, No. 3, 189–205 (1989).

[79] O. E. Gregul and V. V. Kirichenko, "On semihereditary serial rings," *Ukr. Mat. Zh.*, **39**, No. 2, 156–161 (1987).

[80] M. Griffin, "Multiplication rings via their quotient rings," *Canad. J. Math.*, **26**, 430–449 (1974).

[81] J. K. Haack, "Self-duality and serial rings," *J. Algebra*, **59**, No. 2. 345–363 (1979).

[82] B. R. Hardy and T. S. Shores, "Arithmetical semigroup rings," *Canad. J. Math.*, **32**, No. 6, 1361–1371 (1980).

[83] A. Hattori, "A foundation of torsion theory for modules over general rings," *Nagoya Math. J.*, **17**, 147–158 (1960).

[84] C. Herrmann, "On a condition sufficient for the distributivity of lattices of linear subspaces," *Arch. Math.*, **33**, No. 3, 235–238 (1979).

[85] D. A. Hill, "Injective modules over non-Artinian serial rings," *J. Austral. Math. Soc.*, **44**, 242–251 (1988).

[86] Y. Hinohara, "Note on noncommutative local rings," *Nagoya Math. J.*, **17**, 161–166 (1960).

[87] G. Ivanov, "Left generalized uniserial rings," *J. Algebra*, **31**, 166–181 (1974).

[88] G. Ivanov, "Decomposition of modules over serial rings," *Comm. Algebra*, **3**, No. 11, 1031–1036 (1975).

[89] N. Jacobson, *Structure of Rings*, Amer. Math. Soc., Providence, Rhode Island (1968).

[90] C. U. Jensen, "A remark on arithmetical rings," *Proc. Amer. Math. Soc.*, **15**, No. 6, 951-954 (1964).

[91] C. U. Jensen, "Arithmetical rings," *Acta Math. Acad. Sci. Hungar.*, **17**, No. 1, 115–123 (1966).

[92] I. Kaplansky, "Elementary divisors and modules," *Trans. Amer. Math. Soc.*, **66**, 464–491 (1949).

[93] F. Kasch, *Moduln und Ringe*, B.G.Teubner, Stuttgart (1977).

[94] K. H. Kim and F. W. Roush, "Reflexive lattices of subspaces," *Proc. Amer. Math. Soc.*, **78**, No. 1, 17–18 (1980).

[95] K. H. Kim and F. W. Roush, "Regular rings and distributive lattices," *Comm. Algebra*, **8**, No. 13, 1283–1290 (1980).

[96] V. V. Kirichenko, "Generalized uniserial rings," Preprint, 1975.

[97] V. V. Kirichenko, "Generalized uniserial rings," *Mat. Sb.*, **99**, No. 4, 559–581 (1976).

[98] V. V. Kirichenko, "On serial hereditary and semihereditary rings," *Zap. nauchn. sem. LOMI Acad. Nauk SSSR*, **114** 137–147 (1982).

[99] V. V. Kirichenko and O. Ya. Bernik, "Semiperfect rings of distributive module type," *Dokl. Akad. Nauk USSR*, -A, No. 3, 15–17 (1988).

[100] V. V. Kirichenko and P. P. Kostyukevich, "Biserial rings," *Ukr. Mat. Zh.*, **38**, No. 6, 718–723 (1986).

[101] V. V. Kirichenko and Yu. V. Yaremenko, "Noetherian biserial rings," *Ukr. Mat. Zh.*, **40**, No. 4, 435–440 (1988).

[102] I. B. Kozhukhov, "Uniserial semigroup rings," *Uspekhi Mat. Nauk*, **29**, No. 1, 169–170 (1974).

[103] H. Krause, "Maps between tree and band modules," *J. Algebra*, **137**, 186–194 (1991).

[104] H. Kupisch, "Beiträge zur Theorienichthalbeinfacher Ringen mit Minimalbedingung," *J. Reine Angew. Math.*, **201**, 100–112 (1959).

[105] H. Kupisch, "Über eine Klasse von Ringen mit Minimalbedingung, *Arch. Math.*, **17**, 20–35 (1966).

[106] H. Kupisch, "Einreihige Algebren über einem perfekten Körper," *J. Algebra*, **33**, No. 1, 68–74 (1975).

[107] H. Kupisch, "A characterizaton of Nakayama rings," *Comm. Algebra*, **23**, No. 2, 739–741 (1995).

[108] I. Lambek, *Lectures on Rings and Modules*, Blaisdell, Waltham (1966).

[109] D. Latsis, "Localizaton dans les anneaux duos," *C. R. Acad. Sci.*, **A282**, No. 24, 1403–1406 (1976).

[110] Lee Sin Min, "On the constructions of local and arithmetical rings," *Acta. Math. Acad. Sci. Hung.*, **32**, No. 1–2, 31–34 (1978).

[111] T. H. Lenagan, "Noetherian rings with Krull dimension one," *J. London Math. Soc.*, **15**, No. 1, 41–47 (1977).

[112] L. R. Le Riche, "The ring $R\langle X \rangle$," *J. Algebra*, **67**, 327–341 (1980).

[113] G. M. Low, P. F. Smith, "Multiplication modules and ideals," *Comm. Algebra*, **18**, No. 12, 4353–4375 (1990).

[114] R. Makino, "QF-1 algebras with faithful direct sums of uniserial modules," *J. Algebra*, **136**, 175–189 (1991).

[115] V. T. Markov, A. V. Mikhalev, L. A. Skornyakov, and A. A. Tuganbaev," Modules," In: Progress of Science and Technology: Series on Algebra, Topology, and Geometry. Vol. 19, *Itogi Nauki i Tekhn. VINITI*, All-Union Institute for Scientific and Technical Information, Akad. Nauk SSSR, Moscow (1981), pp. 31–134.

[116] V. T. Markov, A. V. Mikhalev, L. A. Skornyakov, and A. A. Tuganbaev," Endomorphism rings and submodule lattices," In: Progress of Science and Technology: Series on Algebra, Topology, and Geometry. Vol. 21, *Itogi Nauki i Tekhn. VINITI*, All-Union Institute for Scientific and Technical Information, Akad. Nauk SSSR, Moscow (1983), pp. 183–254.

[117] J. Martinez, "On commutative rings which are strongly Prüfer," *Comm. Algebra*, **2**, No. 9, 3479–3488 (1994).

[118] R. Mazurek, "Distributive rings with Goldie dimension one," *Comm. Algebra*, **19**, No. 3, 931–944 (1991).

[119] R. Mazurek and E. R. Puczylowski, "On nilpotent elements of distributive rings," *Comm. Algebra*, **18**, No. 2, 463–471 (1990).

[120] R. Mazurek and E. R. Puczylowski, "On semidistributive rings," *Comm. Algebra*, **25**, No. 11, 3463–3471 (1997).

[121] P. Menal," Group rings in which every left ideal is a right ideal," *Proc. Amer. Math. Soc.*, **76**, No. 2, 204–208 (1979).

[122] W. Menzel, "Über den Untergruppenverband einer Abelschen Operatorgruppe. Teil II. Distributive und M-Verbande von Untergruppen einer Abelschen Operatorgruppe," *Math. Z.*, **74**, No. 1, 52–65 (1960).

[123] W. Menzel, "Ein Kriterium für Distributivität des Untergruppenverbands einer Abelschen Operatorgruppe, *Math. Z.*, **75**, No. 3, 271–276 (1961).

[124] A. V. Mikhalev, "Special structural spaces of rings," *Dokl. Akad. Nauk SSSR*, **150**, No. 7, 259–261 (1963).

[125] J. L. Mott, "Equivalent conditions for a ring to be a multiplication ring," *Canad. J. Math.*, **16**, 429–434 (1964).

[126] J. L. Mott, "Multiplication rings containing only finitely many minimal primes," *J. Sci. Hiroshima Univ. Ser. A-I.* **33**, 73–83 (1969).

[127] I. Murase, "On the structure of generalized uniserial rings, I," *Sci. Papers College Gen. ed., Univ. Tokyo*, **13**, 1–22 (1963).

[128] I. Murase, "On the structure of generalized uniserial rings, II," *Sci. Papers College Gen. ed., Univ. Tokyo*, **13**, 131–158 (1963).

[129] I. Murase, "On the structure of generalized uniserial rings, III',
Sci. Papers College Gen. ed., Univ. Tokyo, **14**, 11–25 (1964).

[130] I. Murase, "Generalized uniserial group rings, I," *Sci. Papers College Gen. ed., Univ. Tokyo*, **15**, 15–28 (1965).

[131] I. Murase, "Generalized uniserial group rings, II," *Sci. Papers College Gen. ed., Univ. Tokyo*, **15**, 111–128 (1965).

[132] B. J. Müller, "The structure of serial rings," In: *Methods in Module Theory*, G. Abrams, J. Haefner, and K. M. Rangaswamy eds., Marcel Dekker, New York, 1992, 249–270.

[133] B. J. Müller, "Goldie-prime serial rings," *Contemp. Math.*, **130**, 301–310 (1992).

[134] B. J. Müller and S. Singh, "Uniform modules over serial rings II," *Lect. Notes Math.* **1448**, 25–32 (1989).

[135] B. J. Müller, S. Singh, "Uniform modules over serial rings," *J. Algebra*, **144**, 94–109 (1991).

[136] T. Nakayama, "Note on uniserial and generalized uniserial rings," *Proc. Imp. Acad. Tokyo*, **16**, 285–289 (1940).

[137] T. Nakayama, "On Frobenius algebras. II," *Ann. Math.*, **42**, No. 1, 1–21 (1941).

[138] A. G. Naoum, "The Ohm type properties for finitely generated multiplication ideals," *Period. Math. Hung.* **18**, No. 4, 287–293 (1986).

[139] A. G. Naoum, "On the product of a module by an ideal," *Beiträge Algebra Geom.*, **27**, 153–159 (1988).

[140] A. G. Naoum, "On the dual of a finitely generated multiplication module II," *Beiträge Algebra Geom.*, **27**, 83–88 (1988).

[141] A. G. Naoum and F. H. Al-Alwan, "Dedekind modules," *Comm. Algebra*, **24**, No. 2, 397–412 (1996).

[142] A. G. Naoum and F. H. Al-Alwan, "Dense submodules of multiplication modules," *Comm. Algebra*, **24**, No. 2, 413–424 (1996).

[143] A. G. Naoum and M. M. Balboul, "On finitely generated multiplication ideals in commutative rings," *Beiträge Algebra Geom.* **19**, 75–82 (1985).

[144] A. G. Naoum and M. A. K. Hasan, "The residual of finitely generated multiplication modules," *Arch. Math.*, **46**, 225-230, (1986).

[145] A. G. Naoum and K. R. Sharaf, "A note on the dual of a finitely generated multiplication module," *Beiträge Algebra Geom.*, **27**, 5-11 (1988).

[146] R. Pierce, *Associative Algebras*, Springer, New York (1982).

[147] Z. Pogorzaly, "Representation-finite biserial algebras over a perfect field," *Bull. Pol. Acad. Sci. Math.*, **33**, No. 7-8, 349-353 (1985).

[148] G. E. Puninskii, "Superdecomposable pure-injective modules over commutative valuation rings," *Algebra i Logika*, **31**, No. 6, 655-671 (1992).

[149] G. E. Puninskii, "Endodistributive and pure injective modules over uniserial rings," *Uspekhi Mat. Nauk*, **48**, No. 3, 201-202 (1993).

[150] G. E. Puninskii, "Indecomposable pure-injective modules over uniserial rings," *Trudy Mosk. Mat. Obshch.*, **56**, 1-13 (1994).

[151] G. E. Puninskii, "Pure-injective modules over right Noetherian serial rings," *Comm. Algebra*, **23**, No. 4, 1579-1592 (1995).

[152] G. E. Puninskii, "Serial Krull-Schmidt rings and pure-injective modules," *Fundam. Prikl. Mat.*, **1**, No. 2, 471-490 (1995).

[153] G. E. Puninskii and R. Wisbauer, "Σ-injective modules over right duo and right distributive rings," *J. Pure Appl. Algebra*, **113**, 55-66 (1996).

[154] V. B. Repnitskii, "Arithmetical modules," *Mat. Sb.*, **110**, No. 1, 150-157 (1979).

[155] T. Shores and W. J. Lewis, "Serial modules and endomorphism rings," *Duke Math. J.*, **41**, No. 4, 889-909 (1974).

[156] S. Singh and F. Mehdi, "Multiplication modules," *Canad. Math. Bull.*, **22**, No. 1, 93-98 (1979).

[157] L. A. Skornyakov, "When are all modules serial?" *Mat. Zametki*, **5**, No 2, 173-182 (1969).

[158] A. Skowronski and J. Waschbüsch, "Representation-finite biserial algebras," *J. Reine Angew. Math.*, **345**, 172-181 (1983).

[159] P. F. Smith, "Some remarks on multiplication modules," *Arch. Math.*, **50**, No. 3, 223–235 (1988).

[160] P. F. Smith, T. Ukegawa, and N. Umaya, "On noncommutative Noetherian multiplication rings," *Math. Sem. Notes Kobe Univ.*, **11**, No. 2, 259–276 (1983).

[161] K. I. Sonin, "Regular Laurent series rings," *Fundam. Prikl. Mat.*, **1**, No. 1, 315–317 (1995).

[162] K. I. Sonin, "Regular skew Laurent series rings," *Fundam. Prikl. Mat.*, **1**, No. 2, 565–568 (1995).

[163] K. I. Sonin, "Biregular Laurent series rings," *Vestn. MGU, Mat., Mekh.*, No. 4, 20–22 (1997).

[164] K. I. Sonin, "Krull dimension of Malcev–Neumann rings," *Comm. Algebra*, **26**, No. 9, 2915–2931 (1998).

[165] B. Stenström, *Rings of Quotients: An Introduction to Methods of Ring Theory*, Springer, Berlin (1975).

[166] W. Stephenson, "Modules whose lattice of submodules is distributive," *Proc. London Math. Soc.*, **28**, No. 2, 291–310 (1974).

[167] A. A. Tuganbaev, "Rings all modules over which are direct sums of distributive modules," *Vestn. MGU, Mat., Mekh.*, No. 1, 61–64 (1980).

[168] A. A. Tuganbaev, "Distributive Noetherian rings," *Vestn. MGU, Mat., Mekh.*, No. 2, 30–34 (1980).

[169] A. A. Tuganbaev, "Distributive modules," *Uspekhi Mat. Nauk*, **35**, No. 5, 245–246 (1980).

[170] A. A. Tuganbaev, "Integrally closed rings," *Mat. Sb.*, **115**, No. 4, 544–559 (1981).

[171] A. A. Tuganbaev, "Integrally closed Noetherian rings," *Uspekhi Mat. Nauk*, **36**, No. 5, 195–196 (1981).

[172] A. A. Tuganbaev, "Distributive modules and rings," *Uspekhi Mat. Nauk*, **39**, No. 1, 157–158 (1984).

[173] A. A. Tuganbaev, "Distributive rings," *Mat. Zametki*, **35**, No. 3, 329–332 (1984).

[174] A. A. Tuganbaev, "Rings with distributive lattice of ideals," In: *Abelian Groups and Modules* [in Russian], Tomsk, No. 5 (1985), pp. 88–104.

[175] A. A. Tuganbaev, "Right distributive rings," *Izv. Vuzov, Mat.*, No. 1, 46–51 (1985).

[176] A. A. Tuganbaev, "Rings with flat right ideals and distributive rings," *Mat. Zametki*, **38**, No. 2, 218–228 (1985).

[177] A. A. Tuganbaev, "Distributive rings and endodistributive modules," *Ukr. Mat. Zh.*, **38**, No. 1, 63–67 (1986).

[178] A. A. Tuganbaev, "Rings whose lattice of right ideals is distributive," *Izv. Vuzov, Mat.*, No. 2, 44–49 (1986).

[179] A. A. Tuganbaev, "Rings with distributive lattice of right ideals," *Uspekhi Mat. Nauk*, **41**, No. 3, 203–204 (1986).

[180] A. A. Tuganbaev, "Distributive series rings," *Mat. Zametki*, **39**, No. 4, 518–528 (1986).

[181] A. A. Tuganbaev, "Distributive and semigroup rings," In: *Abelian Groups and Modules* [in Russian], Tomsk, No. 6 (1986), pp. 133–144.

[182] A. A. Tuganbaev, "Series rings and weak global dimension," *Izv. Vuzov, Mat.*, No. 11, 70–78 (1987).

[183] A. A. Tuganbaev, "Hereditary rings," *Mat. Zametki*, **41**, No. 3, 303–312 (1987).

[184] A. A. Tuganbaev, "Distributive rings and modules," *Trudy Mosk. Mat. Obshch.*, **51**, 95–113 (1988).

[185] A. A. Tuganbaev, "Distributive rings with the finiteness condition," *Izv. Vuzov, Mat.*, No. 10, 50–55 (1988).

[186] A. A. Tuganbaev, "On monoid rings," In: *Abelian Groups and Modules* [in Russian], Tomsk, No. 7 (1988), pp. 124–130.

[187] A. A. Tuganbaev, "Modules with distributive lattice of submodules," *Uspekhi Mat. Nauk*, **44**, No. 1, 215–216 (1989).

[188] A. A. Tuganbaev, "Rings with distributive lattice of right ideals," In: *Abelian Groups and Modules* [in Russian], Tomsk, No. 8 (1989), pp. 124–127.

[189] A. A. Tuganbaev, "Distributive rings and modules," *Mat. Zametki*, **47**, No. 2, 115–123 (1990).

[190] A. A. Tuganbaev, "Embeddings into serial modules," In: *Abelian Groups and Modules* [in Russian], Tomsk, No. 9 (1990), pp. 115–120.

[191] A. A. Tuganbaev, "Rings with projective principal right ideals," *Ukr. Mat. Zh.*, **42**, No. 6, 861–863 (1990).

[192] A. A. Tuganbaev, "Serial rings and modules," *Mat. Zametki*, **48**, No. 2, 99–106 (1990).

[193] A. A. Tuganbaev, "Elementary divisor rings and distributive rings," *Uspekhi Mat. Nauk*, **46**, No. 6, 219–220 (1991).

[194] A. A. Tuganbaev, "Rings of quotients of distributive rings," In: *Abelian Groups and Modules* [in Russian], Tomsk, No. 10 (1991), pp. 148–152.

[195] A. A. Tuganbaev, "Distributive monoid algebras," *Mat. Zametki*, **51**, No. 2, 101–108 (1992).

[196] A. A. Tuganbaev, "Quaternion algebras over local rings," *Uspekhi Mat. Nauk*, **47**, No. 3, 179–180 (1992).

[197] A. A. Tuganbaev, "Generalized quaternion algebras," *Mat. Zametki*, **53**, No. 5, 120–128 (1993).

[198] A. A. Tuganbaev, "Quaternion algebras over commutative rings," *Mat. Zametki*, **53**, No. 2, 126–131 (1993).

[199] A. A. Tuganbaev, "Flat modules and distributive rings," *Ukr. Mat. Zh.*, **45**, No. 5, 721–724 (1993).

[200] A. A. Tuganbaev, "On quaternion algebras," *Trudy Sem. im. I. G. Petrovskogo*, Moscow (1994) No. 17, 209–219.

[201] A. A. Tuganbaev, "Endomorphism rings and distributivity," *Mat. Zametki*, **56**, No. 4, 141–152 (1994).

[202] A. A. Tuganbaev, "On semigroup rings," In: *Abelian Groups and Modules* [in Russian], Tomsk, No. 11–12 (1994), pp. 218–225.

[203] A. A. Tuganbaev, "Endomorphism ring of a distributive module," *Uspekhi Mat. Nauk*, **50**, No. 1, 215–216 (1995).

[204] A. A. Tuganbaev, "On left distributivity of right distributive rings," *Fundam. Prikl. Mat.*, **1**, No. 1, 289–300 (1995).

[205] A. A. Tuganbaev, "Endomorphisms of distributive modules," *Uspekhi Mat. Nauk*, **50**, No. 3, 167–168 (1995).

[206] A. A. Tuganbaev, "Noetherian semiprime rings and distributivity," *Fundam. Prikl. Mat.*, **1**, No. 3, 767–779 (1995).

[207] A. A. Tuganbaev, " Right or left distributive rings," *Mat. Zametki*, **58**, No. 4, 604–627 (1995).

[208] A. A. Tuganbaev, "Distributive semiprime rings," *Mat. Zametki*, **58**, No. 5, 736–761 (1995).

[209] A. A. Tuganbaev, "Flat modules and distributivity," *Uspekhi Mat. Nauk*, **50**, No. 6, 221–222 (1995).

[210] A. A. Tuganbaev, "Distributively decomposable rings," *Uspekhi Mat. Nauk*, **51**, No. 3, 215–216 (1996).

[211] A. A. Tuganbaev, "Flat modules and rings that are finitely generated over its centre," *Mat. Zametki*, **60**, No. 2, 254–277 (1996).

[212] A. A. Tuganbaev, "Piecewise domains and nonsingularly prime ideals," *Uspekhi Mat. Nauk*, **51**, No. 4, 171–172 (1996).

[213] A. A. Tuganbaev, "Direct sums of distributive modules," *Mat. Sb.*, **187**, No. 12, 137–156 (1996).

[214] A. A. Tuganbaev, "On distributive rings and modules," In: *First International Tainan-Moscow Algebra Workshop*, Walter de Gruyter, Berlin–New York, (1996), pp. 323–328.

[215] A. A. Tuganbaev, "Semidistributive hereditary rings," *Uspekhi Mat. Nauk*, **52**, No. 4, 215–216 (1997).

[216] A. A. Tuganbaev, "Ideals of distributive rings," *Fundam. Prikl. Mat.*, **4**, No. 2, 791–794 (1998).

[217] A. A. Tuganbaev, *Semidistributive Rings and Modules*, Kluwer Academic Publishers, Dordrecht–Boston–London, (1998).

[218] D. A. Tuganbaev, "Uniserial Laurent series rings," *Fundam. Prikl. Mat.*, **3**, No. 3, 947–951 (1997).

[219] D. A. Tuganbaev, "Uniserial Laurent skew series rings," Preprint (1998).

[220] M. Uematsu, K. Yamagata, "On serial quasi-hereditary rings," *Hokkaido Math. J.*, **19**, 165–174 (1990).

[221] T. Ukegawa, "Some properties of noncommutative multiplication rings," *Proc. Jpn. Acad.*, **54A**, No. 9, 279–284 (1978).

[222] T. Ukegawa, "Left Noetherian multiplication rings," *Osaka J.Math.*, **17**, No. 2, 449–453 (1980).

[223] T. Ukegawa, "Some remarks on M-rings," *Math. Jpn.*, **28**, No. 2, 195–203 (1983).

[224] M. H. Upham, "Serial rings with right Krull dimension one," *J. Algebra*, **109**, 319–333 (1987).

[225] P. Vàmos, "Finitely generated Artinian and distributive modules are cyclic," *Bull. London Math. Soc.* **10**, No. 3, 287–288 (1978).

[226] N. Vanaja and V. N. Purav, "A note on generalized uniserial rings," *Comm. Algebra*, **21**, No. 4, 1153–1159 (1993).

[227] E. M. Vechtomov, "Distributive rings of continuous functions," *Mat. Zametki*, **34**, No. 34, 321–332 (1983).

[228] E. M. Vechtomov, *Rings of Continuous Functions and F-spaces, Selected Themes* [in Russian], Moscow State Teachers' Training University Press, Moscow (1992).

[229] A. Vogel, "Spectrum of a locally cyclic modules over a commutative regular rings," *Comm. Algebra*, **9**, No. 16, 1617–1637 (1981).

[230] B. Wald and J. Waschbüsch, "Tame biserial algebras," *J. Algebra*, **95**, 480–500 (1985).

[231] R. B. Warfield, "Decomposability of finitely presented modules," *Proc. Amer. Math. Soc.*, **25**, 167–172 (1970).

[232] R. B. Warfield, "Serial rings and finitely presented modules," *J. Algebra*, **37**, No. 3, 187–222 (1975).

[233] R. B. Warfield, "Bezout rings and serial rings," *Comm. Algebra*, **7**, 533–545 (1979).

[234] J. Waschbüsch, "Self-duality of serial rings," *Comm. Algebra*, **14**, No. 4, 581–589 (1986).

[235] R. Wisbauer, "Modules locally of serial type," *Period. Math. Hungar.*, **18**, 39–52 (1987).

[236] R. Wisbauer, *Foundations of Module and Ring Theory*, Gordon and Breach, Philadelphia, (1991).

[237] M. H. Wright, "Serial rings with right Krull dimension one, II," *J. Algebra*, **117**, 99–116 (1988).

[238] M. H. Wright, "Krull dimension in serial rings," *J. Algebra*, **124**, 317–328 (1989).

[239] M. H. Wright, "Links between right ideals of serial rings with Krull dimension," *Lect. Notes Math.*, **1448**, 33–40 (1989).

[240] M. H. Wright, "Certain uniform modules over serial rings are uniserial," *Comm. Algebra*, **17**, 441–469 (1989).

[241] M. H. Wright, "Uniform modules over serial rings with Krull dimension," *Comm. Algebra*, **18**, No. 8, 2541–2557 (1990).

[242] M. H. Wright, "Prime serial rings with Krull dimension," *Comm. Algebra*, **18**, No. 8, 2559–2572 (1990).

[243] M. H. Wright, "Right locally distributive rings," In: *Ring Theory. Proc. Bien. Ohio State-Denison Conf., Granville, Ohio, May, 1992. Singapore etc.*, 350–357 (1993).

[244] W. Xue, "Two examples of local artinian rings," *Proc. Amer. Math. Soc.*, **107**, No. 1, 63–65 (1989).

[245] W. Xue, "A note on three artinian rings," *Comm. Algebra*, **18**, No. 7, 2243–2247 (1990).

[246] W. Xue, "Exact modules and serial rings," *J. Algebra*, **134**, 209–221 (1990).

[247] W. Xue, "Rings with Morita duality," *Lect. Notes Math.*, **1523**, 1–197 (1992).

[248] Y. Yukimoto, "Artinian rings whose projective indecomposables are distributive," *Osaka J. Math.*, **22**, 339–344 (1985).

[249] B. V. Zabavskii and N. Ya. Komarnitskii, "Distributive elementary divisor domains," *Ukr. Mat. Zh.*, **42**, No. 7, 1000–1004 (1990).

[250] V. V. Zharinov, "Distributive lattices and their applications in complex analysis," *Trudy Mat. Inst. Akad. Nauk SSSR, Moscow*, **162**, 3–116 (1983).

Index